NON-CONVENTIONAL ENERGY RESOURCES

(Alternative Energy Sources and Systems)

Dr. R.K. Singal

Non-Conventional Energy Resources

(Alternative Energy Sources and Systems)

By

Dr. R.K. SINGAL
B.Sc. Engg., M.E. (Hons.), Ph.D., F.I.E.,
C.Eng, D.S.S.T. (Russia), C.M.P.D., C.F.M.,
Galgotia's College of Engg. & Technology
(Greater Noida)

S.K. KATARIA & SONS®
Publisher of Engineering Books
4885/109, Prakash Mahal, Dr. Subhash Bhargav Lane,
Opposite Delhi Medical Association, Daryaganj, New Delhi–110002 (INDIA)
Phone: +91-11-43551243, Mobile: +91-9871775858, 7678425209
e-mail: katariabook@gmail.com, katariabooks@yahoo.com
Website: www.skkatariaandsons.com
(e-books, subscriptions and bulk buying also available on our website)

Published by:
S.K. Kataria & Sons®
4885/109, Prakash Mahal, Dr. Subhash Bhargav Lane,
Opposite Delhi Medical Association, Daryaganj, New Delhi-110002 (INDIA)
Phone: +91-11-43551243, Mobile: +91-9871775858, 7678425209
e-mail: katariabook@gmail.com, katariabooks@yahoo.com
Website: www.skkatariaandsons.com

First Edition: January, 2005; Third Edition: 2010; Reprint: 2026

Laser Typesetting by : Plus Computers, Delhi

Printed at : G.S. Enterprises

Dedicated

To Future Generations.
All Belongs to Them.

Preface

Human societies require increasing amounts of energy for industrial, commerical, domestic, argiculture, and transport uses. Currently these energy needs are met from short-term, depletable fossil fuel supplies like petroleum, coal and natural gas. Attention has been recently diverted to the use of renewable energy sources which are going to sustain the future societies. Therefore, a large number of universities, colleges and polytechnics are offering a number of courses on alternative, non-conventional energy sources and conversion systems. Such courses like Non-conventional Energy Resources or Renewable Energy Sources or Energy Systems or Direct Energy Conversion Systems are offered at undergraduate and postgraduate levels either as a core subject or as elective subject.

There are plenty of books on research and development or as reference material on the subject. There is a shortage of a text book for the direct and exclusive use of the students. An effort has been made in this direction to compile this exclusive text book covering the courses named differently. Every effort has been made to describe the matter in clear and simple way. The diagrams are very illustrative and simple. The students can understand, remember and reproduce the material and drawings with least effort. The text has been arranged to sustain the interest of the students throughout the book.

Practical data have been included for the reference of practising and research engineers.

Projects and papers have been included to help the students for their projects, term papers and seminars.

The author is grateful to the publisher for bringing out this book in a very short time. Although the author is constantly in touch with the students and the subject teachers, the book has been written keeping in mind their difficulties and requirements. Any constructive criticism on the book will be welcome with gratitude. After all, the present effort is being made to help the student community.

Dr. R.K. Singal

A Tree's Petition

Dear, my dear Prime Minister, Madam*,
Save me from the axe of Adam.

He wants to end my life so dear,
With no conscious, prejudice or fear.

Life of flora and fauna is thy concern,
This is the talk of nations, I learn.

I want to serve the man with dignity,
Why then, he cuts me with futility ?

The gifts of heaven, I give him free,
The food, the medicines and energy.

Important link in life cycle of nature,
I purify the environment for all creatures.

I save earth's surface from denudation,
Floods, famines, furries of the Creation.

I do not encroah his rights and liberties,
Save him from sun, wind and miseries.

I also feel and enjoy life's honey,
Madam, recall Jagdish Bose's testimony.

I appeal thee to save me by legislation,
I also want to live and serve the nation.

— Dr. R.K. Singal

* *Smt. Indira Gandhi*

Contents

PART 1

CURRENT ENERGY SYSTEMS

Chapter 1

ENERGY UTILIZATION

Introduction ; Energy Characteristics : Different Form of Energy : Mechanical Energy ; Heat Energy ; Chemical Energy ; Nuclear Energy ; Electrical Energy ; World Energy Production ; Energy Use ; Role of Energy Consumption.

1.1. INTRODUCTION

Man arrived on this earth and started exploiting different sources of energy for raising his standard of living. Energy has been an important component to meet the day-to-day need of human beings. The degree of civilisation is measured by the energy utilization for human advancement or needs. Some people rate energy as an important yardstick for measuring the success of all human culture. As per them, civilization can be defined as equal to energy plus intelligence. The human societies have continued to increase the amount and form of energy flowing through each individual life in search of more and more comfort. Energy consumption is an absolutely necessary component of industrial society.

Before the Industrial Revolution of the 18th century, most energy used relied on two sources : human and animal muscles, and the energy of the wind and water available in nature. The chemical energy in firewood was the main source of heat and light. The main disadvantages of these natural energy flows of the earth were their intermittency, lack of dependability, and their usually extreme low energy densities. However, these traditional energy resources were renewable, inexhaustible and free for the taking. These were the predominant energy forms used by mankind during its early developmental millenia.

The earth's population grew and nature alone could no longer support life on earth. Man discovered solutions when they were needed. The solutions came in the form of fossil fuels that were concentrated in certain pockets of the earth. Fossil fuels had much higher energy densities. This brought on the industrial revolution and thus the modern way of life as we know it today.

The high energy density of fossil fuels and metallurgical advances brought high temperatures of working cycles and therefore higher efficiencies. The traditional energy resources could not withstand the high efficiencies, abundance and low price of fossil fuels. However, the great economic squeeze of oil producers during early 1970s brought the era of cheap energy to an end. This and environmental concerns reopened the era of traditional, renewable (or alternative) sources of energy. Most renewable resources are aimed at electric production as their main contribution. Technological solutions are being worked out for the economical, sustainable and clean conversion systems based on renewable energy sources.

1.2. ENERGY CHARACTERISTICS

Energy is a fundamental concept such as God, and is very difficult to define. However, energy has been defined as the capability to produce an effect or capacity to do work. Energy has certain characteristics.

1. Energy is available in different forms.
2. Energy can neither be created nor destroyed.
3. Energy can be transformed from one form to another form.
4. Energy can be stored.
5. Energy can be transported from one system to other system or from one place to another.
6. An energy is associated with a potential. For example ; thermal potential (temperature difference) for heat energy or pressure difference for mechanical energy or electrical potential for electrical energy. Free flow of energy takes place only from a higher potential to a lower potential.
7. The energy is measured as N-m or in joules. The energy density is expressed as J/kg. Different sources and forms of energy have different energy densities.
8. Energy does not have absolute value. It is measured relative to a datum. The common data are ground level, atmospheric pressure and a temperature at 0K. Heat energy is assumed to have zero value at 0K. Similarly potential energy is zero at ground level.
9. The transfer of heat energy takes place due to temperature gradient. All other forms of energy exchange may take place due to difference in other intensive properties. These forms of energy can also be called work.
10. The forms of energy are graded as per their availability or exergy content. A high grade energy like electrical energy can be converted into mechanical work without significant loss. On the other hand, whole of low grade energy like heat cannot he converted into mechanical work. Some part of heat is further degraded (into very low temperature heat) and rejected to environment.

1.3. DIFFERENT FORMS OF ENERGY

Broadly speaking there are two main types of energy : heat and work. However, there are many forms of energy.

1. Electrical energy.
2. Mechanical energy including shaft energy, kinetic energy, potential energy flow energy, etc.
3. Chemical energy expressed as calorific value of fuel, heat of combustion, Gibb's free energy.
4. Heat energy including enthalpy and internal energy, solar energy, geothermal energy.
5. Nuclear energy of fission and fusion.

1.3.1. Mechanical Energy

1. Kinetic Energy

The energy possessed by a body or a system by virtue of its mass and velocity of motion is called kinetic energy.

$$K.E. = \frac{mV^2}{2}$$

where m = mass of body [kg]

V = Velocity of the body, [m/s].

$$K.E = \text{kg}\ .\ (\text{m}/\text{s})^2$$

$$= \frac{kgm}{s^2}.m = \text{N-m} = \text{J}.$$

The energy is the capacity to do work and power is the rate of doing work. The total power of a fluid stream (air or water) is equal to the rate of the incoming energy of that stream.

Total power of a stream,

$$P = \dot{m}\,ke$$

$$= \dot{m}\frac{V^2}{2}, \text{[W]}$$

where $\dot{m}$ = mass flow rate, [kg/s]

V = stream velocity [m/s]

ke = kinetic energy of a unit mass [J/kg]

The mass flow rate of stream is given by the continuity equation,

$$\dot{m} = \rho AV$$

where ρ = incoming stream density [kg/m^3]

A = Cross sectional area of stream [m^2]

$$\therefore \quad P = \frac{\rho AV^3}{2}.$$

The kinetic energy of air is utilized in a wind mill and that of water in a hydro-turbine.

2. Potential Energy

It is the energy possessed by a body or system by virtue of its position above the datum (ground) level. The work done is due to its falling on earth's surface. Potential energy,

$$PE = Wh = mgh \text{ [N-m]}$$

Where W = weight of the body [N]

m = mass of the body [kg]

h = distance of fall of body, [m]

g = acceleration due to gravity

$= 9.81\ \text{m}/\text{s}^2$

$$PE = Wh$$

$$= N \times m = \text{N-m}$$

$$PE = mgh$$
$$= kg \times \frac{m}{s^2} \times m$$
$$= \frac{kg - m}{s^2} . m = \text{N-m.}$$

The total power of a stream,

$$P = \dot{m}\, pe = \dot{m}\, gh$$

where $\dot{m}$ = mass flow rate [kg/s]
h = distance of fall of stream, [m]
pe = potential energy of unit mass [J/kg]

$$\dot{m} = \rho \dot{Q}$$

ρ = incoming stream density [kg/m^3]
$\dot{Q}$ = volumetric flow rate, [m^3/s]

$$\therefore \quad P = \rho \dot{Q} g h . \text{ [W]}$$

The potential energy of water stored at a height is utilized in a hydroelectric power plant.

3. Flow Energy

Flow energy or pressure energy is possessed by a flowing fluid by virtue of its pressure. It can be expressed as the work done by the flowing fluid to push the quantity represented by mass *m* into and out of a system. Mathematically, it is equal to the product of pressure *p* and volume *V*.

Flow energy

$$FE = pV = pmv$$

4. Shaft Energy

A rotating shaft has rotational energy. This can be expressed by shaft work.

Shaft energy,

$$W = \int_1^2 T d\theta$$

where T = torque applied to the shaft (N-m)
$d\theta$ = angular displacement of the shaft (rad)

Shaft power,

$$P = \int_1^2 T \frac{d\theta}{d\tau} = Tw$$

where w = angular velocity of the shaft [rad/s]

$$P = 2\pi NT \text{ [W]}$$

where N = Shaft speed, (rps)

The rotating shaft possesses kinetic energy.

Kinetic energy of rotation,

$$K.E = \frac{1}{2} I w^2$$

where I = Mass moment of inertia

$= mk^2$

m = mass of rotating shaft [kg]

k = radius of gyration [m]

The work done by a rotating shaft is equal to change in kinetic energy.

$$W_{1-2} = \text{Initial } KE - \text{Final } K.E.$$

5. Piston Energy or Displacement Work

In a constant mass closed system, a piston is moved in a cylinder from position 1 to position 2 by a fluid. The amount of work done,

$$W_{1-2} = \int_{V_1}^{V_2} p\,dv$$

The amount of work done depends upon the process of change of volume with pressure. The most general relationship is given by

$$pV^n = \text{constant}$$

where n is called the polytropic exponent and varies from zero to infinity. Its value for certain processes is given in Table 1.1.

Table 1.1. Values of *n* for various process.

Process	n	c_n
Constant pressure	0	C_p
Constant temperature	1	∞
Adiabatic reversible	$\gamma = \dfrac{C_p}{C_v}$	0
Constant volume	∞	C_v
Polytropic	0 to ∞	$C_v\left(\dfrac{\gamma - n}{1 - n}\right)$

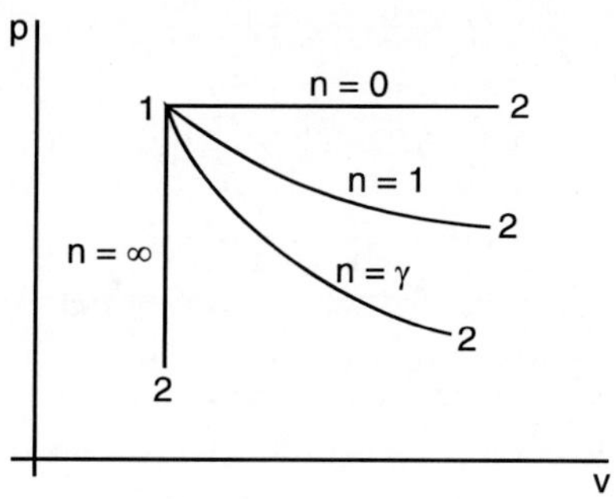

Fig. 1.1. $pV^n = C$.

1.3.2. Heat Energy

Heat is energy transferred across the boundary of a system due to temperature difference between the system and surrounding.

$$Q = m\,C_n\,(T_2 - T_1)\ [\text{kJ}]$$

where m = mass [kg]

T_1, T_2 = Temperature is °C or K.

C_n = Specific heat [kJ/kg-K]

C_n depends upon the process taking place during heat transfer and the values for various processes are given in Table 1.1.

1. Solar Energy

The sun is a large sphere of hot gases. As viewed from the earth, the sun is a solar disc of uniform brightness with a surface temperature of 5762 K. The solar radiations are reaching the earth as heat energy and can be calculated by Stefen-Boltzmann equation. The rate at which solar energy arrives at the top of the atmosphere is called solar constant. This is the amount of heat energy received in unit time on a unit area perpendicular to the sum's direction at the mean distance of the earth from the sun. The solar constant is 1353 watt per square metre. The value of solar radiations as reaching the earth's surface is reduced due to diffusive and scattering effect of the atmosphere gases and dust particles.

2. Internal Energy

The energy can be stored within a system and can be transferred from one system to another. The energy can be stored at microscopic and macroscopic level of the system. The kinetic and potential energy of a system are associated with the co-ordinate frame selected and can be specified by the macroscopic parameters of the mass, velocity and elevation.

The energy processed by a body or a system by virtue of its molecular arrangement and motion of molecules is called internal energy. This energy is associated with thermodynamic state of the system. For perfect gases, internal energy depends only on the temperature of the system whereas for non perfect gas, vapours, liquids the internal energy strongly depends on temperature and weakly on pressure.

3. Enthalpy

In the study of energy, we encounter many processes of heating and cooling at constant pressure. In a constant mass closed system,

$$Q_{1-2} = U_2 - U_1 + W_{1-2}$$

Now
$$W_{1-2} = \int_1^2 p\,dv = p(V_2 - V_1)$$

$\therefore$
$$Q_{1-2} = U_2 - U_1 + p_2V_2 - p_1V_1$$
$$= (U_2 + p_2V_2) - (U_1 + p_1V_1) = (H_2 - H_1)$$

In an open, steady-flow process,

$$U_1 + p_1V_1 + \frac{V_1^2}{2} + gZ_1 + Q_{1-2} = U_2 + p_2V_2 + \frac{V_2^2}{2} + gZ_2 + W_{1-2}$$

If there are no changes in kinetic and potential energy

$$U_1 + p_1V_1 + Q_{1-2} = U_2 + p_2V_2 + W_{1-2}$$
$$Q_{1-2} - W_{1-2} = (U_2 + p_2V_2) - (U_1 + p_1V_1)$$
$$= (H_2 - H_1)$$

The sum of internal energy (U) and product of pressure and volume (pV) or flow energy in called enthalpy (H).

$$H = U + pV$$

For unit mass

$$h = u + pv$$

For an ideal gas,

$$pv = RT$$

$$h = u + (RT)$$

Now $\quad U = U(T)$

$\therefore \quad h = h(T)$

The enthalpy of an ideal gas is a function of temperature only.

The enthalpy of steam or hot water from wells is utilized in a steam turbine to produce power. This enthalpy is called geothermal energy.

1.3.3. Chemical Energy

1. Calorific Value

Fossil fuels in the form of solids, liquids and gases are hydrocarbons or mixture of hydrocarbons which combine with oxygen of air to produce CO_2 and H_2O and exothermic heat. The combustion heat produced per kg of solid and liquid fuel or per m^3 of gaseous fuel is called its calorific value. The calorific value or heating value of a fuel depends upon its composition. The heating values of some important fuels are given in Table 1.2.

Table 1.2. Heating Values of Fuels

Fuel	Heating Value
Bituminous coal	25 to 30 MJ/kg.
Anthracite coal	35 MJ/kg.
Lignite	15 to 20 MJ/kg.
Diesel oil	50 MJ/kg.
Petrol	45MJ/kg.
Ethane gas ($C_2 H_6$)	52 MJ/kg.

2. Gibb's Free Energy

For a pure substance undergoing an infinitesimal reversible process.

$$dG = dH - Tds - sdT$$
$$= Vdp - sdT$$

Where G is called Gibb's free energy or Gibb's function.

The maximum useful work that can be obtained from a steady flow stream when it undergoes a process at the beginning and end of which the pressure and temperature are in equilibrium with the atmosphere.

$$\Delta W_{max} = (\Delta H - T\Delta S) = \Delta G$$

In a fuel cell, a chemical reaction takes place where reactants and products can be converted into a steady flow process and both are in equilibrium with the atmosphere.

1.3.4. Nuclear Energy

1. Nuclear Fusion

Energy is produced in the sun and stars by continuous fusion reactions in which four nuclei of hydrogen fuse resulting in one nucleus of helium and two positrons.

$$4\ {}_1H^1 \rightarrow {}_2He^4 + 2\ {}_{+1}e^{\circ}$$

This results in a decrease in mass of about 0.0276 amu corresponding to 25.7 MeV. The heat produced in these reactions maintains temperatures of the order of several million degrees.

Many problems have to be solved before an artificially made fusion reactor becomes a reality. The most important difficulties are generating and maintaining high temperatures in the plasma and conversion of fusion energy to electricity.

2. Nuclear Fission

There are many fission reactions that release different energy values. On an average, the fission of a U^{235} nucleus yields about 193 Mev.

$${}_{92}U^{235} + {}_0n^1 \rightarrow {}_{56}Ba^{137} + {}_{36}kr^{97} + 2\ {}_0n^1$$

The mass balance

$$235.0439 + 1.00867 \rightarrow 136.9061 + 96.9212 + 2 \times 1.00867$$

$$236.0526 \rightarrow 235.8446$$

$$\therefore \quad Dm = 235.8446 - 236.0526 = -\ 0.2080 \text{ amu}$$

$$\Delta E = 931 \times (-\ 0.2082) = -\ 193.6 \text{ Mev}$$

$$= -\ 3.1 \times 10^{-11} \text{ J}$$

This is called prompt energy. The total energy, produced per fission reaction, therefore, is greater than prompt energy and is about 200 Mev. Burning of 1 g of fissionable material generates nearly 1 MW-day of energy.

3. Nuclear Breading

Nuclear fuel is defined as all uranium, plutonium and thorium isotopes. Reactor fuel contains fissionable nuclei as well as other nonfissionable isotopes of uranium plutonium or thorium. Nuclear breeding is to convert U^{238} to Pu^{239} and U^{234} to U^{235} with the help of fast neutrons.

$${}_{92}U^{238} + {}_0n^1 \rightarrow {}_{94}Pu^{239} + {}_{-1}e^0$$

Pu^{239} is fissionable material and can be used in a fission reactor.

1.3.5. Electrical Energy

When a current flows through a resistance, it can drive a motor, motor can drive a pulley and pulley can raise a weight.

The current flow in amperes is given by

$$I = \frac{dC}{d\tau}$$

where C = change in columns

τ = time in seconds

dC is the charge crossing the boundary during time $d\tau$. The work done for a voltage potential, E is given by :

$$dW = EdC = EId\tau$$

$$W = \int_1^2 EId\tau$$

The electrical power is the rate at which work is transferred and can be expressed as

$$P = \underset{d\tau \to 0}{Lt} \frac{dW}{d\tau} = EI$$

Electrical energy is the most convenient form of energy because it can be

1. easily transported,
2. easily controlled,
3. easily converted into other forms of energy at about 100% efficiency.

The only shortcoming of electrical energy is that it cannot be stored in large quantities.

Alternative energy of future is hydrogen energy which can also be easily stored. But it is highly inflamable.

Different modes of energy discussed are first converted into electrical energy.

1.4. WORLD ENERGY PRODUCTION

The energy consumption from Christ birth to 1750*AD* has been estimated as 2.34×10^{15} kWh, from 1800 AD to 1900 AD 1.17×10^{15} kWh. During the period 2000 *AD* to 2100 *AD*, it may be anything from 29.3×10^{15} kWh to 117×10^{15} kWh due to increase in world population and man's standard of living. This may fall short of total world resources of fossil fuels, estimates of which are given in Table 1.3.

Table 1.3. World Resources in Fossil Fuels.

Fuel	Consumption
Coal	5.27×10^{15} kWh.
Natural gas	0.618×10^{15} kWh.
Oil	0.996×10^{15} kWh.
Uranium	23.40×10^{15} kWh.
Thorium	14.00×10^{15} kWh
Total	44.28×10^{15} kWh

The rate of production of fossil fuels increases with time, reaches a maximum, and then decays. The peaking indicates that the end of its availability is in sight and alternative source of energy must become increasingly available to meet the growing energy requirements. According to present estimates, about 88% of the world's fossil fuels (excluding nuclear fuels) is coal. There is a reasonable agreement among various investigators that the peaking time of coal falls somewhere about the year 2040. It takes somewhere between 20 to 40 years for a new technology to come to fruition and it is therefore prudent to plan for shifting a major share of the total world energy production from fossil fuels to some other plentiful sources as easily as possible.

The following possible New and Renewable Forms of Energy have been recommended by U.N.O.

1. Tidal Energy
2. Solar Thermal Energy
3. Wind Energy
4. Biomass (Wastes)
5. Coal Tar
6. Oil Shales (Oil in rocks)
7. Geothermal Energy
8. Draught Animals
9. Ocean waves
10. Ocean Thermal Energy Conversion (OTEC)
11. Agricultural Residues.
12. Solar photovoltaic.
13. Tar sand
14. Peat (Raw coal)

1.5. ENERGY USE

Energy consumption differs greatly between developed and developing countries. The amount and patterns of energy use differ from one country to the next. In less developed countries, total energy consumption, per capita energy consumption (GJ/a), share of commercial energy sources (fossils) as compared to traditional energy sources (fuel wood, animal dung and agricultural residues) are far less than developed countries. Similarly energy use for different sectors differ from country to country. The basic data of India is shown in Table 1.4 and energy balance in Table 1.5.

Table 1.4. Basic Data of India

1. Population	:	1000m	
2. Category	:	Balanced growth of industry and agriculture	
3. Urbanization	:	23%	
4. GNP/capita	:	195 US $	
5. Share of industry	:	27%	
6. Share of agriculture	:	38%	
7. Growth GNP/capita	:	4%	
8. Oil imports/total exports	:	60%	
9. Energy consumption/capita	:	12.6 GJ	
(*i*) Traditional energy	:	52%	6.6 GJ
(*ii*) Commercial energy	:	48%	6.0 GJ
(*a*) Transport	:	28%	1.7 GJ
(*b*) Industry	:	43%	2.6 GJ
(*c*) Domestic	:	8%	0.5 GJ
(*iii*) Fossil fuels	:	95%	5.7 GJ
(*iv*) Electricity generation	:	17%	

Table 1.5. Energy Balance of India (PJ)

Sector	Commercial Energy					Traditional energy	Total
	oil	gas	coal	electricity	Total		
I. Rural 1. Agriculture 2. Domestic	221	—	—	47	268	3059 3059	3327
II. Urban 1. Domestic 2. Services	276	5	—	50	331	474	805
III. Transport 1. Freight 2. Private 3. Public	830	—	300	8	1138 625 115 400	—	1138
IV. Industry	278	40	1215	175	1708	875	2583
V. Electricity 1. Generation 2. Distribution losses	90		575	180 –80	485 80		485 80
Total	1695	45	2090	180	4010	4408*	8418

* 60% wood.

1.6. ROLE OF ENERGY CONSUMPTION

By the end of 21st century, as per U.N projections, the world population will be about 10 to 12 billion people. It is very important to provide adequate supplies of energy, cleanly, safely and sustainably. There is a tendency among the countries of the world for a correlation between income growth and growth in energy consumption. Although the correlation is not perfect but some relationship holds good for many countries. The ratio of energy consumption to GDP is defined as the energy intensity of the economy. The GDP grows through the increase in population and results in an increase in energy consumption also. Therefore, growth of GDP virtually parallels that of energy consumption.

The per capita consumption of India is only 310 KG OE in comparison to 779 KGOE of China, 4011 KGOE of Japan and 7861 KGOE of USA. A substantial amount of improvement is, therefore, needed to elevate the standard of living of Indian poeple to a greater height and thereby put India into its rightful place in the comity of advanced nations as far as the per capita energy consumption is concerned. India, having a population of 15% of the world population, ranks sixth in terms of energy demand. The energy consumption in India is 3% of the global total. This mismatch represents the negative differential in the quality of life of the Indian people. We have to provide clean, reliable, economic energy at twenty five times the current availability to every one of the billion Indians. So many things have to be provided but every thing requires energy.

TUTORIAL SHEET No. 1.1.

1. Discuss the history of energy resources.
2. What is energy ? Discuss the different characteristics of energy.
3. Classify and discuss various forms of energy.
4. What are different types of Mechanical Energy ?
5. Write notes on :
   (*i*) Heat energy
   (*ii*) Nuclear energy
   (*iii*) Electrical energy.
6. Prove that energy requirement in 21st century will fall short of world energy resources.
7. What is the basic economic data of India ? Draw an energy balance of India.
8. Discuss the role of energy consumption for India.

❑❑❑

Chapter 2

ENERGY RESOURCES

Classification of Energy Resources ; Commercial Primary Energy Resources; Renewable Sources of Energy; New Sources of Energy ; Advantages and Demerits of Renewable Sources of Energy.

2.1. CLASSIFICATION OF ENERGY RESOURCES

Adequate supplies of energy to the growing population of the world desirous of higher standard of living can be cleanly, safely and sustainably met by exploiting various sources of energy. The various sources of energy can be conveniently grouped as follows :

1. Commercial Primary Energy Resources

India's energy use is mostly based on fossil fuels. Although the country has significant coal and hydro resource potential, it is relatively poor in oil and gas resources. Nuclear energy has the potential to meet the future needs of electricity demand in the country. The country has developed the capacity to build and operate nuclear power plants observing international standards of safety.

Coal, lignite, oil and natural gas, hydroelectric and nuclear fuels can be grouped as commercial primary energy resources.

2. Renewable Sources of Energy

India is endowed with abundant natural and renewable resources of energy namely, solar, wind and biomass. The available renewable resources need to be exploited by giving a commercial orientation, wherever possible. Efforts are being made to make the renewable sources cheaper and more efficient. The main attention has to be directed to the following sources of renewables, namely, (*i*) solar photo-voltaic, (*ii*) wind, and (*iii*) hydrogen fuel-cell.

3. New Sources of Energy

There are new sources of energy available for local exploitation. In many cases, autonomous and small power plants can be built to avoid transmission losses. Most prominent new sources of energy as identified by UN are :

Tidal energy, ocean waves, OTEC, peat, tar sand, oil shales, coal tar, geothermal energy, draught animals, agricultural residues, etc.

2.2. AVAILABILITY OF COMMERCIAL PRIMARY ENERGY RESOURCES

The geographical distribution of available primary commercial energy sources in the country is quite uneven. About 77% of the hydropotential is located in the northern and north-eastern region. Similarly, about 70% of the total coal reserves are located in the eastern region. Most of the hydrocarbon reserves (crude oil and natural gas) lie in the West.

The geographical distribution of various energy resources available in the country is given in Table 2.1.

Table 2.1. Primary Commercial Energy Resources

Region of India	Coal (*Bt*)	Lignite (*Bt*)	Crude oil (*Mt*)	Natural gas (*BCM*)	Hydropower (*TWH*)
Nothern	1.06	2.51	0.03	0.00	225.00
Western	56.90	1.87	519.47	516.42	31.40
Southern	15.46	30.38	45.84	80.94	61.80
Eastern	146.67	0.00	2.19	0.29	42.50
North-Eastern	0.89	0.00	166.17	152.00	239.30
Total	220.98	34.76	733.70	749.65	600.00

Note : *Bt* : Billion tonnes ; BCM : Billion Cubic Metres; TWH : Trillion Watt Hours ; *Mt* : Million Tonnes.

Source : Planning Commission of India.

2.2.1. Coal

The geological coal reserves of the country are estimated at 220.98 billion tonnes (*Bt*) as on January, 2001. Out of this, proven reserves are 84.41 *Bt*, while 98.55 *Bt* are indicated reserves and 38.02 Bt are inferred reserves. Coal is the principal source of commercial energy accounting for nearly 50 % of the total supplies. About 70% of the power generated are coal and lignite based.

2.2.2. Lignite

The lignite deposits in the southern and western regions have emerged as an important source of fuel supply for the states like Tamilnadu, Rajasthan and Gujarat. The current estimates of geolagical lignite reserves in the country are 34.76 Bt spread over Tamilnadu and Pandicherry (87.5%), Rajasthan (6.9%), Gujarat (4.9%), Kerala (0.31%) and Jammu and Kashmir (0.37%). The share of lignite for power generation is increasing fast from 24.3 Mt in 2001-2002 to 44.96 Mt in 2006-2007.

2.2.3. Oil and Natural Gas

India has 733.70 *Mt* of crude and 749.95 BCM of natural gas as estimated in 2001. India has about 0.4% of the world's proven reserves of crude oil. The transport sector is the main consumer. Similarly India's natural gas reserves are also 0.4% of world's natural gas reserves. Gas is increasingly being used for power generation, industrial application and transport sector. The share of power generation capacity, based on gas, is about 10% of the total installed capacity.

India has an estimated 1000 BCM of Coal Bed Methane (CBM), which is likely to emerge as a new source of commercial energy in the country.

2.2.4. Nuclear Resources

The current installed capacity of nuclear power plant is 2860 MW accounting for 2.8% of the total installed capacity of the country. It has great potential to meet the future needs of electricity demand in the country.

2.2.5. Hydro Power

The Central Electricity Authority (CEA) has assessed India's hydro potential to be about 148700 MW of installed capacity. The hydro electric capacity currently under operation is about 26000MW and 16083 MW is under various stages of development. India is endowed with economically viable hydro potential of 600 TWH. The key advantage of hydro electric power is the ability to store energy and the flexibility of its used during peak load periods.

2.3. EXPLOITATION OF COMMERCIAL PRIMARY ENERGY RESOURCES

The production of commercial primary energy resources has been shown in Table 2.2 for different periods. The same have been projected for Tenth Five Year Plan terminating in the year 2007.

Table 2.2. Production of Commercial Energy Sources

Source of energy	Unit	Production Periods					
		1960-61	1970-71	1980-81	1990-91	2001-02	2006-07
Coal	Mt	55.67	72.95	114.01	211.73	325.65	405.00
Lignite	Mt	0.05	3.39	4.80	14.07	24.30	55.96
Crude oil	Mt	0.45	6.82	10.51	33.02	32.03	33.97
Natural gas	BCM	—	1.44	2.35	1.79	29.69	37.62
Hydro Power	BkWh	7.84	25.25	46.54	71.66	82.80	103.49
Nuclear Power	BkWh	—	2.42	3.00	6.14	16.92	19.30

Significant hydro and nuclear generation capacity is likely to be added during the Tenth Five-Year Plan period. The capacity addition programme includes 16083 MW from hydel power plants and 1300 MW from nuclear power plants.

2.4. ECONOMIC GROWTH AND ENERGY DEMAND

Primary commercial energy demand grew almost three fold at an annual rate of 65% between 1981 and 2001 to reach 314.7 Mt of oil equivalent (MTOE). India's commercial energy demand for the next decade will be among the highest in the world for sustained economic growth, rise in income levels, increased availability of goods and services, raising the standards of living.

The average annual world economic growth in the period 1997-2020 is projected at 3.2% while the energy growth rate is estimated at 2.1% per annum.

The projected requirement of commercial energy in India is expected to grow at an average rate of 6.4% Estimated energy demand for 2006-07 and 2011-12 are given in Table 2.3.

Table 2.3. Estimated Energy Demand

Primary Fuel	Unit	Demand (in original units)		Demand (MTOE)	
		2006-07	2011-12	2006-07	2011-12
Coal	Mt	460.50	620.00	190.00	254.93
Lignite	Mt	57.79	81.54	15.51	22.05
Oil	Mt	134.50	172.47	144.58	185.40
Natural gas	BCM	47.45	64.00	42.70	57.60
Hydro Power	BkWh	148.08	215.66	12.73	18.54
Nuclear Power	BkWh	23.15	54.74	6.04	14.16
Total				411.91	1 553.68

The difference in demand and supply can be met by imports, efficient management and exploitation of renewables.

2.5. RENEWABLE ENERGY SOURCES POTENTIAL

There is a growing consensus that renewable energy sources will play a very important part in the solution of energy needs. There is growing concern over the pollution, resource depletion and possibility of climate change due to continuing use of conventional fossil fuels and nuclear fuels. Recent technological developments have improved the cost-effectiveness of many of the renewables. It is expected that half of the world's energy needs may be met by renewables by about the middle of this century. The energy requirements of rural areas, particularly, remote villages can be met by locally available renewables as it is difficult to service these areas through the conventional power grid. Table 2.4. gives the available potential and the actual potential exploited till August, 2001 for various renewable sources of energy.

Table 2.4. Renewable Energy Sources Potential

Source/Technology	Units	Available Potential	Actual Potential Exploited
Blo-gas	Million	12	3.22
Bio-mass-based power	MW	19500	384
Efficient wood stove	Million	120	33.86
Social energy	MW/km^2	20	1.74
Small Hydro	MW	15000	1398
Wind Energy	MW	45000	1367
Energy recovered from wastes	MW	1700	16.2

2. 5.1. Solar Energy

The earth and its atmosphere is continuously receiving 1.7×10^{17} W of radiations from the sun. A world population of 10 billion with a total power need per person of 10 kW would require about 10^{11} kW of energy. It is thus apparent that if irradiation on only one percent of earth's surface could be converted into useful energy with 10% efficiency, solar energy could fullfil the energy needs of the entire world population. Another interesting comparison is that the total energy on earth is less than the solar energy incident upon the earth in one year.

Solar energy is, however, very diffuse, cyclic and often undependable. It, therefore, needs systems and components that can gather and concentrate it efficiently for conversion to any of uses and that can do the conversion as efficiently as possible.

The solar energy resource can be divided into natural and technological collection system. Further sub-division is shown in Fig. 2.1.

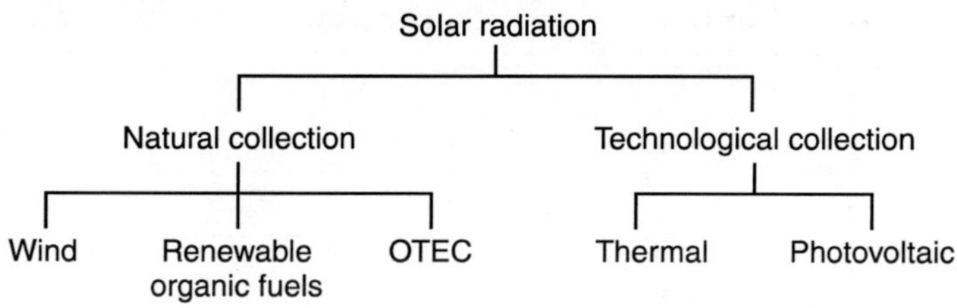

Fig. 2.1. Solar Energy Resources.

2.5.2. Thermal Utilization

Flat plate collectors of solar radiation are available for operation over a temperature range of temperature upto approximately 365 K for producing hot water, space heating, cooking and operation of absorber type air conditioning systems. This thermal low-temperature utilization is also economically attractive.

Operating temperatures as high as 4000K have been achieved by the use of focussing devices for solar furnaces. Generation of steam to operate pumps for irrigation purposes and conventional steam turbines for generation of electricity has also proved technologically feasible. The cost of solar power in favourable locations will be compatible with that of nuclear power when the development of these plants has been completed.

2.5.3. Photo Voltaic Convertors

Photovoltaic solar cells utilize energetic photons of the incident solar radiations directly to produce electricity at high conversion efficiencies of about 24% (theoretical) and 15% (achieved). The technology of photovoltaic conversion is well developed as a part of satellite and space-travel technology, but large scale application is hampered by the high price of photocells. Efforts to reduce the production costs of solar cells are directed towards drawing thin monocrystalline materials, development of concentrating devices to reduce surface areas of solar cells required and highly innovative fabrication techniques. These efforts have brought down the cost of solar photovoltaic (*PV*) module from $27/watt in 1978 to $ 3.33/watt in July 2003. The total cost of solar energy including the cost of balance of system, installation, etc., is $ 21.26 cent (US).

2.5.4. Wind Energy

The wind power technology is well developed and cost estimates in favourable regions of the world are fairly close to those of other energy sources. Rotor diameters of 50m and heights of 100m are dimensions typically for wind generation to produce appreciable amount of power. The unreliability of power production due to large variability in wind velocity has hamphared the large-scale applications even in favourable regions.

In 1978, the generation cost of electricity from wind turbine was around \$1.0/ unit which came down to 5 cents in 1998, and in January 2003, it has become only 3.6 cent. The available potential of wind energy in India is 45000 MW out of which 1367 MW have been exploited till August 2001. Table 2.5 provides the top six wind energy generator-countries in the world.

Table 2.5. Wind Generation Countries

Country	Generation Capacity (MW)
Germany	10,650
United states	4329
Spain	4039
Denmark	2515
India	1507
Italy	755

2.5.5. Solar Energy Conversion by Oceans

Oceans forming almost 71% of the world's surface serves as a tremendous storehouse of solar energy due to the temperature differences produced by the sun as well as kinetic energy stored in the waves. There are a number of places in the ocean where temperature differences of the order of 20-25K exist at depths of less than 1000 m, and these temperature differences could be used to operate low pressure heat engines. Although the thermodynamic efficiencies are low, the available amount of thermal energy is very large. The cost of using the temperature gradients in the ocean for practical solar energy utilization depends largely on the development of efficient and cheap heat exchangers which form the bulk of the plant cost.

2.5.6. Biological Conversion

Bioconversion is expected to provide an increasing amount of our total energy needs in the future. It will be possible to cultivate appropriate plants exclusively for the purpose of generating power either by direct bioconversion or by pyrolytic conversion into liquids or gaseous fuels. Cultivation of special plants like sugarcane can improve the conversion efficiency from the present 1% to 2.5 %. The re-cycling processes are within the realm of possibility to convert the waste products of food and animals to methane and methanol.

2.6. NEW ENERGY SOURCES

These sources are mostly small in quantities and very site specific. These can be exploited for local applications.

2.6.1. Geothermal Energy

Geothermal energy is primarily energy from the earth's own interior. Vast natural heat stores lie beneath the earth's crust. It is recoverable in the form of steam or hot water. These are used for electric generation, space heating, cooking, and medicinal purposes. The main locations of geothermal energy are USA, Italy Japan, New Zealand.

The world projections for geothermal electric production are given in Table 2.6.

Table 2.6. World Geothermal Energy Utilization

Country	Electricity (MW)		Space and Process Heat (MW)
	Installed	Under construction	
USA	773	641	30
Italy	421	—	1
New Zealand	203	150	70
Japan	166	100	37
Mexico	150	30	—
El Salvador	60	35	—
Iceland	33	30	475
USSR	6	58	300
Phillipines	60	605	—
Turkey	0.5	—	—
Hungary	—	—	363
France	—	—	5
Total	1872	1649	1281

It is seen that while geothermal energy is not the sought-after sole and long-range solution to our energy problems, it never-the-less represents a not insignificant factor if its resources are developed in a careful and efficient manner. India does not have high-temperature hydrothermal areas.

2.6.2. Tidal Energy

The tides are yet another source of energy from the oceans in addition to OTEC. This energy can be tapped from coastal water by building dams that entrap water at high tide and release it at low tide back to the sea. Power can be obtained by turbines from both in-out flows of water. The amount of energy is very large but only in a few parts of the world.

The total power of tides is estimated at (2 to 5) × 10^6 MW, but difficulties associated with finding suitable locations, high construction costs and non-uniform availability of power have hindered the progress of harnessing the tidal energy. Development work is going on in different countries like France, Russia Canada, U.K, USA and Sweden to harness this energy.

Three possible locations have been identified for the development of tidal power in India : Gulf of Kutch and Gulf of Cambay in Gujarat, Sunderbans in west Bengal. A feasibility of tidal power scheme for 600 MW installed capacity in Gulf of Kutch is under study.

2.6.3. Wave Energy

Ocean waves are caused by the wind, which in turn is caused by uneven solar heating and subsequent cooling of earth's crust and the rotation of the earth. The harnessing of wave energy requires development of special power conversion devices. The main advantage of power from waves, like most alternative energy sources, is free and renewable energy source. Wave-power devices unlike solar or wind devices, donot use up large land masses. It is pollution free and leaves the water in a placid state in the water. A staggered array of power devices can produce electricity, protect coastlines from the destructive action of waves, minimize erosion and even help create artificial harbours.

Like other alternative energy sources, waves lack dependability, and there is relative scarcity of accessible sites of large wave activity. Economic factors like capital investment, costs of maintenance, repair and replacement are hindering the development.

The suitable world sites for wave energy harnessing are Molakai and Alenuihaha channels in the Hawaiian islands, Pacific Coasts of North America, the Arabian Sea off India and Pakistan, the North Atlantic coast of Scotland, the coast of New England, etc.

2.6.4. Peat

Anthracite and Bituminous coals are used for power generation world over. Because of high moisture content and low heating value (15 to 20 MJ/kg), lignite is not economical to transport over long distances and it is usually burned by power plants at the mining site. Peat is a heterogeneous material consisting of decomposed plant matter and inorganic mineals. It contains upto 90 % moisture. Although not attractive as a utility fuel, it is abundant in many parts of the world. Therefore, it can be used in electric generating plants and in district heating. It is available in some states of USA, Ireland, Finland and Russia.

2.6.5. Oil Shale

Shale is a fine-grained rock formed by stratification under pressure. One ton of oil shale can be decomposed to yield 95 to 130 L of low-sulphur synthetic crude oil at about 500°C. This oil makes a good refinery feedstock that can be processed further to various useful liquid and gaseous fuels.

There are vast amounts of oil-shale deposits in USA, France, Russia and China. Several pilot plants have been operated but with difficulties of large amount of water requirement, environmental concerns, disposal of spent rock, large amount of energy requirement for mining, transportation and processing. Oil-shale plants will become competitive in future when cost and availability of natural crude oil will become adverse.

2.6.6. Tar Sands

Tar sand contains 2 to 3 times the oil reserves of all Middle East. It is found in Alberta, Canada. There are many problems surrounding the commercial exploitation of oil from tar sands.

2.6.7. Agricultural Residues

The estimated production of agricultural residues in India is 200 million tonnes per year and that of wood is 130 million tonnes. At an average heating value of 18 MJ/kg db, a total potential of 6×10^{12} MJ/year or approximately

75×10^7 MJ/hour exists. At a power conversion rate to 25% total useful potential is about 75000 MW. This can supply all our villages with power at a rate of 30,000 kWh per day per village against the present meagre consumption of only 130 kWh per day per village.

The versatility of fluidised-bed combustion technology can be utilized for the efficient combustion of forestry and agricultural waste materials such as saw dust, wood chips, hog fuel, rice husk, straws, nutshells and pits. The steam generated can be used for power generation. Another method of utilization is pyrolysis of biomass in a fluidised-bed gasifier to produce combustible gas for cooking, heating or power generation.

2.6.8. Small Hydel Projects

Hydropower, available free of cost in the form of flow water, is an important, pollution free, renewable source of energy. The utilization of energy in flowing water in rivers and irrigation canals through micro-hydro schemes is a good example of appropriate technology for rural development.

China has developed 1200 MW of power by small hydel projects by creating water heads of about 5m along the rivers. It is estimated that 15000 MW capacity of power can be created in India with the help of small hydro turbines of 200 kW to 1000 kW capacity each. Till August, 2001, 1398 MW of power in terms of installed capacity from small, mini and micro hydel schemes has been exploited.

2.6.9. Draught Animals

Draught animal is an excellent energy device readily available in the rural scene and within the means and technical competence of the farmer for ploughing, water lifting and small scale cart operators for transportation. In India, there may be about 130 million draught animals including young stock; and 80 million work animals ploughing two-thirds of land and carry two -thirds of rural transportation. These are move economical than tractors for ploughing farms of sizes less than four hectares and more competent than trucks for handling vehicles for small scale transportation (about one tonne) over short distances (below 20 km).

The animal power is 9650 kJ/Bullock-hour and total draught animal energy of working animals in India is 215,000 MW. Five pairs of bullock can replace one tractor and therefore, eight million tractors can be replaced by 80 million working animals, saving huge amount of capital investment in addition to savings in imported diesel/kerosene oil.

2.7. ADVANTAGES OF RENEWABLES

1. These sources of energy are renewable and there is no danger of depletion. These recur in nature and are inexhaustible.

   Fossil fuels are also renewable, but on a geologic time scale of hundreds of millions of years. They however, are consumed much faster than they are renewable. For all practical purposes, they have to be considered as finite.
2. Renewables are free for the taking. The power plants based on renewable donot have any fuel cost and hence negligible running cost.
3. Renewables are more site specific and are used for local processing and

application. There is no need for transmission and distribution of power. Hence there is no economic and technological losses of transmission and distribution.

4. Renewables have low energy density and more or less there is no pollution or ecological balance problem.
5. Most of the devices and plants used with the renewables are simple in design and construction which are made from local materials, local skills and by local people. The use of renewable energy can help to save foreign exchange and generate local employment.
6. The rural areas and remote villages can be better served with locally available renewable sources of energy. There will be huge savings from transporting fuels or transmitting electricity from long distances. This can also serve the development of far flung areas and arrest the migration of population which is creating many social problems.

2.8. DEMERITS IN THE USE OF RENEWABLE SOURCES OF ENERGY

1. Low energy density of renewable sources of energy need large sizes of the plant resulting in increased cost of delivered energy.
2. Intermittency and lack of dependability are the main disadvantages of renewable energy sources. The recurrence is often periodic ranging from daily (sun) to a few short years (biomass).
3. Low energy density also results in lower operating temperatures and hence low efficiencies.
4. Although renewables are essentially free, there is definite cost effectiveness associated with its conversion and utilization . Whereas for conventional energy sources, the processing cost has traditionally been borne by large industries which borrow money from a bank and then charge the consumer for each unit of energy used. In renewable energy, installations are total responsibility of the user in most cases. As a result there is a shift in the economic risk. The user has to make a huge additional investment before deriving any benefit from it.
5. *Energy Intensive :* Much of the construction materials used for renewable energy devices are themselves very energy intensive. For example, large quantities of aluminium, steel, copper, concrete, glass and plastics are needed for construction which require large quantities of energy for conversion from ore to finished products.

   The photovoltaic convertors require the extreme energy intensive processing of sand to crystalline silicone. The energy consumed in the processes for producing the finished products and in constructing the systems can be large fraction of the net energy generated during the life of the plant.
6. *Environment Pollution* : Renewable energy systems are not pollution free as usually expected. The mining of large quantities of mineral ores, the processing of these ores to the finished material, the manufacturing of the equipment, and the construction of the plants are the processes that generate their own pollution and health and occupational safety risks.
7. The low efficiency of these plants can result in large heat rejections and hence thermal pollution.
8. The renewable energy plants use up larger land masses.

2.9. RENEWABLE ENERGY PROGRAMME OF INDIA

Government of India has established an important Ministry of New and Renewable Energy Sources to plan and implement programmes in renewable energy sources to achieve the following goals.

1. Add substantial capacity of renewable energy sources in the next few decades.
2. Renewable energy sources will reduce dependence on fossil fuels.
3. There will be expansion and diversification in the energy supply mix.
4. There will be greater sustainability of national development.
5. India can meet environmental and social responsibility in a better way.
6. Degree of national energy security will increase.
7. India can emerge as world leader in the development and utilization of renewable energy sources.
8. India has one of the largest renewable energy programme in the world.
9. It is not a programme. It is a movement as clear from the following data.

2.9.1. Installed Capacity of Renewable Energy Systems

The installed capacity of renewable energy systems at present is given below based on data as given by Ministry of New and Renewable Energy Sources, Government of India.

S.No.	Programmes	Capacity
A.	Grid Connected Systems (installed capacity in MW)	
1.	Wind Power	2002
2.	Small Hydro Power	1530
3.	Biomass/Cogeneration Power	571
4.	Urban and Industrial Waste Power	26
5.	Solar Photovoltaic Power	3
	Total	**4132**
B.	**Decentralized Systems**	
6.	Biogas Plants (Nos. in lakh)	35.5
7.	Community/Institutional/Night-Soil based Biogas Plants (Nos.)	3,902
8.	Improved Chulha (Nos. in lakh)	350
9.	Solar Home Lighting Systems (Nos.)	2,60,000
10.	Solar Street Lighting Systems (Nos.)	43,470
11.	Solar Lanterns (Nos. in lakh)	4.42
12.	SPV Pumps (Nos.)	6,400
13.	Solar Water Heating Systems (lakh sq. meter collector area)	7

2.9.2. Major Achievements and Thrust Areas

• Over 4200 MW grid power from Wind, Small Hydro, Biomass and Solar Energy. • 3600 remote villages/hamlets, including those in Sunderbans, Bastar, Ladakh and the North East electrified through solar energy. • Deployment of 7.5 lakh Solar Photovoltaic Systems for lighting, communication and related uses. • 7 lakh sq. meter collector area Solar Water Heating Systems installed. • Largest solar-steam cooking system for 15,000 persons/day set up at Tirupati Tirumala Devasthanam. • 35 lakh improved chullas in rural homes. • Demonstration of the use of Ethanol, Bio-Diesel and Hydrogen for vehicles and stationary applications. • Zero emission vehicles including two, three and four wheelers and passenger buses being promoted under research and development and demonstration programmes. • Fuel cell vehicle developed and demonstrated for the first time outside the USA, the European Union and Japan. • Bangalore declared solar thermal city with Thane in Mumbai to follow soon. • 30 MW capacity Solar Photovolatic products exported to various developed and developing countries. • Constitution of Hydrogen Energy Board for drawing up and implementing the National Hydrogen Energy Roadmap. • 280 Educational Energy Parks setup for demonstration of renewable energy systems and devices. Integrated Rural Energy Programme implemented in 860 blocks. • Rs. 2500 crore direct subsidy given so far to beneficiaries/users of renewable energy systems and devices, including subsidy for grid connected renewable power projects. • Rs. 3200 crore loan provided so far by Indian Renewable Energy Development Agency Limited for 1600 renewable energy projects.

2.9.3. Specialised R&D Institutes

• Sardar Swaran Singh National Institute for Renewable Energy setup as a premier Research and Development Center near Jalandhar in Punjab. • Centre for Wind Energy Technology setup as a scientific and industrial research organization for wind resource assessment, equipment certification and R&D at Chennai in Tamil Nadu. • Solar Energy Centre set up for development of solar energy systems and devices at Gurgaon in Haryana.

2.9.4. Major Goals

• Electrification of over 20,000 remote and unelectrified villages and hamlets through renewable energy by 2007 and all households in such villages and hamlets by 2012. • Around 10 percent additional power generation capacity through renewables by 2012. • Research and development in new technology areas, viz., Hydrogen, Fuel Cells, Hybrid Vehicles, Geothermal, Tidal etc. • Benchmarking renewable energy systems and devices to international standards, specifications and prices.

TUTORIAL SHEET No. 2.1.

1. Give a brief classification of various energy resources. What is the future of non-conventional energy sources in India.
2. Describe in detail the availability of primary commercial energy resources.
3. Write notes on :
   (*i*) Exploitation of primary commercial energy sources.
   (*ii*) Projection of energy demand for economic growth.
4. What are renewable energy sources potential available and exploited in India ?
5. Describe the following renewables.
   (*a*) Thermal solar energy (*b*) photovoltaic convertors.
6. Write notes on :
   (*a*) Wind energy (*b*) OTEC
   (*c*) Biological conversion of solar energy.
7. How new energy sources are different from renewables? Explain geothermal energy, tidal energy and wave energy.
8. Justify the use of the following sources of energy in Indian context.
   (*a*) Small hydel projects (*b*) Draught animals.
   (*c*) Agricultural residues.
9. What are the advantages of use of renewable sources of energy ?
10. Explain the limitations of the use of renewable sources of energy.

❑❑❑

Chapter 3

ENERGY CONVERSION TECHNOLOGIES

Introduction ; Electricity Generation : Transport Energy; Rural Energy ; Limitations of Current Power Generation Systems ; Direct Energy Conversion Devices.

3.1. INTRODUCTION

The energy available in natural form is not directly usable. The fossil energy is chemical energy of fuels or nuclear energy of fissile materials. The hydro energy and wind energy are available as kinetic energy or potential energy of water and air streams. Solar energy is available as photon energy or thermal energy. The usable forms of energy are electricity for heating, lighting and motive power for industries, transport, domestic and commercial activities. Therefore, the natural forms of energy are to be converted into usable forms with the aid of suitable plants and technologies.

3.2. ELECTRICITY GENERATION

Electrical energy is most convenient form of usable energy and is produced in a power plant. There are many types of power plants.

1. Utility plants or Central power plants

These are large capacity power plants usually located near the natural source of energy. The power generated is transmitted over long distances to the consumers. Depending upon the input natural energy sources these plants may be thermal power plants, nuclear power plants or hydro power plants.

The flow diagram of thermal power plants is shown in Fig. 3.1. The flow diagram of a nuclear power plant is shown in Fig. 3.2. along with various stages of energy conversion. Similarly, different stages of hydro power plant are shown in Fig. 3.3.

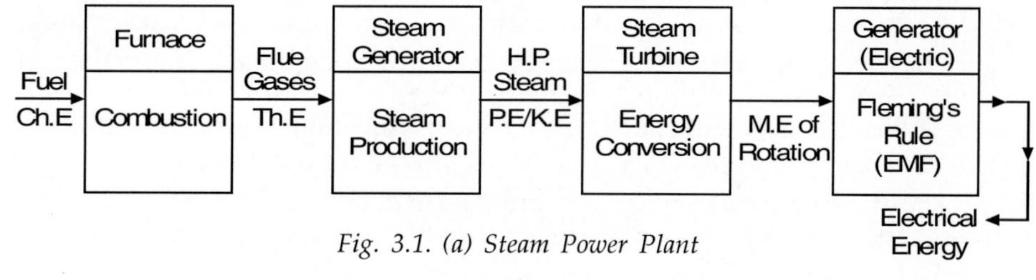

Fig. 3.1. (a) Steam Power Plant

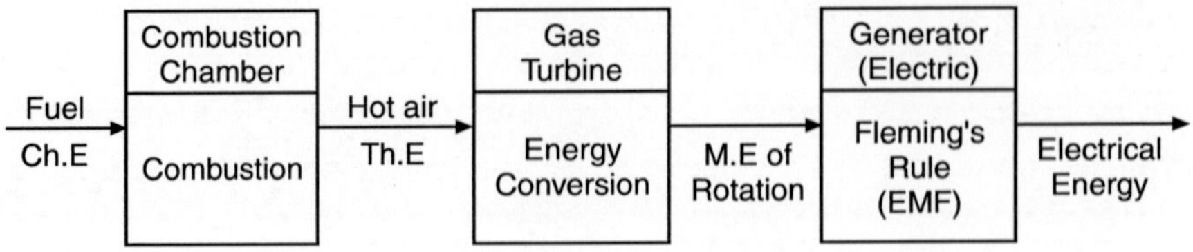

Fig. 3.1. (b) Gas Turbine Plant.

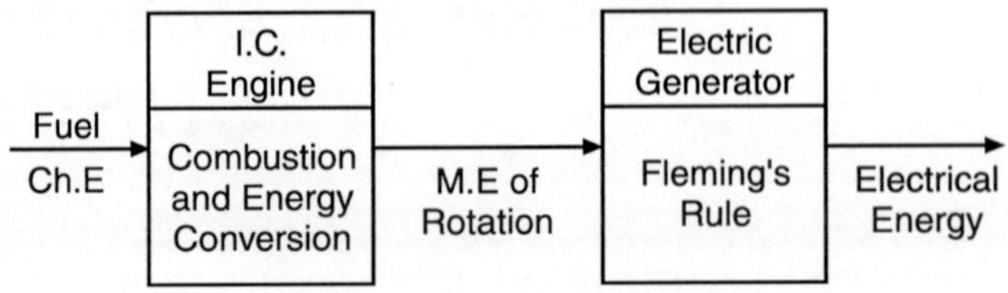

Fig. 3.1(c) Diesel Engine Plant.

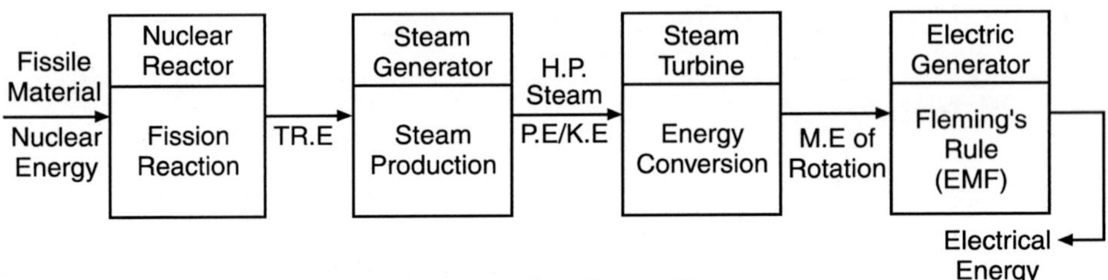

Fig. 3.2. Nuclear Power Plant.

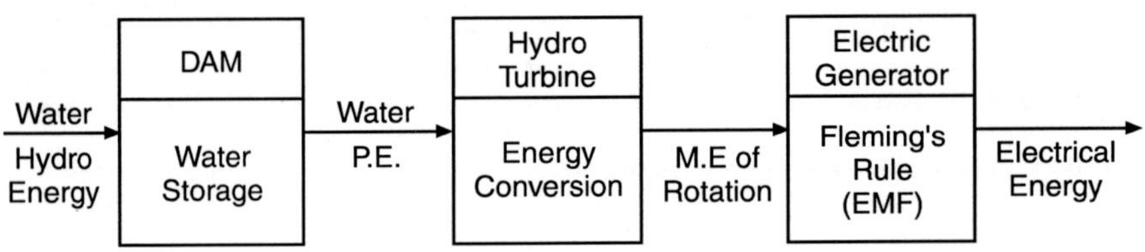

Fig. 3.3. Hydro Power Plant.

In a steam power plant, pulverised coal or lignite is burned in the furnace of a steam generator. The chemical energy of fuel is released as thermal energy during combustion and hot flue gases are used to generate steam. The enthalpy of high pressure steam is converted into mechanical energy in a steam turbine which in turn rotates an electric generator. The electricity thus produced is transmitted over long distances and distributed to the consumers for industrial, domestic and commercial applications.

In a gas turbine plant, natural gas or furnace oil is combusted in a combustion chamber to heat compressed air. The enthalpy of hot air is converted into mechanical energy is a gas turbine, which in turn rotates an electric generator. The electricity thus produced is transmitted over long distances and distributed to the consumers. In a diesel engine plant both combustion and energy conversion of heat into mechanical energy take place inside the cylinder of the engine. The electrical generator is rotated by the engine to produce electricity.

The nuclear power plant is similar to steam power plant where nuclear fission takes place in a nuclear reactor and thermal energy produced is used to raise steam in a steam generator which is further used in running a steam turbine as in a steam power plants.

In a hydro-electric power plant, water streams in hills/mountains are collected behind a dam. The potential energy of stored water is conduited to a hydro turbine where potential energy of water is used to rotate the turbine. The mechanical energy of rotation of turbine is used to run an electrical generator to produced electricity. The electricity generated is transmitted over distances and distributed to consumers of electric energy.

2. Captive Power Plants

Certain metallurgical and other industries are very sensitive to even short interruption in power supply. Similarly, hospitals and hotels cannot afford power failure even for a very short duration. These manufacturing and service industries are allowed to install their own power plants to meet their total power requirement. Such plants are called capture power plants. These are thermal power plants of smaller capacity. Long transmission lives are not required.

3. Cogeneration Plants or Total Energy Plants

There are many industries especially process industries like refineries, fertilizer plants, pharmaceuticals, organic and inorganic chemical plants, sugar mills, paper mills, etc. which require bulk of heat energy along with electrical power. Similarly, hospitals, hotels and other community centres need electricity as well as heat energy for various processes. The present system of buying power from the grid and installing a low pressure boiler is not efficient. The total energy requirements can be met from a single plant called co-generator plant at a very high thermodynamic efficiency.

The most common cogeneration plants use a high efficiency, high pressure boiler to generate steam for running a steam turbine. The required heat energy is supplied by medium or low pressure steam extracted from the turbine at suitable points A schematic diagram is shown in Fig. 3.4.

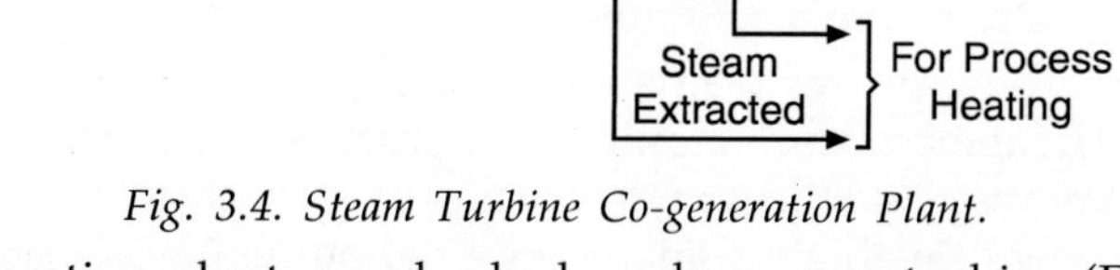

Fig. 3.4. Steam Turbine Co-generation Plant.

The cogeneration plant can also be based on a gas turbine (Fig. 3.5) where air exhausted from the turbine is utilized for process heating. Similarly in a diesel engine co-generation plant, exhaust gases from the engine and jacket cooling water can be used for meeting the heat requirements.

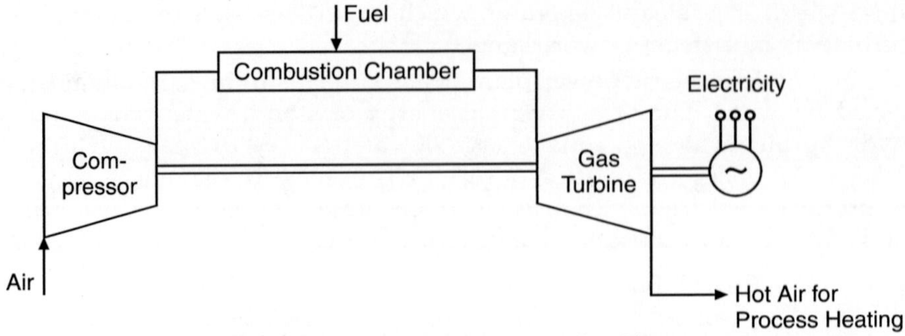

Fig. 3.5. Gas Turbine Co-generation Plant.

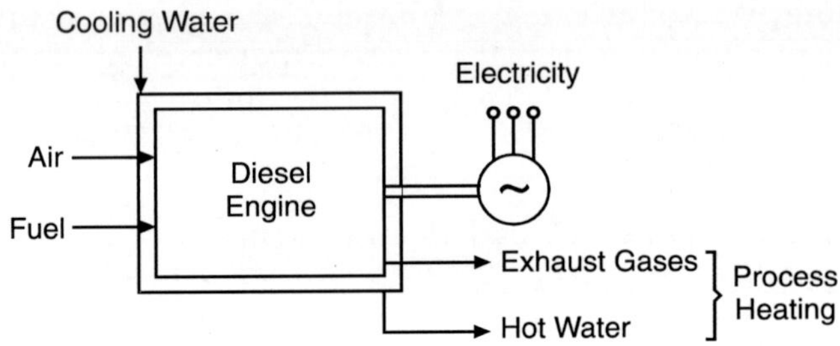

Fig. 3.6. Diesel Engine Co-generation Plant.

4. Autonomous Power Plants

These are small capacity power plants which normally work on local sources of energy. These plants generate electricity for local use and transmission losses are avoided. These plants can be :

(*i*) Wind Power Plants.
(*ii*) Solar Power Plants.
(*iii*) Biomass Power Plants.
(*iv*) Mini and Micro Hydro Power Plants.

These plants are normally conceived as hybrid system. The intermittency of solar energy and wind energy is compensated by standby power plants. The plant uses the solar energy or the wind energy as the main source of energy when available and sufficient, a diesel engine or small hydro power plant is used at other times. In such hybrid schemes, there is no need of storage of energy.

5. Combined Cycle Power Plants

The gas turbine power plants working on Brayton cycle are high temperature plants. The temperature at inlet to gas turbine may be of the order of 1200-1300 K. The exit temperatures of air are also very high of the order of 700–900K. In order to utilize this high grade energy the outlet gases may be used to raise steam in a boiler to run a steam turbine power plant. Such arrangements are called combined cycle power plants. Due to high temperature range of utilization of heat for the combined plant, the thermal efficiencies of gas turbine plant and steam turbine plant are almost added resulting in very high thermal efficiency of the combined cycle power plant. There can be many arrangements but a schematic diagram of a typical combined-cycle plant is shown in Fig. 3.7.

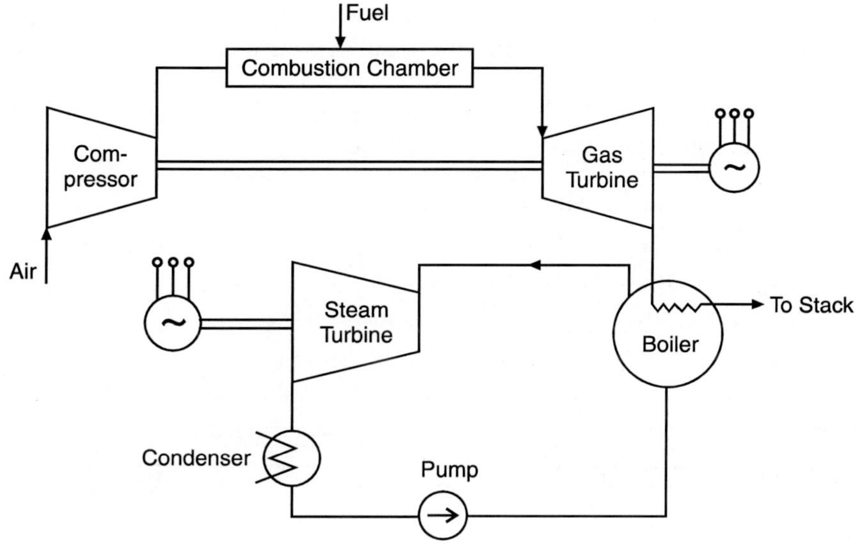

Fig. 3.7. Combined Cycle Power Plant.

3.3. TRANSPORT ENERGY

There are more than 500 million vehicles in the world. Cars and trucks alone consume one-third of world's oil production. There are aero-engines, marine engines, automobile engines, engines for earthmoving and material handling machines. The total capacity of these mobile engines or power packs is many times the total installed capacity of the power plants used for electricity generation.

The engines used for transport vehicles are gas turbines for ships and aircrafts and piston-cylinder internal combustion engines using petrol, diesel and compressed natural gasas fuel for other vehicles. In addition to consumption of these hydrocarbons, these engines emit green house gas causing thermal pollution and adversely affecting, the global climate. In addition, nitrous oxides are also emitted which are primary components of tropospheric ozone. The emission of carbon by vehicles cannot be absorbed by the present forests.

3.4. RURAL ENERGY

India is essentially an agricultural country with 78% of its population living in rural countryside and 70 % of its gross domestic product emanating from rural agricultural sector. There are more than 5 lakh villages scattered all over the country. About 95% of these villages are small villages with population less than 2000. On the average, the villages consume 150 kWh of energy per day for energising irrigation pumps, for lighting, for cooking, etc. As the demand factor is very low and because these villages are widely dispersed, it is most uneconomical to electrify them from power grids. Transportation of commercial fuels like coal, diesel and petrol over long distances is very difficult.

The development, welfare and prosperity of rural villages depend upon the availability of the following services.

1. Cooking energy is 80% of the basic rural energy needs. This amounts to 0.5 to 1 kWh per person per day.

2. Water for irrigation, drinking and washing : 25 m^3/hectate-day for drip irrigation of corn, wheat, cotton, millets ; 50–60 m^3/hectare-day for flood irrigation for rice and sugarcane ; 40–50 liters/head-day for livestock.
3. Domestic amenities : Cooking, drinking and wash water, lighting and heating as per Table 3.1.
4. Community services : Radio and T.V. sets, street lighting and cooperative milk chilling as per Table 3.2.
5. Utilities for village industries : Power, water and steam as per Table 3.3.
6. Food supply system.

Table 3.4. For Imperishable Food supply system

Table 3.5. For Perishable Food supply system

Table 3.6. For Animal Husbandry Food supply system.

Table 3.1. Domestic Amenities

S.No.	Amenities	Common mode of energy input (present)	Potential alternative sources (proposed)
1.	Cooking	Wood, cowdung	Bio-gas, solar cooker, biogas plate
2.	Drinking water.	Animal power human labour	Wind, water current energy, solar plumps, electricity.
3.	Wash water	Open surface water	Water pumps run by wind, solar, micro-hydro, electricity.
4.	Lighting	Kerosene	Bio-gas lanterns, solar lights, electricity.
5.	Heating	Wood, cowdung	Solar, biogas heaters.

Table 3.2. Community Services

S.No	Services	Common mode of energy input (present)	Potential alternative sources (proposed)
1.	Radio, T.V. sets	Nil	Solar, electricity
2.	Street lighting	Nil	Solar, biogas, electricity.
3.	Milk chilling	Nil	Solar refrigeration, electricity

Table 3.3. Utilities For Industries

S.No	Utilities	Common mode of energy input (present)	Potential alternative sources (proposed)
1.	Electricity	Nil	Micro-hydro, solar, wind, biogas generators.
2.	Water	Open ponds, canal system	Wind, solar, microhydro, biogas pump sets,
3.	Hot, water, steam	Wood, cowdung	Solar heaters/ boilers on biogas, biomass

Table 3.4 Imperishable Food Supply System.

Food Chain Activity	Common Mode of Energy Input	Potential solar device
Soil preparation (ploughing, harrowing)	Draught animals (Diesel tractor)	—
Sowing	Draught animals, Human labour, (Diesel tractor)	—
Weeding, pest control	Human labour (Diesel)	—
Irrigation, drainage	Draught animals.	Solar pumps
Fertilising, manuring	Draught animals, Human labour (Diesel tractor)	
On-farm transport	Draught animals, Human Labour (Diesel tractor)	
Threshing	Draught animals, Human labour (Electricity)	
Drying	Open sun drying	Solar dryers.
Storage	Natural ventilation	
Milling	Human labour, water power (electricity)	
Cooking	Wood, dung, agricultural residues.	Solar cookers

Note : Sources shown in brackets are available only in some villages.

Table 3.5. Perishable Food Supply System.

Food Chain Activity	Common Mode of Energy Input	Potential Solar Device	Remarks
Soil preparation (Ploughing)	Human labour, Draught animals (Diesel Tractor)	Green house	Rural activity
Sowing, planting	Human labour		-do-
Weeding, pest control	Human labour (Diesel)		-do-
Irrigation, drainage	Draught animals (electricity, diesel)	Solar pump	-do-
Fertilizing, manuring	Draught animals (Diesel)		-do-
Harvesting	Human labour		-do-

Contd...

Food chain Activity	Common Mode of Energy Input	Potential Solar Device	Remarks
Transport to markets	Draught animals (Diesel Tractor)		-do-
Storage (Refrigeration)	Electricity, diesel	Solar Cooling/ refrigerator	Presently urban activity can be shifted as rural activity.
Processing (Agro Industry	Electricity, coal, diesel	Solar dryers, boilers, heaters, refrigerators.	-do-
Marketing transport	Human labour (diesel)		Urban
Consumption (cooking, refrigerator)	Wood, gas, electricity	Solar cooker, refrigerator	Rural and urban.

Note : Sources shown in brackets are available only in some villages.

Table 3.6. Animal Husbandry Food Supply System

Food chain Activity	Common mode of energy input	Potential solar device	Remarks
Food collection	Milk, meat		Rural Activity
Transport	Human labour, draught animals, diesel vehicles		Rural
Storage, chilling	Refrigeration (electricity)	Solar milk cooler.	Presently urban activity can be shifted as rural activity
Processing (chilling, heating)	Electricity, diesel, coal	Solar refrigerator heater, boiler,	-do-
Marketing, distribution	Human labour, diesel		-do-
Consumption, refrigeration, cooking	Wood, gas, electricity	Solar cooker, refrigerator	Rural & Urban

3.4.1. Local Energy Resources in Villages

The following renewable sources of energy are generally available in the villages which can be tapped and converted into usable energy.

1. Solar energy : 4.5-7.0 kWh/m^2 for 250-300 days/year.
2. **Water current energy** in the rivers and irrigation canals : Low/zero head turbines of 200 kW – 1000 kW for power generation and lift irrigation. Total potential of Indian villages : 5000 MW.
3. **Biomass :** Agricultural residues 200 million tonnes/year plus wood 130 million tonnes/year for power generation of 75,000 MW by combustion or gasification.

4. Animal Wastes : 13.5 kg/day cowdung per cattle to produce 0.46 nm^3 biogas which can be used to generate 1 kWh of electricity.
5. Wind energy : 100 kWh/m^2 to 500 kWh/m^2 of turbine swept area to generate 20,000 MW of electricity or use for irrigation water.

3.4.2. Alternative Available Technologies

The following conversion devices have been developed and are available for use in rural areas.

1. Solar Cookers : Box type solar cookers ; Hot plate type solar cookers, solar ahara, bio-gas plates.

2. Water Pumping :

(*i*) NPL Abhimanyu solar water pump of 1 kW rating using freon R–114 priced at Rs. 37000/-

(*ii*) CEL solar photovoltaic water pumping system. 30–40m^3/day of water capacity, total head 10m, priced at Rs 40,000/-.

(*iii*) NAL wind mill water pumping system of 5m diameter to lift 23m^3/day of water from a depth of 20m at wind speed 10m/s, priced at Rs. 11,750.

(*iv*) Bharat Biogas engine of 3.7 kW/1500 rpm, 100 × 100 mm centrifugal pump, priced at Rs. 16500/-.

(*v*) Jyoti 5 kW/1500 rpm, biogas engine with 100 × 100 mm centrifugal pump priced at Rs 37,000/-.

(*vi*) Micro-hydro pumping system as per site conditions.

3. Solar Devices : Photovoltaic solar system for street lighting TV sets,. Solar dryers, kilns, air and water heaters, distillation plants, solar refrigerator, milk cooler, vaccine freezer.

4. Hot water, steam and electricity generation plants.

Thermal solar, photovoltaic, biogas engines ; biomass plants, windmills ; micro-hydro plants.

3.5. LIMITATIONS OF CURRENT POWER GENERATION SYSTEMS

The current power generation systems of large capacities suffer from some serious disadvantages and price of electricity from the power grid is high. The power supply at the same time is not reliable.

1. Carnot Limitation

The steam turbine power plants based on coal or nuclear fuels, gas turbine plants and diesel engine plants work on the principle of conversion of thermal energy into electrical energy. The maximum possible cycle efficiency is limited by Carnot efficiency of $\frac{T_{max} - T_{min}}{T_{max}}$. Specific heat rates of most modern thermal power stations is in the range of 8400 – 10,000 kJ/kWh generated and about 65–70% of the heat available from the fuel is lost to the cooling water in the condensing plant and boiler stacks. The overall achievable efficiency of 40% of a modern central power station is called Carnot limitation.

2. Matallurgical Limitation

T_{min} is site specific and depends upon the temperature of cooling water available. T_{max} is limited by the availability of technical materials for boiler tubes and turbine blades. This is called metallurgical limitation.

3. Reliability and Mechanical Links

There are many mechanical links in modern thermal and hydro power plants in the form of boilers, penstocks, turbines, pumps, generators. There is loss of reliability and mechanical losses get added up with every additional component specially rotating systems. The energy output in India is only 3600 kWh/kW installed capacity.

4. Ecological Balance

The modern power plants are large size plants and disturb the ecological balance of the site. The need of a large dam in a hydro power plant results in submergence of large fertile areas under the water reservoir, displacement of huge population, loss of crops and forests, adverse effect of high water table of very large areas damaging the corps. The thermal discharges from a thermal power plant into rivers or lakes serving as source of cooling water and fly ash discharges from the stacks can adversely change the environment.

5. Plant Costs

Huge capital requirement (Rs 5000/kW installed) on boilers, turbines, generators, cooling water treatment and supply system, cooling towers, etc, and availability of fuel and cooling water are main constraints in siting thermal power plants. The construction costs of a dam for a hydro power plant are very high (Rs. 40,000-50,000/kW installed). The gestation period is many years.

6. Transmission and Distribution

One of the most expensive form of energy transmission is that of electric power from power plants. A huge investment in procuring rights of way, installing towers, erecting lines and providing protection against lighting and other problem is required. The erection and repair costs for overhead lines are very high and overhead lines are prone to breakdown disrupting power supply on a large scale.

In addition to high investment costs, considerable power losses occur during transmission. In India, the transmission losses are as high as 18 to 20% against 8 to 10% in developed countries. This is due to high pilferages and voltage drops. The investment costs on transmission lines in India constitute 30 % of power generation plant costs whereas in developed countries it is 100% of power generation plant cost.

The power distribution system can be very complex for a metropolitan city or industrial estate. In addition to huge investments, the erring distribution system causes frequent disruptions in power supplies resulting in production losses.

3.6. DIRECT ENERGY CONVERSION DEVICES

Many of the disadvantages of current power generation system can be overcome by small power packs not subject to Carnot limitation, free from rotating prime movers and generators installed locally at the consumer points not needing transmission and distribution systems or cooling water.

3.6.1. Fuel Cell

If an electric current is passed through a dilute solution of an acid or an alkali by means of two platinum electrodes, hydrogen is produced at the cathode and oxygen is evolved at the anode. It this process is reversed by removing the power supply and connecting the two electrodes through a suitable resistance, the presence of hydrogen at one electrode and oxygen at the other will produce a small current in the external circuit, water being produced as a by-product. This reverse process of electrolysis is the essence of the fuel cell technology as chemical energy stored in hydrogen and oxygen have been combined to produce electricity. The fuel cells donot have moving components and it is quieter, require less maintenance and attention in operation. Fuel cells convert chemical energy directly to electrical energy at room temperatures. These are very efficient and are not subject to Carnot limitation.

Hydrogen and oxygen were selected as reactants for the fuel cells used for spacecraft power supplies because of relatively high reactivity of hydrogen. The reactants for commerical fuel cells can be cheap and readily available natural gas or petroleum derivatives as fuel and air as oxidants.

The fuel cells can be installed at individual apartments, commercial buildings as on-site power units. The local trains and buses, trucks or personal cars can be propelled by methanol fuel cells.

3.6.2. Solar Cell

The solar cell utilizes energetic photons of the incident solar radiations directly on a p-n junction to produce electricity at high conversion efficiencies. The technology of solar cell is well developed as a part of satellite and space-travel technology. It is also a direct conversion technology and does not use a working fluid like steam or gas and are not subject to Carnot limitation. It is simple, convenient and, lacking moving parts, dependable. The solar cell is modular, so that arrays of identical modules can be assembled to meet various power needs ranging from small residential systems installed on roof tops to relatively large central systems.

Efforts at reducing costs include search for lower-cost cell-manufacturing techniques and new and cheaper materials than silicon.

3.6.3. Thermoelectric Power Units

When two dissimilar materials are joined together to form a loop and the two junctions are maintained at different temperatures, an emf will be developed around the loop. This is called Seeback effect. This principle is used in thermocouples to measure temperatures. The electricity produced by the device is proportional to the temperature difference between hot junction and cold junction and therefore is subject to Carnot limitation. But the device is very simple, without any moving part and hence very dependable without much need of operation and maintenance attention. Due to material constraints, the present conversion efficiencies are very low of the order of 3%. With the development of new materials in the field of ceramics and semi-conductors, the thermoelectric generator is projected with high potential in the field of base load and peak load power generation. The present applications are restricted to small units operating in remote areas as signalling and navigational device.

The cold junction is maintained either by cooling water or atmospheric air and the source of heat can be oil or gas burner, direct solar radiators by paraboloidal concentrator, heat of radio-active decay of isotopes.

3.6.4. Thermionic Converter

A Thermionic generator converts heat directly into electricity by using thermionic emission. An emitter which is a metallic cathode is heated till electrons cross the surface Fermi barrier and are collected at the opposite colder anode. The container holding the two electrodes is filled with ionized cesium vapour to minimize energy losses of the electrons crossing the gap between the electrodes. The emitter is positively charged and the collector is negatively charged. It is a low-voltage high current device limited by Carnot's law. Thermal efficiencies of 10 – 20% have been realized and higher efficiencies are possible in future.

This device is being designed for space power applications. For the time being it is used to supply power to boats, power tools and irrigation pumps.

3.6.5. Magneto-Hydro-Dynamic Generator

MHD is a direct heat-to-electricity conversion technique based on Faraday Law that when an electric conductor moves across a magnetic field, a voltage is induced in it which produces an electric current. Here the conductor is an ionized gas which is passed at high velocity through a powerful magnetic field, a current is generated and can be extracted by placing electrodes in a suitable position in the stream. It produces d.c power directly.

MHD is the most promising direct conversion technology where the mechanical link can be by passed. It can overcome some of the limitations of conventional power generation. By hybridization of MHD and thermal power plants, the efficiency can be raised from 40% to 55%, thus better utilizing the fuel resources and reducing the conventional pollution.

TUTORIAL SHEET No. 3.1.

1. Discuss various types of power plants currently used for bulk electricity generation.
2. Write a detailed note on rural energy, requirements and resources.
3. Discuss the following aspects of rural energy.
   (*i*) Local energy resources available in villages
   (*ii*) Alternative technologies.
4. Draw the flow charts for the following food chains. Also indicate where solar energy can be fruitfully utilized.
   (*i*) Imperishable food supply system
   (*ii*) Perishable food supply system
   (*iii*) Animal husbandry food supply system.
5. What are the main limitations of current power generation system ? How these limitations can be overcome by direct energy generators ?
6. What are the basic principles of the following technologies ?
   (*i*) Solar cell
   (*ii*) Fuel cell
   (*iii*) Thermoelectric generator
   (*iv*) Thermionic generator
   (*v*) MHD generator.

PART 11

RENEWABLE ENERGY SYSTEMS

Chapter 4

WIND ENERGY

Introduction : Wind Power Plant ; Classification of Wind Power Plants ; Principles of Power Generation ; Power Coefficient of Wind Turbines : Forces Acting on Blades ; Wind Turbines Operation and Control : Site Selection, Specification of Wind Turbines ; Off-shore Wind Farms ; New Developments, Wind Diesel Hybrid System, Novel Wind Turbines.

4.1. INTRODUCTION

The winds account for a power of 2×10^{13} W. Though this is only a small portion of the incident solar energy when compared, but it is more than the total energy consumption rate of the world. It is estimated that just one percent of potential wind power is equivalent to 2×10^{11} W or approximately three percent of the current world energy consumption rate. If wind power were used for the generation of electricity, its thermal equivalent would be as high as 8–9 per cent of total. Technical and economic problems tend to limit the attractiveness of wind energy conversion systems, but the potential extractable wind power is not insignificant.

Following a better understanding of aerodynamics of rotating aerolifts, there is a revival of interest in harvesting wind energy. The importance of wind as alternative source of energy is increasing with the depletion of fossil fuels and the need to maintain our ecosystem. The Government of India has formulated and launched a special wind energy programme. At a conservative assessment, wind power potential in India is around 20,000 MW. At present, an aggregate wind power capacity of about 2002 MW has been established in the country and it is proposed to establish additional capacity in the coming years. Although in many countries and some parts of India, electric power is being commercially produced from windmills, wind energy can be more appropriately utilized for pumping water for irrigation due to high variation of power density of wind over time and location. In India, windmills have been tried for pumping water for drinking, irrigation for fodder crops, forest nurseries, and agro-forestry plantations. But, so far, wind mills have not been tried among individual farmers and in the irrigation of annually cultivated crops. There is a large potential of ground water, especially to operate sprinkler systems.

4.2. WIND POWER PLANT

The wind energy is the kinetic energy of the airflow. The total power of a wind stream is equal to the rate of the incoming kinetic energy of the stream.

$$P_{tot} = \dot{m}\ kE_i$$

$$= \dot{m}\frac{V_i^2}{2}$$

where P_{tot} = total wind power [W]
$\dot{m}$ = mass-flow rate [kg/s]
V_i = incoming velocity [m/s]
KE_i = kinetic energy of incoming stream

The mass flow rate is given by the continuity equation,

$$\dot{m} = \rho A V_i$$

where ρ = incoming wind density [kg/m³]
A = cross-sectional area of stream [m²]

$$\therefore \quad P_{tot} = \frac{\rho A V_i^3}{2}\ [W]$$

The total power of a wind stream is directly proportional to its density, area and the cube of velocity.

The wind velocity (V_i) increases with the height above the earth surface. If V_i is measured at 10m height where air density is 1.205 kg/m^3 (at 20°C and 1.013 bar), the power density will be as follows :

V_i	P_{tot}
3 m/s	16.27 W/m^2
6 m/s	130.14 W/m^2
7 m/s	439.22 W/m^2

Due to comparatively low power density of air stream, wind power plants are required to be built with a correspondingly large rotor size for large power outputs. Presently, a wind power plant of 2 MW capacity with a rotor diameter of 60 m is in operation in Denmark.

Wind speeds increase with height because of reduction in the drag effect of the earth's surface. The tower height is usually 1 to 2 times the rotor diameter.

The wind velocity across the turbine decreases since kinetic energy is converted into mechanical work. Two types of forces act on the blade of the wind turbine rotor. The lift force acts perpendicular to the direction of air flow as circumferential force in the direction of the turbine rotation that provides the torque. The drag force acts perpendicular to lift force as axial force in the direction of the wind stream that provides an axial thrust. This must be counterbalanced by proper mechanical design.

The main components of a wind power plant are :

1. **Wind mill** of a suitable size and shape. The blades are mounted on the hub of the rotor which has a provision inside to adjust the pitch of the blades to suit the wind velocity.
2. **Wind direction follow up device :** The horizontal axis wind turbines are rotated about a vertical axis with the help of a motor so that maximum swept area of the rotor faces the wind direction. In small turbines, yaw action is controlled by a tail vane.

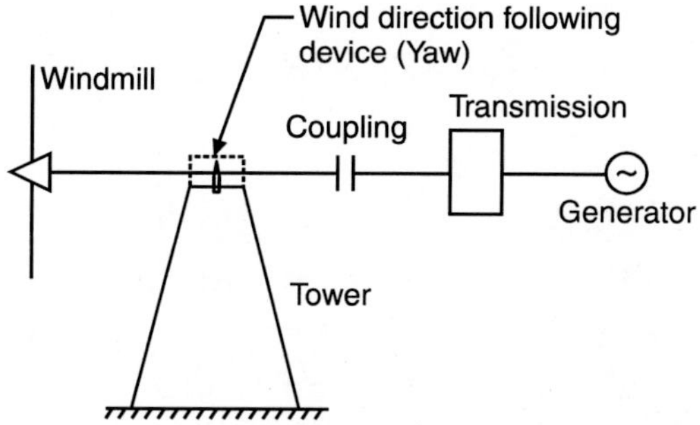

Fig. 4.1. Wind Power Plant.

3. **Coupling:** The turbine rotor is connected to the generator through a hydraulic transmission system.
4. **Hydraulic Transmission :** The speed of wind turbines is controlled by varying the pitch of the rotor blades. The normal speed of rotation is 40 to 50 rpm. It has to be increased to match the generator speed of 1800 rpm. A mechanical gear box is used for top mounted equipment. For bottom mounted equipment hydraulic transmission is used which gives a high degree of design flexibility and large savings in cost.
5. **Electric generator** : For large systems, synchronous generators are used
6. **Tower** : Truss tower is widely used for wind mills.

4.3. CLASSIFICATION OF WIND POWER PLANTS

The wind power plants can be classified as per different parameters.

1. Axis of Rotation

There are vertical axis and horizontal axis wind turbines. They use either lift or drag forces to harness the wind energy. The majority of modern wind turbines are the horizontal axis lift machines.

Fig. 4.2. shows simple diagrams of different types of wind turbines.

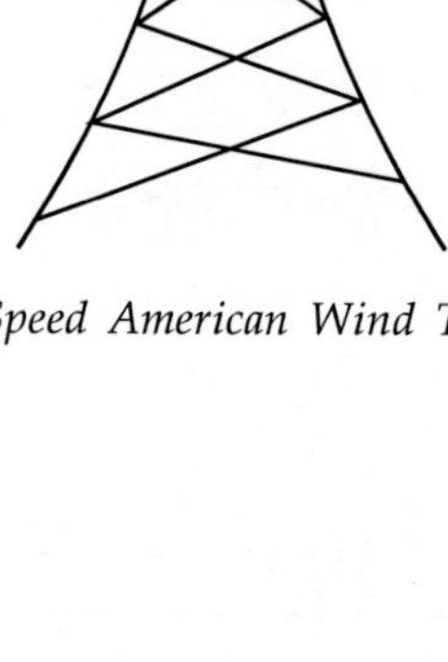

(a) Low Speed American Wind Turbine.

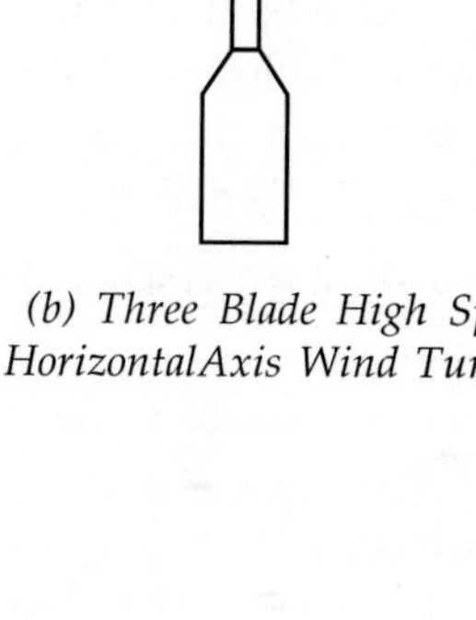

(b) Three Blade High Speed HorizontalAxis Wind Turbine.

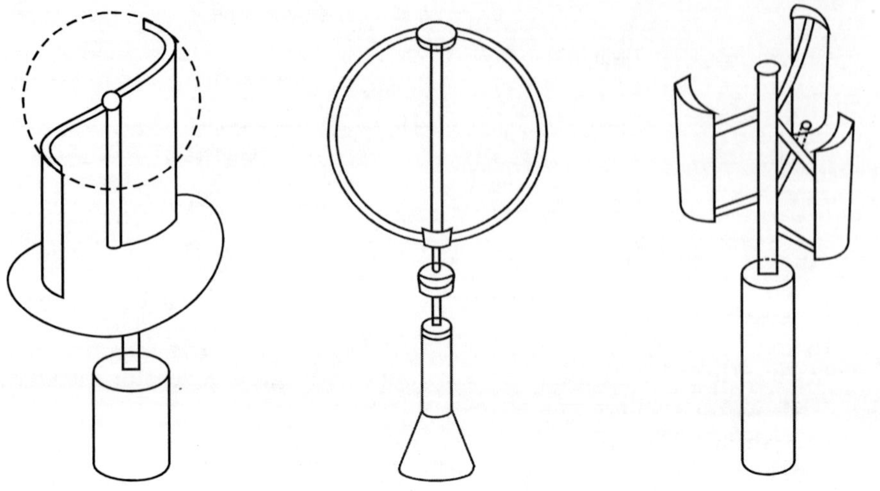

Fig. 4.2. (b) Vertical Axis Wind Turbines

2. Size of Machine

According to power output, wind power plants are divided into the following classes :

1. Small Wind Power Plants. The power output of 10 to 50 kW and a rotor diameter of 1 to 16m.

2. Medium Wind Power Plants. The power output of 50 to 500 kW and a rotor diameter of 16 to 50 m.

3. Large Wind Power Plants. The power output of 500 to 5000 kW and a rotor diameter of 50 to 130 m.

The tower (hub) height is usually 1 to 2 times the rotor diameter.

3. Applications

The main applications of wind plants are :

1. Grinding grains
2. Drainage
3. Pumping
4. Saw milling
5. Electrical generation

The wind power plants are also classified as

1. Small Machines. For autonomous power generation for local use, wind farms, homes and rural use.

2. Large Machines. Power generation for central grid system.

4.4. PRINCIPLES OF POWER GENERATION

The wind mill works on the *principle of momentum.* The work done by the turbine rotor is the difference between the kinetic energies of incoming and outgoing streams through the rotor.

A horizontal-axis, propeller type wind mill is shown in Fig. 4.3. The thickness of wheel is *a-b*. The pressure and velocity changes are also plotted.

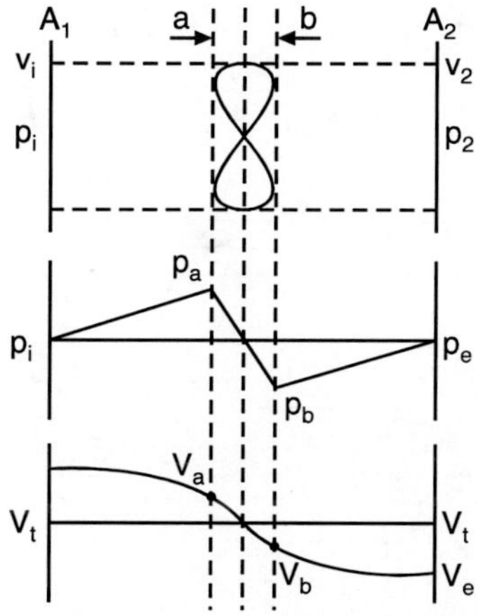

Fig. 4.3. Pressure and Velocity Profiles for Air Flow through a Wind Turbine.

The exit velocity V_e is less than V_i because kinetic energy is extracted by turbine. The pressure p_e is almost equal to p_i. Applying total energy equation, and taking air density ρ = constant.

$$PE_1 = PE_2$$

$$p_i v + \frac{V_i^2}{2} = p_a v + \frac{V_a^2}{2}$$

$$p_i + \rho\frac{V_i}{2} = p_a + \rho\frac{V_a^2}{2}$$

where $$v = \frac{1}{\rho}$$

For exit area,

$$p_e v + \frac{V_e^2}{2} = p_b v + \frac{V_b^2}{2}$$

$$p_e + \rho\frac{V_2^2}{2} = p_b + \rho\frac{V_b^2}{2}$$

Now $$V_i > V_a$$

and $$V_b > V_e$$

$$p_a > p_i$$

$$p_b < p_e$$

$\therefore$ $$p_a - p_b = p_i + \rho\left(\frac{V_i^2 - V_a^2}{2}\right) - \left(p_e + \rho\frac{V_2^2 - V_b^2}{2}\right)$$

As $$p_e = p_i$$

$$V_t = V_a = V_b$$

[velocity does not change within the turbine]

$\therefore$ $$p_a - p_b = \rho\left(\frac{V_i^2 - V_e^2}{2}\right)$$

If A is the projected area of windmill perpendicular to the wind stream, the axial force,

$$F_x = (p_a - p_b)\,A$$

$$= \rho A\left(\frac{V_i^2 - V_e^2}{2}\right)$$

The axial force is also equal to change of momentum.

$$F_x = \Delta(\dot{m}\,u)$$

$$\dot{m} = \rho A\,V_t$$

$$\therefore \quad F_x = \rho A\,V_t\,(V_i - V_e)$$

where

$$V_t = \left(\frac{V_i + V_e}{2}\right)$$

Now steady flow work,

$$W = KE_i - KE_e = \frac{V_i^2 - V_e^2}{2}$$

The power produced is rate of work.

$$\therefore \quad P = \dot{m}\left(\frac{V_i^2 - V_e^2}{2}\right)$$

$$P = \frac{1}{2}\rho\,AV_t(V_i^2 - V_e^2)$$

$$P = \frac{1}{4}\rho A(V_i + V_e)\,(V_i^2 - V_e^2)$$

For maximum power,

$$\frac{dP}{dV_e} = 0$$

$$\therefore \; 3V_e^2 + 2V_i\,V_e - V_i^2 = 0$$

$$V_{e_{opt}} = \frac{V_i}{3}$$

$$\therefore \quad P_{\max} = \frac{8}{27}\rho AV_i^3$$

The maximum efficiency,

$$\eta_{\max} = \frac{P_{\max}}{P_{\text{tot}}} = \frac{8}{27} \times 2$$

$$= \frac{16}{27} = 0.5926. \approx 60\%$$

Therefore actual power,

$$P = \eta P_{\text{tot}} = \eta\,\frac{AV_i^3}{2}$$

where η varies between 30 to 40% for real turbines.

Example 4.1. *A 10m/s wind is at 1 standard atmosphere and 15° C. Calculate :*

1. *The total power density in the wind stream.*
2. *The maximum obtainable power density.*
3. *A reasonably obtainable power density.*
4. *Total power produced if the turbine diameter is 120 m.*

Solution : The air density,

$$\rho = \frac{p}{RT}$$

$$= \frac{1.01325 \times 10^5}{287\,(15 + 273)} = 1.226 \text{ kg/m}^2$$

1. Total power density

$$\frac{P_{\text{total}}}{A} = \frac{\rho V_i^3}{2}$$

$$= \frac{1.226 \times (10)^3}{2} = 613 \text{ W/m}^2$$

2. Maximum power density,

$$\frac{P_{\text{max}}}{A} = \frac{8}{27}\rho V_i^3$$

$$= \frac{8}{27} \times 1.226 \times (10)^3 = 363 \text{ W/m}^2$$

3. Assuming $\eta = 40\%$

Actual power density,

$$\frac{P}{A} = 0.4\left(\frac{P_{\text{tot}}}{A}\right)$$

$$= 0.4 \times 613 = 245 \text{ W/m}^2$$

4. Total power produced,

$$P = \left(\frac{P}{A}\right)\frac{\pi D^2}{4}$$

$$= 0.245 \times \frac{\pi (120)^2}{4} = 2770 \text{ kW.}$$

4.5. POWER COEFFICIENT OF WINDMILLS

The wind velocity changes along the length of a blade depending upon the blade inlet angle and blade velocity. The coefficient of performance or power coefficient C_p is defined as the actual power to the maximum wind power.

$$C_p = \frac{P}{P_{\text{max}}} = \frac{\left(1 + \frac{V_e}{V_i}\right)\left(1 - \frac{V_e^2}{V_i^2}\right)}{2}$$

The ratio of blade tip speed to wind speed,

$$\lambda = \frac{U}{V}$$

Maximum value of C_p is equal to 0.593 and is called Betz limit. The value of C_p for most of the turbine rotors lie between 0.25 to 0.45 and is plotted in Fig. 4.4. for various types of rotors.

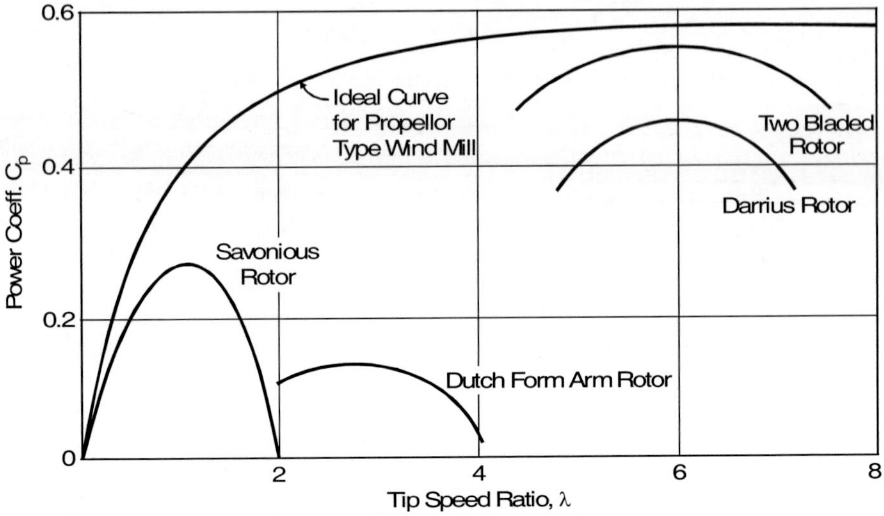

Fig. 4.4. Power Coefficient C_p vs λ.

4.6. FORCES ON THE BLADES AND TORQUE OF WIND MILL

Aerofoil Design

The wind turbine rotor has an aerofoil shape. There is a pressure difference on the two surfaces of the blade as air flows causing the force of lift F_L. There is also force of drag F_D due to obstruction to the air flow. The torque causing the rotation is due to force of lift. The drag force causes the axial thurst on the blade which has to be overcome by the bearing.

The torque causing the rotation of the wind turbine shaft depends upon the turbine rated power output and rotor angular velocity.

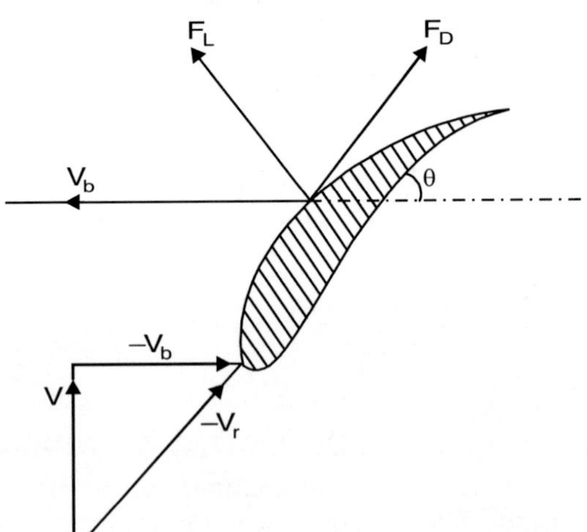

Fig. 4.5. Cross-section of wind turbine blade.

$$T = \frac{P}{\omega}$$

$$= \frac{P}{\pi D N}$$

where T = Torque, [N-m]

ω = Angular velocity [rad/s]

D = Diameter of turbine wheel

$$= \sqrt{\frac{4A}{\pi}} \text{ [m]}$$

N = Wheel speed [S^{-1}]

$$\therefore \quad T = \eta \frac{1}{8} \frac{\rho D V_i^3}{N}$$

$$T_{max} = \frac{2}{27} \cdot \frac{\rho D V_i^3}{N}$$

where $\eta_{max} = \frac{16}{27}$

The axial force or axial thrust,

$$F_x = \frac{1}{2} \rho A \left(V_i^2 - V_e^2\right)$$

$$= \frac{\pi}{8} \rho D^2 (V_i^2 - V_e^2)$$

$$F_{x_{max}} = \frac{\pi}{9} \rho A V_i^2$$

$$= \frac{\pi}{9} \rho D^2 V_i^2$$

Example 4.2. *A propeller wind turbine has a diameter of 120m and runs at 40 rpm. The wind at 1 standard atmosphere and 15°C has a velocity of 15m/s. Calculate:*

(i) the total power density in the wind stream,

(ii) the maximum obtainable power density,

(iii) a reasonably obtainable power density,

(iv) the total power, and

(v) the torque and the thrust.

Solution : The air density,

$$\rho = \frac{p}{RT}$$

$$= \frac{1.01325 \times 10^5}{287(15 + 273)} = 1.226 \text{ kg/m}^3.$$

(*i*) Total power density,

$$\frac{P_{total}}{A} = \frac{\rho V_i^3}{2}$$

$$= \frac{1.226 \times (15)^3}{2} = 2068.87 \text{ W/m}^2$$

(*ii*) Maximum power density,

$$\frac{P_{max}}{A} = \frac{8}{27}\rho V_i^3$$

$$= \frac{8}{27} \times 1.226 \times (15)^3 = 1226 \text{ W/m}^2.$$

(*iii*) Assuming efficiency, $\eta = 35\%$

Actual power density,

$$\frac{P}{A} = 0.35 \times 2068.87 = 724 \text{ W/m}^2$$

(*iv*) Total power produced,

$$P = \text{Power density} \times \text{Area}$$

$$= 724 \times \frac{\pi}{4}(120)^2$$

$$= 8184 \times 10^3 \text{ W} = 8184 \text{ kW}$$

(*v*) Torque at maximum efficiency

$$T = \frac{2}{27}\frac{\rho D V_i^3}{N}$$

$$= \frac{2}{27} \times \frac{1.226 \times 120 \times (15)^3}{(40/60)} = 55170 \text{ N}$$

(*vi*) Maximum axial thrust

$$F_{x_{max}} = \frac{\pi}{9}\rho D^2 V_i^2$$

$$= \frac{\pi}{9} \times 1.226 \times (120)^2 \times (15)^2$$

$$= 1385870 \text{ N}$$

4.7. WIND TURBINE OPERATION AND CONTROL

The power output of a wind mill is proportional to square of its diameter and cube of the wind velocity.

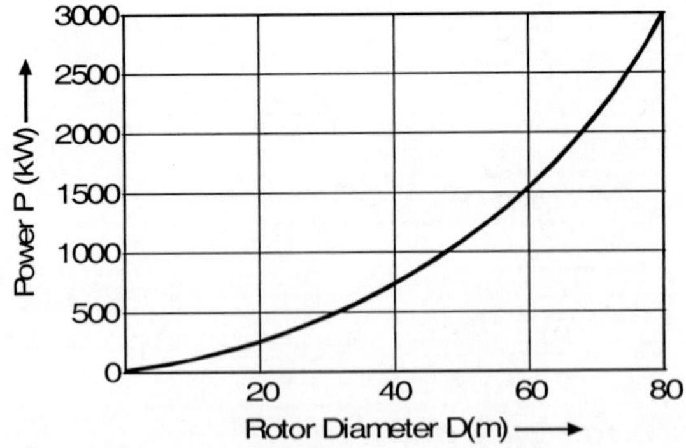

Fig. 4.6(a) Power Output

A small fluctuation in wind velocity will lead to large fluctuations in power. For example, a 20% drop in wind velocity would result in the loss of half the power. The maximum machine efficiency occurs over a relatively narrow power range.

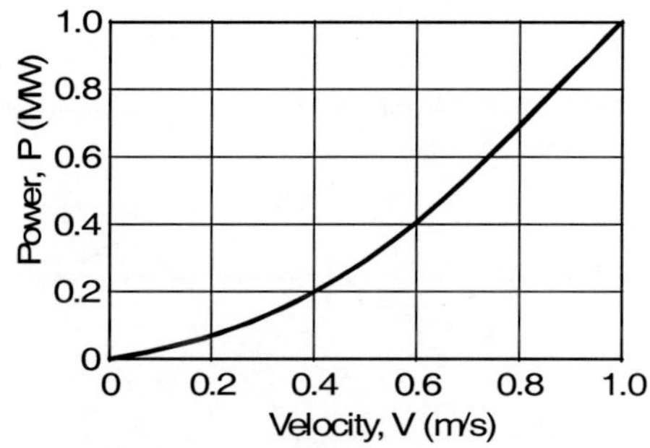

Fig. 4.6.(b) Power-velocity Characteristics.

1. Severe fluctuations in power due to variations in wind velocity result in
   (*a*) Power oscillations on the grid.
   (*b*) Severe strains on the windmill hardware.
2. If a wind turbine is designed corresponding to maximum prevailing velocity at a given site, then
   (*a*) Low powers are produced most of the time,
   (*b*) Capacity of the turbine and generator are unutilized.

3. Flat Rating

It is more cost effective to design a windmill to produce rated power at less than the maximum prevailing wind velocity *i.e.*, using a smaller turbine and generator and to maintain a constant output at all wind speeds above the rated capacity. This is called flat rating.

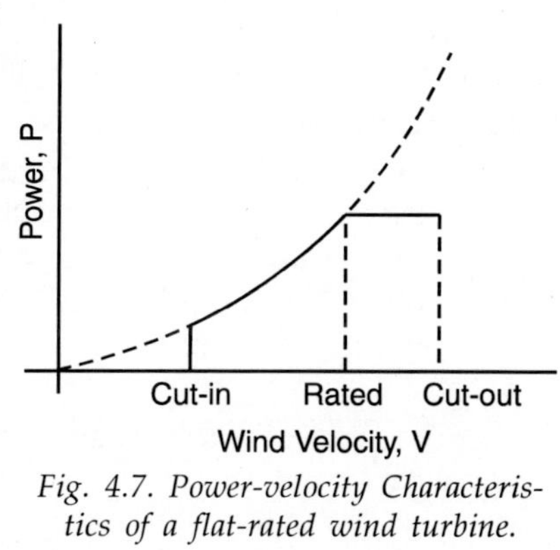

Fig. 4.7. Power-velocity Characteristics of a flat-rated wind turbine.

4. Cut-in velocity

Because of severe loss in efficiency and power at low wind speeds, a wind turbine is designed to come into operation at a minimum wind speed called the cut-in velocity.

5. Cut-out velocity

To protect the turbine wheel against damage at very high wind velocity, it is designed to stop operation (by feathering the blades) at a cut-our velocity.

6. The wind turbine operates with variable load over a narrow range between the cut-in velocity and rated velocity and at constant power between the rated velocity and cut-out velocity. The turbine ceases to operate below cut-in velocity and above cut-out velocity.

Example : Performance data of a wind turbine.

1. Type : Three bladed propeller type

2. Rate power output : 600 kW.
3. Rated wind velocity : 14.5 m/s.
4. Cut-in velocity : 4.5 m/s
5. Cut-out velocity : 25 m/s.

7. Load Curve

The performance of a wind power plant is characterized by the annual energy production for which annual load curves are plotted as shown in Fig. 4.8.

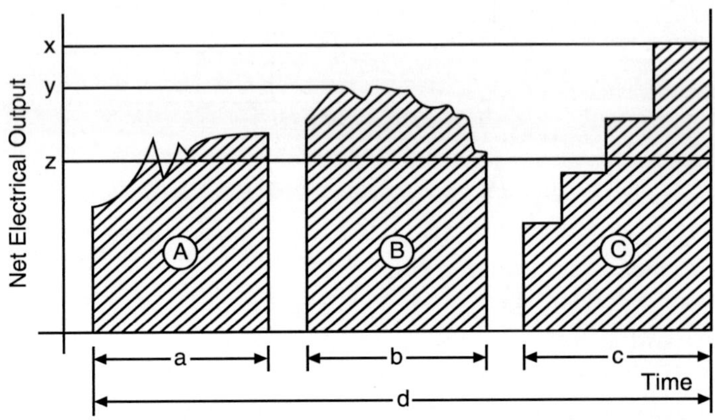

Fig. 4.7. Annual Load Curve of a Wind Power Plant.

8. Availability Factor

It is the fraction of time, during a given period, that the turbine is actually on line. For wind power plants, availability factor is usually 90%.

9. Overall Load Factor

It is the ratio of the total energy generated during a given period of time to the total rated generation capacity during the same period. This factor takes into account operations at less than rated wind velocity, nonoperation below cut-in and above cut-out velocities, and power outages caused by various situations such as repairs, maintenance, etc. The overall load factor is of the order of 30-40%.

The overall load factor is also called the plant operating factor or plant capacity factor. The wind turbines don't always operate at rated power because of changing wind velocities, the overall load factor is much lower than the availability factor.

Refering to Fig. 4.8.

X = rated net capacity of plant.

Y = *peak load during period, d*

Z = average load during period, d.

Plant overall load factor or plant operating factor

$$= \frac{(A+B+C)}{X.d}$$

$$\text{Plant load factor} = \frac{Z}{Y} = \frac{\text{average load}}{\text{peak load}}$$

Plant availability factor $= \dfrac{a+b+c}{d}$

where d is usually taken as 1 year.

4.8. SITE SELECTION

Siting of wind mills depends upon its application and the wind characteristics of the location. Small wind turbines of less than 100 kW capacity are usually used for local power generation and water pumping. Large wind turbines of 100 kW capacity or greater are used in wind farms to generate power for distribution in central power grids. The objective is to minimize the cost of power production.

Suitable sites for large wind mills around the world depend upon favourable wind activity. Maps showing annual average wind-power density, W/m^2, are available.

4.8.1. Mean Wind and Energy Velocities

The wind-power potential can be estimated from the mean velocity, $\overline{V}$, which is based on measurement over a period of time.

Mean wind velocity,

$$\overline{V} = \frac{\sum_i^n V_i}{n}$$

where $\sum_i^n V_i$ = sum of all velocity observations

n = No of observations.

Power plant sizing would be grossly under estimated if it were rated at the mean wind velocity. The power depends upon the cube of the wind velocity. The power plants are sized on mean energy velocity.

The mean energy velocity,

$$\overline{V_E} = \left(\frac{\sum_i^n V_i^3}{n}\right)^{\frac{1}{3}}$$

Example : If $\overline{V_E}$ is greater than $\overline{V}$ by 25%, the power potential would be $(1.25)^3$ or 2 times less if estimated on $\overline{V}$.

4.8.2. Total Energy Production

The total energy production of a wind mill depends upon the following factors :

1. Turbine rating.
2. No. of hours of operation.
3. Availability factor.
4. Variation of wind velocity.

The total energy production can be estimated by Weibull Distribution Model given below

$$\rho\,(V_i > V) = e^{-\left(\frac{V}{C_r}\right)^{K_2}}$$

where $\rho\,(V_i > V)$ = probability that incoming wind velocity exceeds a value V at reference height H (9.1 m).

$$K_r = 1.09 + 0.20\ \text{V}$$

$$C_r = \frac{\overline{V}}{\Gamma\left[1+\left(\frac{1}{k_r}\right)\right]}\ [\text{m/s}]$$

$$\Gamma = \text{gamma function}$$

4.8.3. Exceedance or Wind-Speed Distribution Curves

The wind-distribution curves are used to estimate the annual power output in MWh/year.

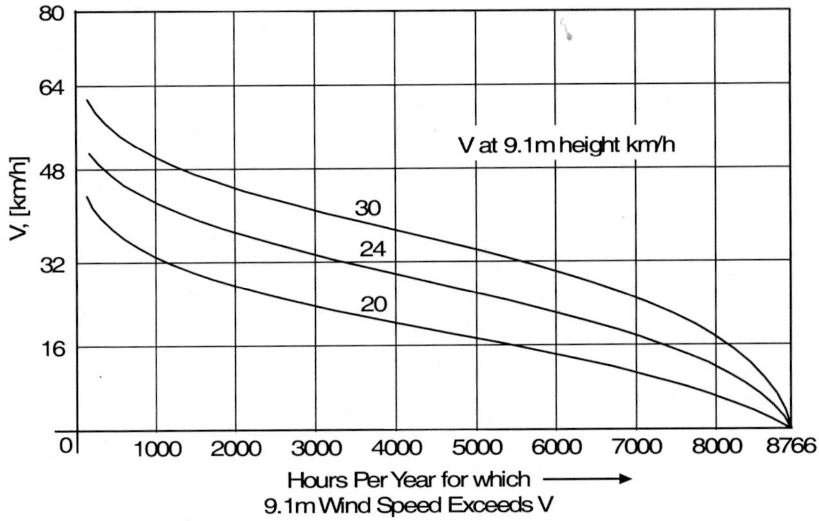

Fig. 4.9. Wind-Speed Distribution Curve

4.8.4. Indian Example

For a wind speed of 10m/s, the effective power obtained is 0.25 kW/m^2 of swept area. As wind velocity of 10m/s occurs only for a few hours a year, the mean annual available power is very less. Depending upon local wind conditions, the annual available energy per m^2 of swept surface may vary between 10 kWh for calm areas and 500 kWh for very windy areas.

Atleast 25 percent of Indian land especially coastal areas are situated in reasonably windy areas. The scope of small water pumping mills is quite significant.

The wind energy data for Veraval situated on the windy west coast of India is as follows :

Max usable energy in wind for 30m^2 swept area	: 20,000 kWh.
Machine power ratio	: 0.20
Annual power output	: 4,000 kWh
Water depth	: 10m
Average daily output	: 273 m^3 of water
Max, daily output	: 2428m^3 of water

4.9. SPECIFICATIONS OF WIND TURBINES

4.9.1. Wind Turbine for Water Pumping

Specific speed or tip-speed ratio of a windmill rotor is the ratio of its peripheral speed to wind speed. Rotors of low tip-speed ratio are preferred for driving piston pumps, which require a high starting torque and low rotational speed. The rotor of low tip-speed ratio also gives high efficiency and low cost. The other important characteristics of windmill rotors are cut-in speed and cut-out speed. Wind turbine for water pumping should have the following characteristics.

1. It should be water pumping, low tip-speed ratio wind mill preferably of vertical axis design.
2. It should be suitable for installation on open wells, canals, river beds, ponds, lakes, etc.
3. The total head over which water has to be lifted should be 15-20m with suction head 10.5 m, if installed on ground level.
4. It should be suitable for rated wind speed of 2 m/s with cut-in speed of 1.5 m/s.
5. Discharge at 2m/s of wind speed and total head of 15 m should be 0.5 to 1 litre/s.
6. The system design should be suitable for operating a sprinkler system for irrigation.

The wind mill WP-2, developed at National Aeronautical Laboratory, Bangalore can pump 2.9m^3/hour against a head of 10m at a wind speed of 3.5 m/s. A sail windmill developed at the same laboratory uses six triangular canvas sails on a rotor of 7m diameter and pumps 5.4 m^3/hour of water at a head of 10m at wind speed of 3.5m/s.

The specifications of a propeller type windmill pump-set, which has been standardised for installation in India, are listed below :

Tower height = 7m
No. of propellers = 12
Rotor diameter = 5m
Machine power ratio = 0.36
Water depth = 20m
Maximum daily output = 23m^3 at a wind speed of 10m/s.

The total cost of the machine including transportation and erection is Rs. 11750 and a similar expenditure is to be incurred on well boring and construction of storage tank. The Government provides a subsidy of Rs. 7500.

The windmills installed so far are not only of small size (about 5m diameter), but also require high maintenance and support. Research and Development (*R* and *D*) efforts are needed so as to reduce the maintenance expenditure of windmills to occasional oiling and changing the pump washer only. Somewhat larger windmills have to be developed than the presently available.

4.9.2. Wind Turbine for Electric Power Generation

The technical data of wind power plants with an electric power output in the range of 500 kW to 2MW are listed in Table 4.1.

Table 4.1. Specifications of Windmills for Power Generation.

1. Rated output, kW	500	1000	1500	2000
2. Rated wind speed m/s	13	14 – 15	15	15
3. Cut-in speed, m/s	4	3	4	3
4. Cut-out speed, m/s	25	25	25	21
5. Rotor diameter, m	36 – 40	53 – 55	63 – 65	60
6. No. of blades	2/3	2/3	3	2
7. Rotor speed, min^{-1}	20 – 30	17 – 25	14 – 20	30.
8. Power control	stall/pitch	stall	stall/pitch	stall
9. Tower hub height, m	40 – 50	58	60	66

The annual energy production of a wind power plant depends upon the wind speed and the rotor diameter as shown in Fig. 4.10.

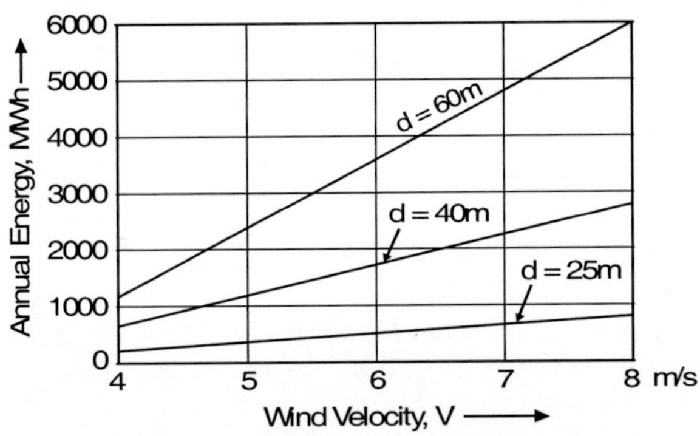

Fig. 4.10. Wind Power Plant Output

Important trend in the development of wind energy technology is the application of high capacity wind turbines.

4.10. OFF-SHORE WIND FARMS

Large capacity wind farms are being set up off the coasts of Northern Europe. These farms offer many advantages as compared to on-land wind farms.

1. The wind speeds are higher over the open sea surface especially at a height of 60m above the water surface.
2. There is less concern about the noise emission.
3. There is no landscape view deterioration.

A 40 MW farm comprising 20 × 2 MW turbines with a rotor diameter of 76 m and hub height of 60 m has been installed near Copenhagen, Denmark. This farm generates 89 GWh electricity annually which is sufficient to feed 25000 households. A 100 MW offshore wind farm is planned for installation near Rostock is Germany. Another 100 MW offshore wind farm is planned off the Netherlands's coast in the North sea. These large capacity off-shore wind farms will be using wind turbines of 2 to 6 MW capacity, rotor diameter of 75 to 100 m and hub height 60 m or more above water surface.

Example 4.3. *Determine the rotor diameter for a wind power plant of 1 MW rated output. The design wind speed is 12m/s, the turbine power coefficient is 0.43 and projected annual duration of utilization is 2400 hours per year.*

Also calculate the turbine power density at the design speed, annual power generation and torque experienced by the rotor at a speed of $50s^{-1}$*. The air density at 20°C is 1.205* kg/m^3*.*

Solution : 1. The power density per m^2 of swept area

$$P = \frac{1}{2} \times 0.43 \times 1.205 (12)^3$$

$$= 447.68 \text{ W/m}^2.$$

2. Rotor area required

$$A = \frac{1 \times 10^6}{447.68} = 2233.74 \text{ m}^2$$

3. Rotor diameter,

$$D = \sqrt{\frac{4A}{\pi}} = \sqrt{\frac{4 \times 2233.74}{\pi}} = 53.33 \text{ m}$$

4. Annual power generation

$$= 1 \times 2400 = 2400 \text{ MWh/year.}$$

5. The torque on the rotor

$$T = \frac{1 \times 10^6}{2\pi \times 50} = 3183 \text{ N-m}$$

4.11. NEW DEVELOPMENTS

The present windmill technology is inadequate for the low wind speed regions in the plains. Special development projects in the following areas must be taken up so that wind energy can also be used in the low wind speed regions.

4.11.1. Artificial Winds

Generation of artificial winds to drive windmills by heating large surfaces with favourable thermodynamic properties is technically feasible. A project report has been prepared to heat a large surface in which case the resulting current (artificial wind) can drive turbines. The efforts needed to pursue the project in the form of money, manpower and time are huge.

4.11.2. Aeroelectric Plant

The low wind velocity in the plains can be augmented by the use of diffusers at intake to wind mills. Besides the propellers, Madaras and Darrieus, there has been a plethora of designs for wind machines. One intriguing power plant design, called the aeroelectric plant, uses the flow up a tower that looks like a cooling tower as shown in Fig. 4.11. Its walls are heated by solar radiation. Since the walls are circular, the sun's rays need not be tracked as it changes position in the sky during the day. The heated walls, in turn, heat the inside air and a flow up the tower is established. This air flow is made to drive a number of air turbines located near the top of the tower. The driving pressure causing air flow is given by the well-known chimney effect.

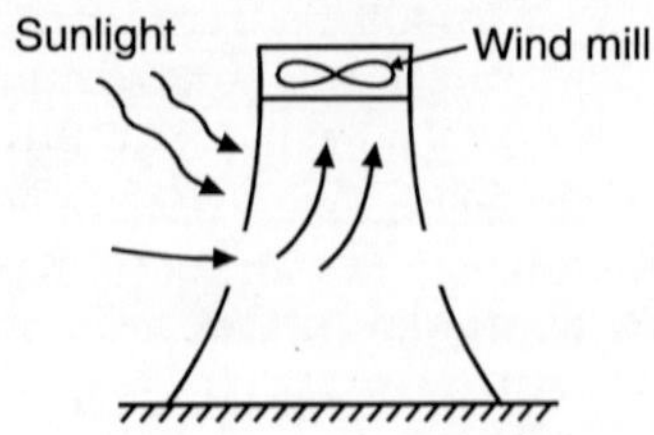

Fig. 4.11. Aeroelectric Plant.

P. Carlson of Californica has proposed a slightly modified form in which, the interior air in a very tall tower would be cooled by pumping water to the top. The water evaporates in the low pressure air there, causing a downward flow of cooled air. The driving pressure can be calculated in a manner similar to that for wet cooling towers.

A conceptual design of such a plant called for 2.4 km high, 300 m diameter tower located in a hot desert and 10 wind turbines surrounding the tower periphery at the bottom producing 2500 MW.

4.11.3. Low Wind Speed Turbines

The turbines available in India and abroad are suitable for a rated wind speed of 3.5m/s or more whereas low wind speed turbines for rated values of 1.5 – 2m/s are needed for plain areas. Special efforts are, therefore, needed to develop cheap and simple rotors, which can cut-in at low wind speeds available in the plains.

The Savonious rotor and American multi-blade type windmills have optimum power coefficients at a very low tip-speed and can therefore be used as starting point to develop windmills suitable for low wind speeds.

4.11.4. Energy Storage Schemes

A combination of a wind mill and a micro-hydro turbine can be used for storage of energy as potential energy of water as shown in Fig. 4.12.

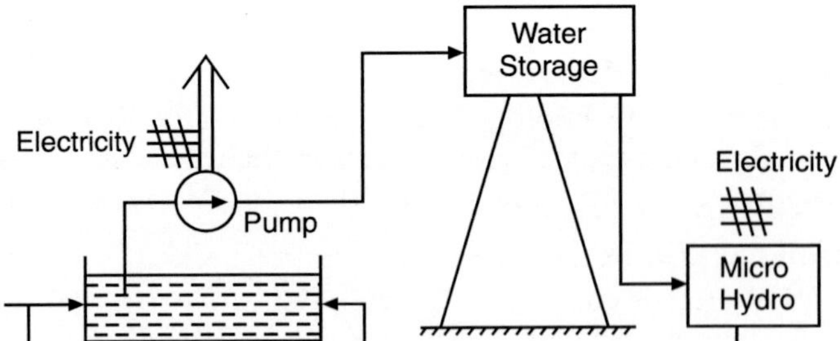

Fig. 4.12. Integrated Energy System.

Water could be pumped to high reservoir during periods of high winds and the potential energy could then be used in a micro hydro power plant during periods of low winds, ensuring a constant output from the wind-water generating system as the basis for sizing of water generator and storage system.

4.11.5. Windmill with Hydraulic Transmission

Hydraulic transmission consisting of a split fluid coupling/ torque convertor can be used to run a centrifugal pump to lift deep ground water for the operation of a sprinkler system for irrigation as shown in Fig. 4.13. The windmill is directly coupled to hydraulic pump part of coupling/torque convertor while the hydraulic turbine part, which is directly coupled to the centrifugal water pump, is lowered deep in the open well. The variable speed/torque hydraulic pump sends pressure fluid through a pipe to drive the hydraulic turbine to operate the centrifugal pump.

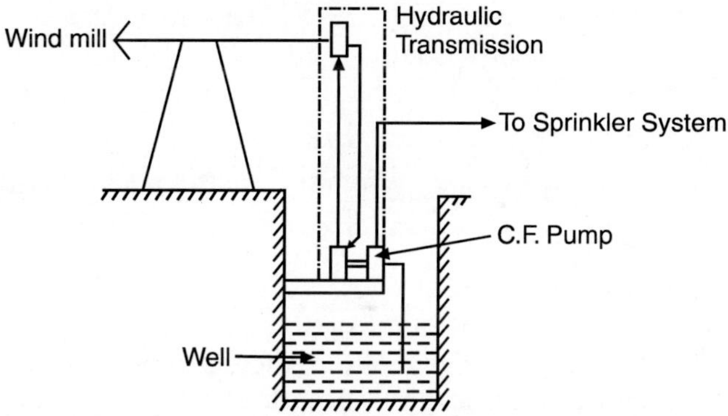

Fig. 4.13. Windmill with Hydraulic Transmission.

The use of hydraulic transmission has many advantages such as optimum adoption of rotor speed to guarantee a maximum output ; no gear box ; no controlling and switching equipments as no electrical generator is involved ; simple maintenance as no reciprocating pumps and allied crank mechanism ; and the centrifugal pump can be installed on ground level or lowered into the deep well to suit the suction lift available.

4.12. WIND DIESEL HYBRID SYSTEMS

A hybrid system is conceived to compensate for the intermittency of wind energy, not by storage but by using the wind as the source of energy when available and sufficient and diesel engine at other times.

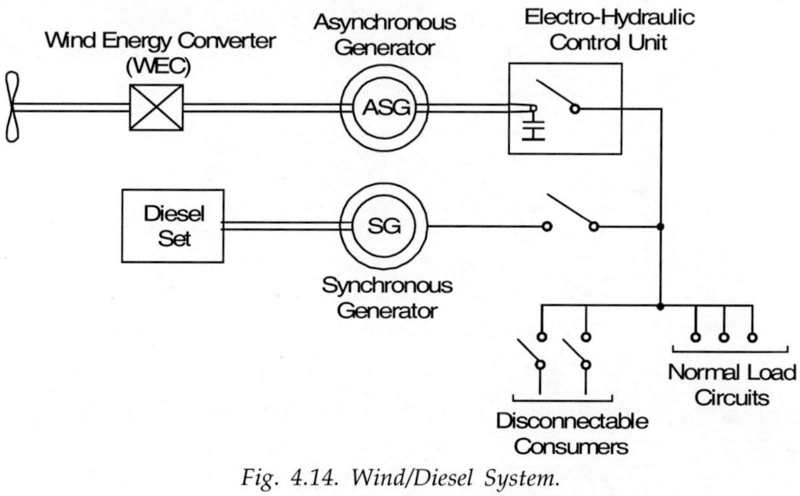

Fig. 4.14. Wind/Diesel System.

A simple combination of a wind energy plant with a diesel power plant is shown in Fig. 4.14. The wind energy plant is always operated in parallel with the diesel engine. It supervises the grid conditions automatically and switches itself on/off depending upon the wind speed. The WEC works as a normal grid connected machine, and frequency and voltage control is taken over by the Diesel-Generator set.

The load circuits of lower priority atleast should be disconnectable to make possible a simple load adaption.

Advantages

1. Simple and robust conception.
2. No additional electronic control unit.
3. Easily extendable.

Disadvantages

1. Operation of the wind energy plant possible when the diesel set is on.
2. Uneconomical part-load operation of the diesel engines cannot be avoided.

4.13. NOVEL WIND TURBINES

Propeller-type wind turbines which work on aerofoil design are the main wind turbines used for wind power generation, there are new designs being investigated which may offer technical and economic advantages in future over the existing designs. : Some of the new turbines are :

1. Magnus rotor wind turbine.
2. Madaras rotor wind turbine.
3. Darrieus Machine
4. Carlson Aeroelectric Tower.

4.13.1. Magnus Effect

A uniform nonviscous air stream when flows past a vertical cylinder, stream lines as shown in Fig. 4.15. are generated. The velocity at any point on the surface of the cylinder.

$$V_\theta = 2\ V_i \sin\theta \qquad ...(1)$$

where V_θ = air velocity on the surface of the cylinder at angle θ.

V_i = incoming uniform air velocity, far from the cylinder

θ = polar angle, measured from the stagnation point.

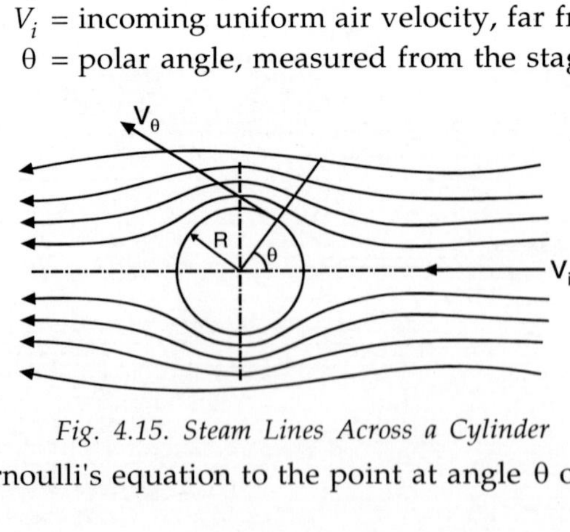

Fig. 4.15. Steam Lines Across a Cylinder

Applying Bernoulli's equation to the point at angle θ on the surface of the cylinder,

$$\frac{p_i}{\rho} + \frac{V_i^2}{2} = \frac{p_\theta}{\rho} + \frac{V_\theta^2}{2} \qquad ...(2)$$

From equations (1) and (2),

$$p_\theta = p_i + \frac{1}{2}\rho V_i^2 (1 - 4\sin^2\theta)$$

$$p_\theta^* = \frac{p_\theta - p_i}{\frac{\rho V_i^2}{2}} = 1 - 4 \sin^2\theta \qquad ...(3)$$

where p_i = pressure of incoming stream

p_θ = πρεσσυρε ον τηε χψλινδερ συρφαχε ατ θ

ρ = air density considered constant

p_θ^* = non-dimensional pressure on cylinder surface at θ.

1. At stagnation point, $\theta = 0$

$$p_\theta - p_i = \frac{\rho V_i^2}{2}$$

$\frac{\rho V_i^2}{2}$ is the dynamic pressure of the undisturbed stream.

2. At $\theta = \pm 30°$ and $\pm 150°$

$$p_\theta = p_i$$

3. At $\theta = \pm 90°$

$$p_i - p_\theta = \frac{3\rho V_i^2}{2}$$

Plotting these values on a cylinder, there is a symmetrical distribution of pressure and there is no resultant force. Therefore, there is no lift or drag on the cylinder for ideal nonviscous flow as shown in Fig. 4.16.

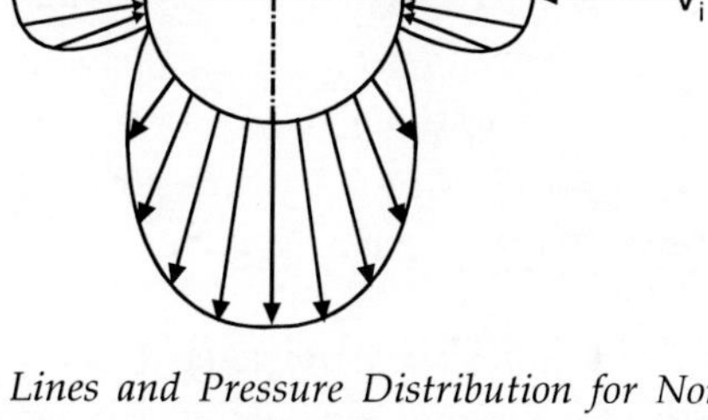

Fig. 4.16. Stream Lines and Pressure Distribution for Nonviscous Flow.

If a cylinder is rotated about its axis in still air, a circulatory flow is produced around the cylinder where the velocity is inversely proportional to the distance from cylinder axis.

$$2\pi r V_r = 2\pi R V_p = \Gamma$$

$$\therefore \quad V_r = \frac{\Gamma}{2\pi r}$$

$$V_p = \frac{\Gamma}{2\pi R}$$

where V_r = velocity at distance r from the cylinder axis.

V_p = cylinder peripheral velocity = $2\pi RN$

r = radial distance from cylinder axis.

R = cylinder radius.

Γ = a constant

$= 2\pi R V_p$

$= (2\pi R)^2 N$ and is called a circulation constant.

N = number of revolutions per unit time.

If a cylinder is rotating in a stream of air, the above two cases will be superimposed in such way that at cylinder top the two velocities will reinforce each other and at the bottom of horizontal cylinder, the two velocities will oppose each other.

$$V = 2V_i \sin\theta + \frac{\Gamma}{2\pi R}$$

$$= V_\theta + V_p$$

The pressure on the cylinder will be higher where the velocities are low.

The Lift Force

Using Barnaullis equation,

$$p = p_i + \frac{1}{2}\rho\left[V_i^2 - \left(2V_i \sin\theta + \frac{\Gamma}{2\pi R}\right)^2\right]$$

where p = πρεσσυρε ον χψλινδερ συρφαχε ατ θ

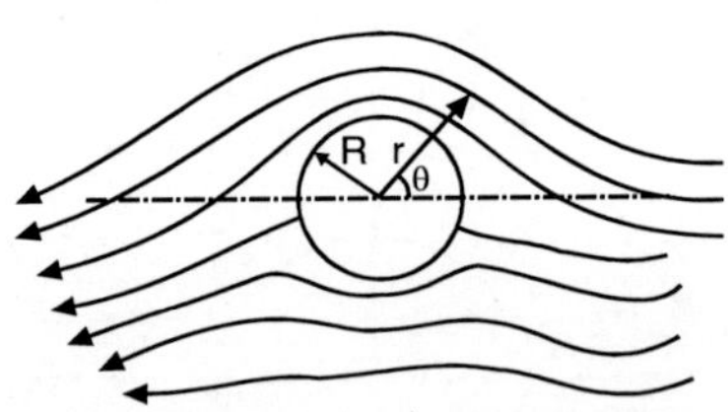

Fig. 4.17. Velocity Stream Lines Around a Rotating Cylinder an Air Stream

The lift force can be obtained by integrating the pressure components perpendicular to free stream.

$$F_L = \int_{-\pi}^{+\pi} -p \sin\theta \, R d\theta H$$

$$= 2\pi\rho \, RHV_p V_i$$

$$= 4\pi^2 \rho R^2 HNV_i$$

where F_L = lift force on cylinder

H = length of cylinder

$$F_L = C_L A\rho \frac{V_i^2}{2}$$

where $C_L = 2\pi \frac{V_p}{V_i}$ = lift coefficient (dimensionless)

The Drag Force

The imbalance in pressures results in drag force in the direction of free stream.

$$F_D = C_D A\rho \frac{V_i^2}{2}$$

where C_D = drag coefficient (dimensionless)

C_D is a function of Reynolds number.

The force acting on the cylinder is the resultant of lift and drag forces.

4.13.2. The Madaras Rotor Wind Machine

A number of tall cylinders are rotated about their axes in the presence of wind. The resultant of lift and drag forces propels the cylinders horizontally along a track as shown in Fig. 4.18.

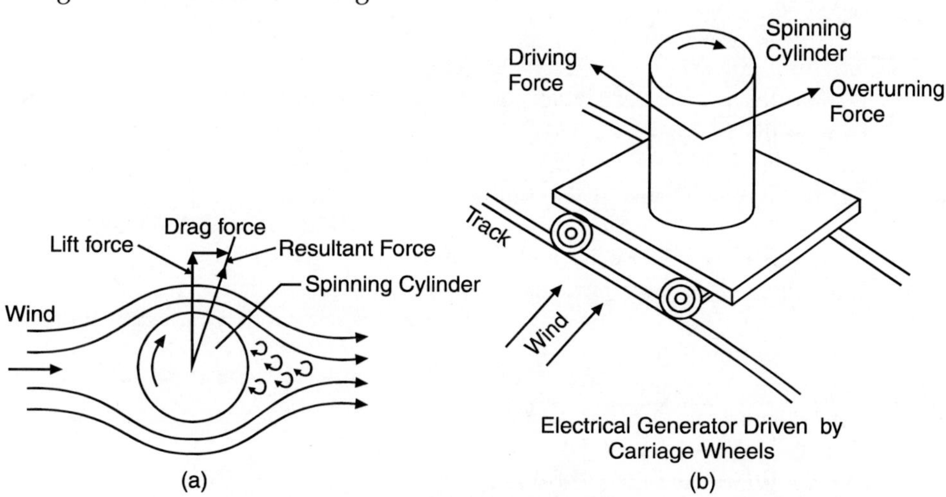

Fig. 4.18. Madaras Rotor.

A continuous train of four-wheel flat cars move on a circular train track. Each flat car has 30m high cylinder rotated about its vertical axis at 120 rpm by a small electric motor. The flat cars move on a circular track by the resultant force on the cylinder. The carriage wheels drive electrical generators and the electricity generated is transmitted to the mains by a trolley.

4.13.3. The Darrieus Machine

It is a vertical-axis machine and looks like an egg beater with two or three slender blades looped over the top and joined at top and bottom to the vertical axis : See Fig. 4.19.

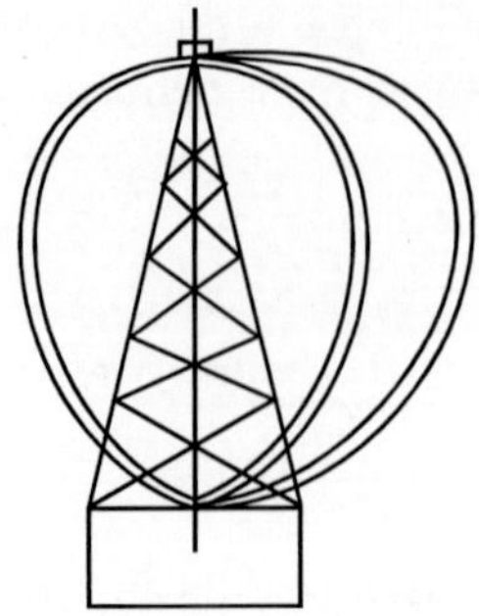

Fig. 4.19. Darrieus Machine

Darrieus machine can spin in any wind direction. It does not require yatch control. It is a vertical axis machine and electrical generator or pump can be installed at ground level.

TUTORIAL SHEET No. 4.1.

1. What is wind energy? Describe the main components of a wind power plant.
2. Give classification of wind turbines on the following basis
   (*i*) Axis of Rotation
   (*ii*) Size of Machine
   (*iii*) Applications.
3. Explain the principle of wind power generation and theory of momentum.
4. Explain the following :
   (*i*) Forces acting on the blades of a wind turbine.
   (*ii*) Power coefficients of windmills.
5. Discuss in details the operation and control of a wind turbine. How are the variations of wind velocity and its directions are taken care?
6. What factors are taken into consideration in site selection ?
7. Give typical specifications of a wind turbine for the following applications.
   (*a*) Water pumping
   (*b*) Power Generation
   (*c*) Off-shore wind farms.
8. Describe an Aeroelectric Power Plant.
9. Explain Wind-Diesel Hybrid system.
10. What is Magnus Effect? How this effect has been used in the development of Madaras Rotor Wind Machine?
11. Discuss wind characteristcs. How these characteristics are used for the following:
    (*a*) site selectcion
    (*b*) selection of type of wind rotor.
    (*c*) performance analysis of wind power plant

❑❑❑

Chapter 5

Small Hydro Power Plants

Introduction, Indian Situation, Hydro Energy; Current Energy; Types of Small Hydro Power Plants ; Hydraulic Machines ; Floating type Micro Hydro Power Plants ; An Alpine Hydro Village, Economic Considerations.

5.1. INTRODUCTION

The utilization of hydro energy to operate agricultural and industrial devices is one of the oldest and widespread techniques of human achievement. Hydraulic rams were used to raise water for lift irrigation and drinking purposes. Water mills had been used to generate mechanical power to drive irrigation pumps, machine tools and small electric generators. The Himalayan Mill is well known in the mountain areas from Afghanistan to Burma which is mostly used for grinding grains.

China has developed 1200 MW of power by small hydel projects by creating water heads of about 5m along the rivers. Chinese model is difficult to develop in India due to different topography. The current number of operational small hydro power plants is 90,000 in China, 10500 in USA, 5300 in Japan, 2600 in Switzerland, 1300 in Austria.

5.2. INDIAN SITUATION

Western Ghats of Maharastra and regions of Uttar Pardesh, Himachal Pradesh, Jammu and Kashmir, Sikkim and North Eastern states located in the Himalayas are suitable sites for small hydel projects where medium to high heads can be available. Equipment costs are low in the high head schemes.

As we come down the hills, the available head decreases and equipment costs go up. Most of our rivers flow through flat terrains which can be harnessed to produce hydropower. The large potential of hydro energy available from flowing water in the rivers and irrigation canals, however, still remains untapped. Generating power on a small scale from low and ultra low head drops of 3 to 20 meters has not been an economic proportion because of the high civil and equipment costs as well as high operating cost to be incurred per unit of power generated. The situation is further deteriorated by very low load factors in rural areas.

The modern high technologies donot adopt to rural ecology. An energy production technique based on renewable source and appropriate technology which can be taken care by trained local people for its operation and maintenance problems is ideally suited. Only appropriate technology involving local resources, manpower and raw materials can ensure environmental harmony and sustainability of the development.

Hydropower, available free of cost in the form of flowing water, is an important, pollution free, renewable source of energy. Contrary to large size hydroelectric power stations, small hydro power plants have a very low, if any, impact on the environment. These can be integrated with irrigation, fish breeding and drinking water supply schemes. A potential of 15000 MW capacity has been identified out which 1398 MW of actual potential has been already exploited.

5.3. HYDRO ENERGY

Hydroelectric power stations harness the kinetic energy of water flow. This energy is produced from the potential energy of water flowing down due to height difference created by a dam or a slope. The net electric power output of a hydroelectric power plant is given by :

$$P = \rho QgH\eta_o$$
$$= \rho QgH\eta_T\eta_g\eta_{Tr}$$

where P = Electric power output in [W]

ρ = Density of water, [kg/m^3]

Q = water flow rate in turbine , [m^3/s]

g = Gravitational acceleration [9.81m/s^2]

H = net useful water head, [m]

η_o = overall efficiency of hydropower plant,

η_T = Turbine efficiency [0.85 – 0.95]

η_G = Generator efficiency [0.95 – 0.99]

η_{Tr} = Transformer efficiency [0.92 – 0.98]

Example 5.1. *Determine the net power output of a small hydroelectric power plant using a Pelton turbine if water flow rate is 2 m^3/s and the plant has an overall efficiency of 82%. Water is delivered to the turbine from a height of 300 m. The density of water is 1000 kg/m^3.*

Solution :

$$\rho = 1000 \text{ kg/m}^3$$
$$Q = 2 \text{ m}^3\text{/s}$$
$$H = 300\text{m}$$
$$\eta_o = 0.82$$
$$P = \rho QgH\eta_o$$
$$= 1000 \times 2 \times 9.81 \times 300 \times 0.82 \times 10^{-6} \text{ MW}$$
$$= 4.8265 \text{ MW.}$$

5.4. CURRENT ENERGY

Special 'zero' head turbines operating on kinetic energy of flowing water in rivers and irrigation canals can be used to harness huge river current energy. The current energy of flowing water may be expressed as :

$$P = \frac{1}{2} AV^3 \eta$$

where P = Power generated (W)

A = Embrassing area of runner (m^2)

V = Water current speed [m/s]

η = Efficiency of runner [0.5 to 0.85]

The efficiency depends upon the type of runner. The relation between river current speed for a turbine with swept are of $4m^2$ and rotor efficiency of 25% is shown in Fig. 5.1.

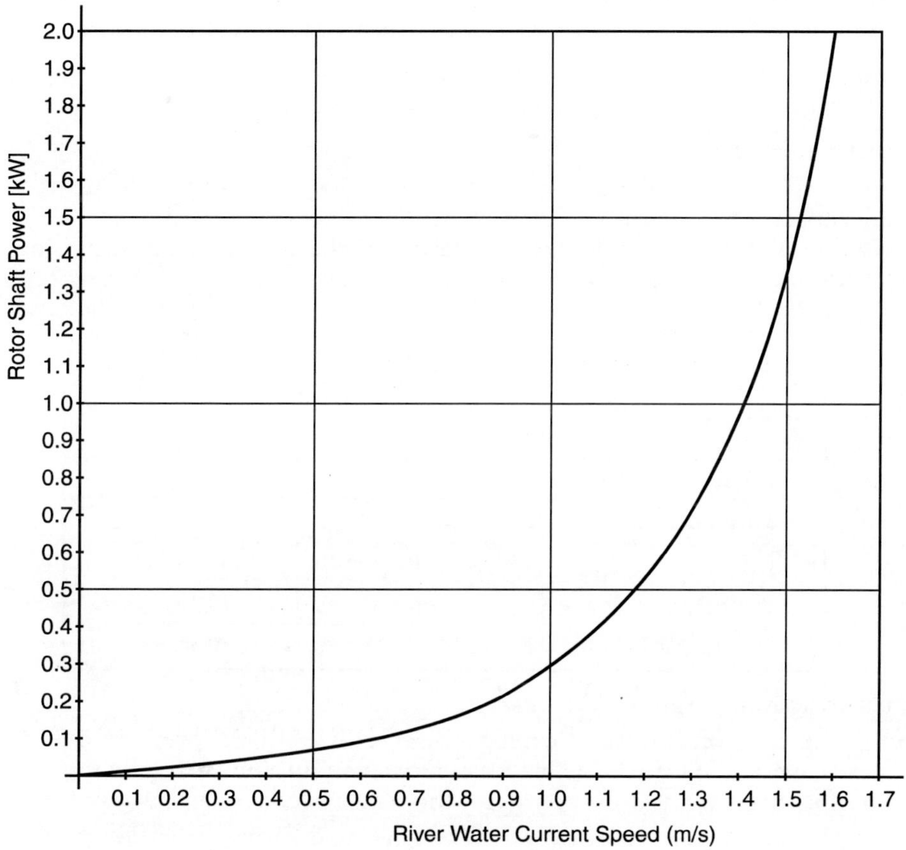

Fig. 5.1. Current Energy.

5.5. TYPES OF SMALL HYDRO POWER PLANTS

The small hydro power plants can be classified according to the size of the hydro turbine.

1. Micro Hydel Projects of capacity 50 to 100 kW.

   Example : A micro hydro power plant with a propeller type turbine with a power output of 50 kW at a head of 10m and water flow rate of 0.8 m^3/s.

2. Mini Hydel Projects of capacity range 100 kW to 1000 kW.

   Example : A mini hydro power plant with Francis turbine with a power output of 750 kW at a head of 25m and water flow rate of 0.4 m^3/s.

3. Small Hydel Projects capacity range of 2MW to 15 MW.

   Example : A small hydro power plant with Pelton turbine with a power output of 3 MW at a head of 400 m and water flow rate of $0.9m^3/s$.

Small hydel projects can be completed in four to five years, mini and micro projects in three years provided civil works of irrigation projects are already completed. Water released for irrigation only is used and power can be generated for 6 to 7 months a year.

5.6. HYDRAULIC MACHINES

Various hydraulic machines used for small hydro power plants are :

5.6.1. Hydro Turbines

Two types of hydro turbines are used in power plants

(*i*) Impulse Turbine

It utilizes the kinetic energy of a high velocity water jet which strikes the buckets on blades mounted on a shaft. The kinetic energy of water jet is converted into mechanical energy of rotation of shafts. The impulse turbine with tangential water flow (Pelton turbine) is used for very high water heads as shown in Table 5.1.

Table 5.1. Operating Range of Hydro Turbines.

Type of Turbine	Operating head range (m)	Output range (kW)
Pelton	100 to 400	50 to 6000
Turgo Impulse	50 to 200	25 to 3000
Francis	20 to 200	50 to 3000
Propeller/Kaplan	2 to 30	10 to 6000
Horizontal Tubular	2 to 25	10 to 6000

(*ii*) Reaction Turbines

The reaction turbines develop power utilizing both pressure energy and kinetic energy of water. The blades of a reaction turbine are totally immersed in the flow of water and the energy of the water is converted into shaft work.

The reaction turbines are Francis turbines with radial water flow and Kaplan turbines with axial water flow.

The schematic representation of various hydro turbines is given is Fig. 5.2. and the efficiency curves are given in Fig. 5.3. The type of turbine for a site is

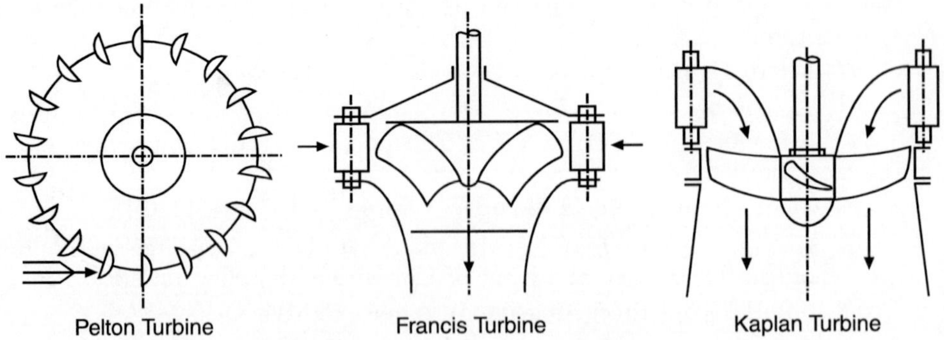

Fig. 5.2. Types of Hydro Turbines.

selected on the basis of available head. In case of overlapping the type of turbine selection is decided from efficiency curves. Pelton turbine has more efficiency but is more suitable for constant load operation. For variable loads, Francis and Kaplan turbines are suited as these have flat efficiency curves.

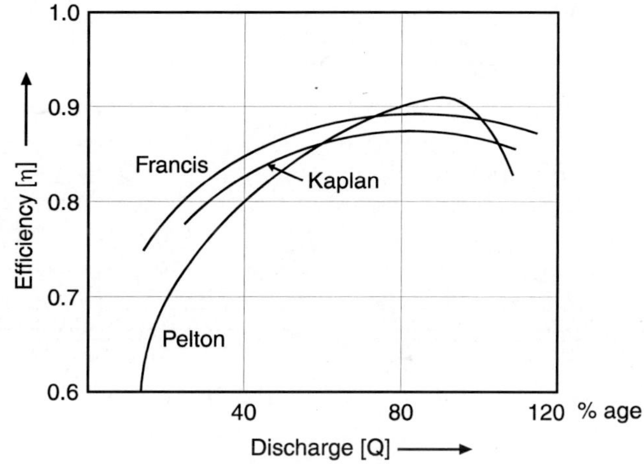

Fig. 5.3. Efficiency Curves.

5.6.2. Himalaya Mill

It is a vertical shaft, horizontal runner hydraulic turbine used for driving grain mills. The wooden radial blades are attached to a boss fitted into a wooden or steel shaft. The water is fed through open wooden channel. It produces about 0.35 kW of shaft power with a water head of 1 m. There are about 10,000 such mills scattered in the hilly areas from Afghanistan to Burma producing power equivalent of about 45,000 kWh daily.

The horizontal water wheel can be improved in performance to raise its efficiency substantially. The improved wheel can be used in a small scale industrial park as a prime mover for flour mills, thrashers, rice husking machines, vertical lathe, circular saw, trip hammer, small electrical generator, carpet loom, etc. The vertical lathe can be used by the craftsmen to produce wooden vessels and trip hammers for embossed copper vessels. The saw mill can be used by carpenters. All these machines can be employed together to meet the local needs for production of domestic and agricultural implements and packing boxes for the export of fruits and other products.

The Himalaya Mill can also be developed into an efficient and modern Pelton turbine using metallic buckets filted to a vertical shaft supported or roller bearings. The water can be directed onto the buckets through adjustable nozzle from a piping system collecting water from natural streams existing abundantly in the hilly areas. The stopping of turbine can be achieved through manual brakes and load/speed regulation by the nozzle. The output of the machines can be standardised at 5 to 10 kW. The mill can be designed and manufactured locally in the small-scale industry park.

The tail race water can be recirculated by lifting it with a hydraulic ram for irrigation purposes.

5.6.3. Hydraulic Ram

Hydraulic ram is a contrivance to raise a part of large amount of water available at some height, to a greater height. This can be employed in hilly areas where some natural source of water like a spring or a stream is available at some altitude. Work done by a large quantity of water in falling through a small height is used to raise a small part of it to a greater height. Action of water hummer makes it feasible.

No external power is, therefore, required to work this machine. Other attractions are : negligible amount of maintenance and supervision costs, continuous operation, high efficiency, quiet operation and possibility of automatic adjustment of water supply. The liftable volume diminishes asymptotically with lifting height. In case of medium lifting height, the hydraulic ram operates with an efficiency absolutely comparable to a piston pump of the same performance.

As the hydraulic ram does not need a driving unit it is to be considered as ideal for lift irrigation and drinking water for hilly areas where the supply in fossil energy carriers or electricity is problematic. The hydraulic ram operates with water streams of 1 to 40 m^3/s and fall heads of 1.5 to 30 m and with lifting heights upto 300 m.

Hydraulic ram is a simple and rugged device for operating irrigation schemes. This can be easily assembled from local materials.

5.6.4. Poncelet Water Wheel

In ancient times, floating mills were used in Germany on large rivers such as Rhine to drive electric generators and machine tools. A vertical undershot water wheel consisting of a horizontal shaft and radial blades fixed to it, was fitted on the side of a barge which was either moored to the bank or firmly anchored in the stream. A smaller barge was used to support the other end of the shaft. The hulls of the barges were built with local materials. The efficiencies realised were very low, in the range of 20 to 30% because of bad blade geometry (straight radial blades).

Poncelet water wheel is an improvement on the straight blades of undershot water wheel, which have suitable curvature. Water strikes the vanes practically without shock (inlet vane angle 15°) and drives it by impulse. Water is discharged from the blades almost vertically downwards. The efficiencies achieved are 55 to 65%.

The main drawback of such machines is low power output for a given size and weight because at any instant only a small part of it is actually in the water being driven by the current while the rest of the machine is idle.

5.6.5. Darrieus Turbine Rotor

Darrieus turbine rotor is similar to modern vertical-axis wind turbine which was first introduced in France by G.J.M. Darrieus in 1920's. Its configuration consists of four hydrofoil blades which rotate at much higher speed than the water current speed. The speed of rotation is primarily dependent upon the machine overall dimensions. It has relatively low starting torque. Because of simplicity of the blade design and because they are relatively thin, blade fabrication costs are low. They require no pitch control for synchronous applications.

Two prototypes have been developed by Intermediate Technology Development Group, London as shown in Fig. 5.4. The blades can be made out of timber or fabricated from steel, ferrocement or glass fibre. The expected performance is shown in fig. 5.1.

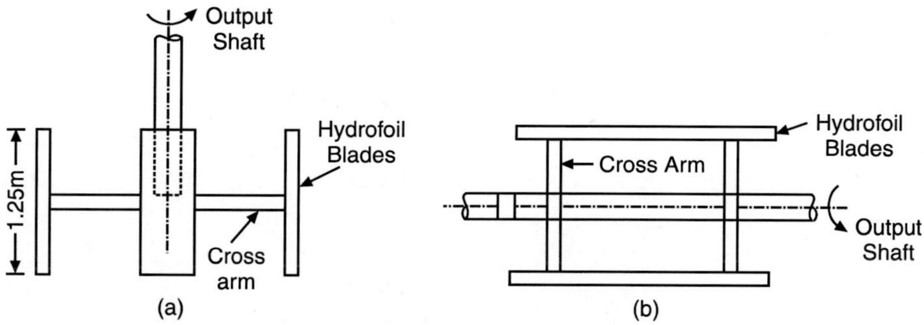

Fig. 5.4. Darrieus Turbine Rotor.

Prototype (*a*) : Vertical axis rotor 13.5 rpm in 1 m/s current. Power take off above surface. Swept area 3.75 m^2.

Prototype (*b*) : Horizontal axis rotor 32 rpm in 1 m/s current. Power take off under water. Swept area 3.75 m^2.

5.7. FLOATING TYPE MICRO HYDRO POWER PLANT

The floating type micro hydro power plants are suitable for use in situations where there is no "head" of water available and is designed to extract kinetic energy from a river or canal water current. Therefore, appropriate sites are along the banks of large rivers or on irrigation canals.

As the energy flux depends upon the cube of water velocity, the latter must be as large as possible. The depth of water required for a 4m^2 runner is atleast 2.5m.

5.7.1. Hydraulic Machines

The Darrieus turbine rotor is a low speed machine. It is a low solidity device which is completely submerged during operation and this, together with its high efficiency (above 50%) means that a rotor of moderate size and modest material content can extract useful amounts of energy from current speeds as low as 1 m/s (2 knots).

The mechanical energy obtained can be used for running irrigation pumps, small electric generators and small compressors for cold storages and ice mills. The small cold storages in hilly areas are necessary to store potatoes and other perishable goods to get better returns. Ice mill can be used to chill and store milk before it is sent to bigger collection centres.

The electricity generated is required to light the houses, schools, community centres and to operate TV sets and other domestic devices. All these measures can improve the wealth of the village multifold and raise the standard of living of the local population which will arrest their migration to big cities.

A turbine with swept area of 4m^2 and rotor efficiency of 25% can give an output of 8 kW at a water current speed of only 2 m/s.

In plain areas, irrigation canals are expected to be an important area of application for floating type micro hydro power plants. In a large canal, like in a river the turbine rotor could be mounted on a pontoon, whereas on a small canal it might be cheaper to mount the rotor under a bridge or beam over the canal. The power output from the rotor could be increased by shaping the canal banks to form a venturi in order to increase the local water velocity through the rotor.

5.7.2. Installation

The appropriate sites for floating type micro hydro power plants are along the banks of large rivers or on irrigation canals. The pontoon carrying the turbine, generator and transmission system can float on two rows of barges or oil drums under which wooden keels are affixed. The site requirements are shown in fig. 5.5.

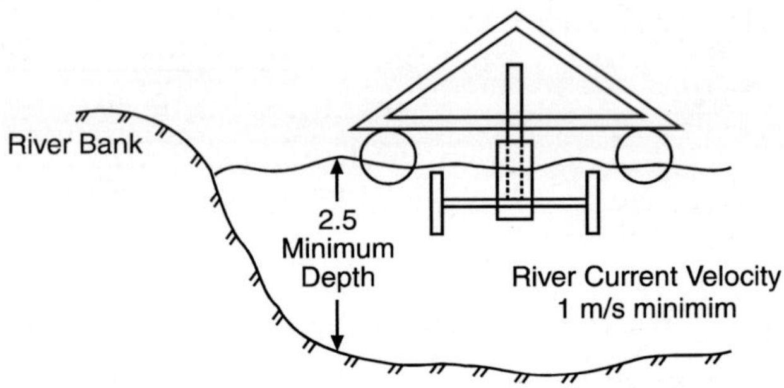

Fig. 5.5. Site requirements for turbine with $4m^2$ swept area.

The mooring details are shown in Fig. 5.6. The mooring cable arrangement protects the rotor against submerged or semi-submerged objects striking the blades. The cable can be adjusted with a winch to ensure sufficient angle of the keels to the current direction so that water side thrust on the keels keeps the pontoon at a sufficient distance from the bank.

5.7.3. Auxiliaries and Controls

The rotational speed of Darriens turbine is very low : 13.5 rpm for vertical axis and 32 rpm for horizontal axis turbines. For Poncelet turbine, the peripheral velocity of runner is about half of water current speed. Therefore, multistage belt drives or gear boxes are needed to increase the speed of generator shaft to about 950 rpm.

Permanent magnet alternator or induction generators are preferred over synchronous generators. In order to reduce equipment costs, the conventional mechanical or hydromechanical speed governor should be dispensed with the inertia of rotor being small, there is a problem of instability in power supply quality.

The above problems can be solved by the use of a flywheel which can ensure stable power supply and can shave-off power demand fluctuations. A flywheel designed for this purpose can be accelerated to extremely high speeds without any risk of breaking

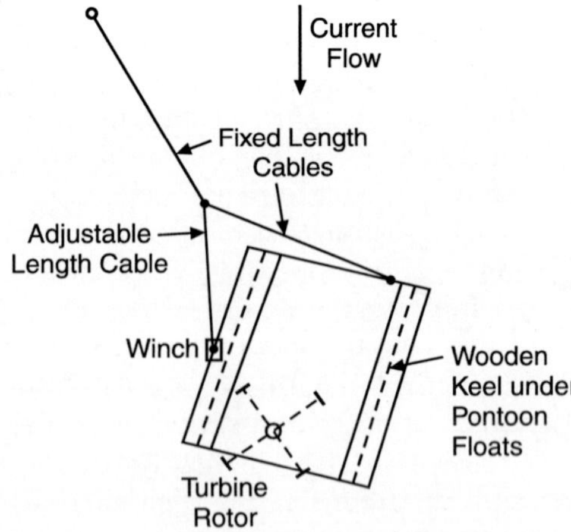

Fig. 5.6. Mooring System.

apart. It can be made from concentric rings of quartz fibre as windings, the rings being fitted over one another and close gaps filled with an elastic substance to keep the plies of the rims together. The flywheel is coupled to a generator and enclosed in a sealed evacuated casing to reduce windage losses. The device operates as a generator when power demand on the system grows, and as an electric motor when it is time to accumulate energy.

5.8. AN ALPINE HYDRO VILLAGE

India is a country of villages. These villages vary in size and population and are scattered all of the country. Some of these are located on the hills and in dense forests with very little connectivity. The population of these villages may be one hundred to two hundred persons. Most of them cultivate land, grow fruits and some of them are craftsmen weaving carpets, baskets and make earthern, wooden and brass pots and articles. Life is hard and incomes are very low.

The high technology industry cannot penetrate these areas and there is no electricity and commercial fuels. The population of these areas felt neglected and many of them are flocking to new economy centres in search of their livelihood.

The development, welfare and prosperity of hilly villages depend upon the availability of cheap and abundant energy for cooking, heating and lighting, community development, irrigation water, mechanised agriculture and small industrial parks based on local materials. A model of an alpine village based on utilization of locally available hydro energy is described below. The model can meet the productivity requirements of the village for self sufficiency.

The layout of an alpine hydro village is shown in fig. 5.7. The capacity and number of various hydro machines will depend upon the size of the village and characteristics of available water stream. The hydro energy is available in the form of a large river and many riverlets and water falls.

The following units have been included in the scheme.

1. A small cold storage and ice mill run by a floating water mill. The capacity of the plant can be around 10 kW. Such cold storages can be built along the flow of water stream to store potatoes, fruits and other perishable products. This will help to avoid spoilage of such products and realize better returns by disposing off uniformly over the year. Similarly ice mills attached to the cold storage can be used to chill and store the milk cooperatively and can be drawn for disposal to collection centres and self use efficiently.
2. Electrical generator run by a floating water mill. The capacity will match the village requirements of electricity for lightening, heating, cooking and for operating radios, TV sets and other small electrical gadgets in the houses, schools, dispensaries and community centres
3. Irrigation pump run by a floating water mill. The capacity will suit the irrigation water requirements.
4. Industry Park consisting of flour mills, rice mills, thrashers, saw mills, vertical lathes, trip hammers, carpet looms, etc. run by improved Himalaya mills individually. The capacities can be standardized at 5 to 10 kW each.

The industry park will meet all the domestic and agricultural requirements and make the village independent of imports of implements and spare parts. This will help to produce the craft goods like wooden, brass and other vessels, toys, beauty carvings, woolen carpets for exports. All packing boxes for export of forest, agricultural and animal yields will be prepared in the industry park.

5. Hydraulic ram to lift the used water from tail race for irrigation and domestic water consumptions.

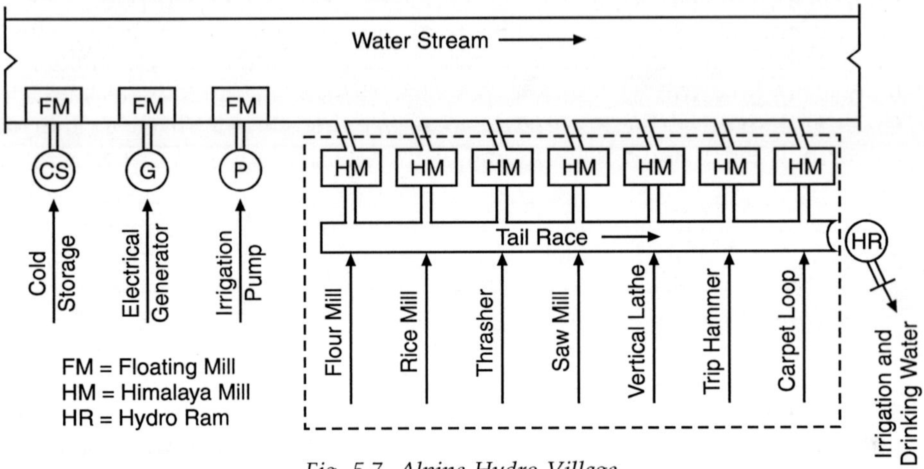

Fig. 5.7. Alpine Hydro Village.

5.9. ECONOMIC CONSIDERATIONS

Bharat Heavy Electricals Ltd. Haridwar and Bhopal plants, Jyoti Limited, Baroda have the capacity to manufacture small hydro turbines as per Table 5.1. Himalaya Mills, Hydraulic Rams and Floating Mills can be designed, manufactured and installed using local materials and skills. Independent small hydel projects become very costly due to heavy civil costs involved. The cost may be Rs. 18000 per kW installed. Therefore such projects become viable if planned as multiple purpose use of water including irrigation and power generation. As we come down the hills, the available head decreases and equipment costs go up. For example, the cost of one hydro set of horizontal tubular design of 3.2 kW capacity operating at a head of 3 m may be Rs. 1.20 lakhs, *i.e.*, Rs. 40,000 per kW installed. The civil costs are site specific but taking 30% of equipment cost, the total installed cost will be a high value of Rs. 52,000 per kW installed. Even though the schemes are costlier but the cost of 1 kWh will be less than from pumped hydroelectric power plant as water is essentially free and other running cost components of operation and maintenance are very low.

TUTORIAL SHEET NO. 5.1.

1. Discuss specific Indian situation to adopt small hydro electric power plants.
2. What is the difference between hydro energy and water current energy ? How is power output estimated ?
3. Give a classification of small hydro power plants.
4. What are various types of hydraulic machines which can be used to exploit hydro energy at small level ?
5. Compare impulse and reaction type of hydraulic turbines.
6. Write notes on
   (*a*) Himalaya Mill
   (*b*) Hydraulic Ram
   (*c*) Poncelet water wheel
   (*d*) Darrieus turbine rotor
7. With the help of suitable sketches, explain "Floating Type" Micro Hydro Power Plant.
8. Discuss the concept of an Alpine Hydro Village. Give a suitable layout.
9. Discuss the economic consideration of small hydro power plant. How can it be justified ?

❑❑❑

Chapter 6

BIOMASS AND BIOENERGY

Introduction ; Availability of Biomass ; Biomass Conversion ; Fluidised Bed Combustion of Biomass ; Circulating Bed Boiler System ; Steam Turbine Cycle ; Gas Turbine Cycles ; A Cooperative Rural Energy Power Plant ; Biomass Gasification ; Biogas Generation, Biofuel Production ; Landfill Gas Production ; Energy Plantations.

6.1. INTRODUCTION

All organic materials such as plants, trees and crops are potential sources of energy and are collectively called biomass. The plants may be grown on land (terrestrial plants) or grown on water (aquatic plants). Biomass also includes forest crops and residues, crops grown especially for their energy content on "energy farms", animal manure, wood waste and bagasse. Coal, oil and natural gas may take million of years to form, but biomass can be considered renewable energy source because plant life renews and adds to itself every year. It can also be considered a form of solar energy as the latter is used indirectly to grow these plants by photosynthesis by the following reaction.

$$6XO_2 + 6H_2O + \text{Light (Minimum 8 photons)} \longrightarrow$$
$$C_6H_{12}O_6 \text{ (glucose)} + 6O_2$$

The biomass sources are highly dispersed and bulky and contain large amounts of water (50 to 90 percent). Thus, it is not economical to transport them over long distances, and conversion into usable energy must take place close to source, which is limited to particular regions. However, biomass can be converted to liquid or gaseous fuels, thereby increasing its energy density and making transportation feasible over long distances.

6.2. AVAILABILITY OF BIOMASS

The total terrestrial crop alone is about 2×10^{12} metric tonnes. This includes

1. Sugar crops such as sugarcane and sweet sorghum ;
2. Herbaceous crops, which are non-woody plants that are easily converted into liquid or gaseous fuels ; and
3. Silviculture (forestry) plants such as cultured hybrid poplar, sycamore, sweetgum, alder, eucalyptus, and other hardwoods.

The terrestrial crops have an energy potential of 3×10^{22} joules. The efficiency of solar energy utilization in natural photosynthesis is only 0.1 to 2%. At present only 1% of world biomass is used for energy conversion.

Current research focuses on the screening and identification of species that are suitable for short-rotation growing and on the optimum techniques for planting, fertilization, harvesting, and conversion. Fast growing trees, sugar, starch and oil containing plants can be cultivated which have about 5% efficiency of solar energy utilization.

The estimated production of agricultural residue in India is 200 million tonnes per year and that of wood is 130 million tonnes. At an average heating value of 18 MJ/kg db, a total potential of 6×10^{12} MJ/year or approximately 75×10^{7} MJ/hour exists. At a power conversion rate of 35%, total useful potential is about 75,000 MW. This can supply all our villages with power at a rate of 30,000 kWh per day per village against the present meagre consumption of only 150 kWh per day per village. With the electrification of the irrigation pumps, there will be a boost in agricultural production resulting in availability of more biomass for energy recovery and hence the process is self adjusting.

Aquatic crops are grown in fresh, sea, and brackish waters both submerged and emergent plants. These include seaweeds, marine algae, etc.

Animal and human waste are an indirect terrestrial crop from which methane for combustion and ethylene can be produced while retaining the fertilizer value of the manure. The daily produce of cowdung is 13.5 kg per cattle which can be used to produce 0.46 m^3 of bio-gas in a Gobar Gas plant. This gas is sufficient to produce 1 kWh of electricity in a bio-gas engine. In India, there are sufficient number of cattles in each village, the animal waste can be used to meet the total energy requirements via above technology.

The human waste can be also be used for production of bio-gas. Community latrines can be planned in the villages for collection of night soil for feeding to biogas plants. Wastes of 200 persons can be used to produce about 5m^3 of gas per day to extract 12 kWh of equivalent energy by running a biogas engine.

6.3. BIOMASS CONVERSION

Biomass can either be utilized directly as a fuel, or can be converted into liquid or gaseous fuels, which can also be used as feedstock for industries. Most biomass in dry state can be burned directly to produce heat, steam or electricity. On the other hand biological conversion technologies utilize natural anaerobic decay processes to produce high quality fuels from biomass.

Various possible conversion technologies for getting different products from biomass are shown in Fig. 6.1. These technologies can be grouped as :

1. Direct combustion, such as wood waste and baggage,
2. Thermo chemical conversion, and
3. Biochemical conversion.

6.3.1. Combustion of Biomass

Various combustion techniques are available for burning biomass, the most suitable depending upon actual application. The selection of most suitable combustion equipment will depend upon properties, such as chemical analysis, volatile content, calorific value, particle size and ash characteristics, as well as the specific application. However, the versatility of the fluidised bed combustion makes it a strong candidate in many cases. This technology may be used for the efficient combustion of forestry and agricultural waste material such as sawdust,

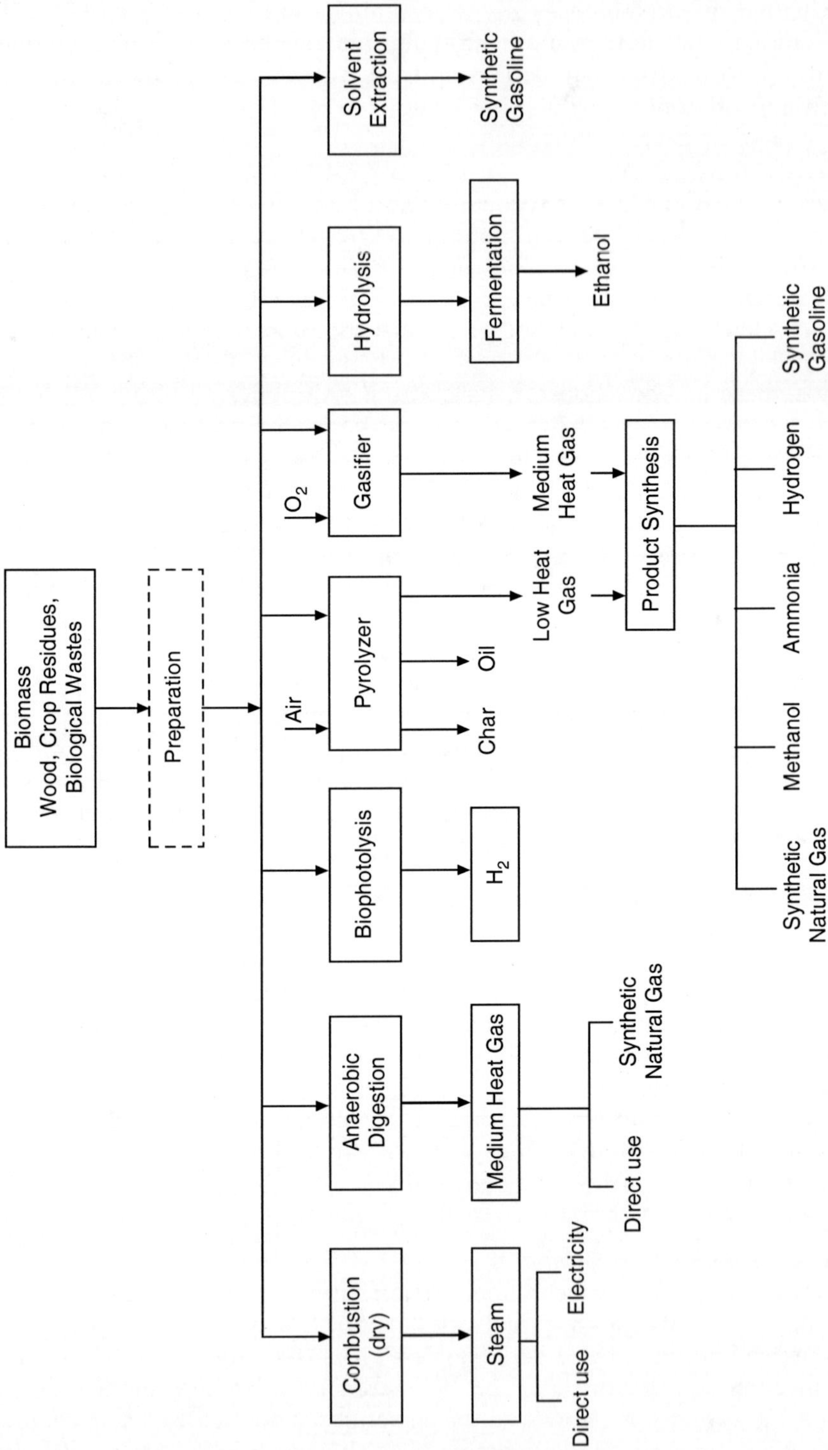

Fig. 6.1. Biomass Conversion Technologies and Products.

wood chips, hog fuel, rice husks, straws nutshells and chips. It offers interesting possibilities in efficient energy recovery.

The fluidised bed may be used in an air heater, liquid phase heater or steam generator. When fired directly into a boiler, steam can be generated for process or power. The ash is clean and inert and may be used as landfill or in concrete.

6.3.2. Thermochemical Conversion

There are two forms of thermochemical conversion : gasification and liquefaction. Gasification takes place by heating, the biomass with limited oxygen to produce low-heat gas or by reacting it with steam and oxygen at high pressure and temperature to produce medium-heat gas. The latter may be used as fuel directly or used in liquefaction by converting it to methanol (methyl alcohol, CH_3OH) or ethanol (ethyl alcohol, $CH_3\,CH_2\,OH$), or it may be converted to high-heat gas. Because the production of high-heat gas is more complex and expensive than low-heat gas, it is intended for use in lieu of natural gas in domestic and industrial applications. Low and medium-gases are considered for use as utility fuels.

6.3.5. Biochemical Conversion

There are two types of biochemical conversion : anaerobic digestion and fermentation.

Anaerobic Digestion

An anaerobe is a microorganism that can live and grow without air or oxygen. It gets its oxygen by the decomposition of matter containing it. Anaerobic digestion therefore, involves the microbial digestion of biomass. It has been used on animal manure (cowdung) and can be extended to other biomass. The process takes place at low temperatures upto 65°C, and requires a moisture content of atleast 80 per cent. It generates a gas consisting mostly of carbon dioxide and methane with minimal impurities such as hydrogen sulphide. The gas can be burned directly or upgraded to synthetic natural gas by removing the CO_2 and the impurities. The residue may consist of protein-rich sludge that can be used as animal feed and liquid effluents that are biologically treated by standard techniques and returned to the soil.

Fermentation

It is the breakdown of complex molecules in organic compounds under the influence of a ferment such as yeast, enzymes, etc. Fermentation is a well-established and widely used technology for the conversion of grains and sugar crops into ethanol. It can be mixed with gasoline to produce gasohol (90% gasoline, 10% ethanol). Gasohal is used as fuel in the internal combustion engines.

6.4. FLUIDISED BED COMBUSTION OF BIOMASS

Biomass is fed into a bed of hot inert particles, such as sand kept in fluidised state with air at sufficient velocity from below. The operating temperature is normally controlled within the range 750-950°C. Ideally it is kept as high as possible in order to maximize the rate of combustion and heat transfer but low enough to avoid the problem of sintering of the bed particles. The rapid mixing and turbulence within the fluidised bed enables efficient combustion to be achieved with high heat releases, as well as effective transfer, than in a conventional boiler. This can result in more compact boiler with less tubing.

A major advantage of fluidised bed combustion is that, because of the low temperatures involved, there is great potential for reducing and controlling atmospheric pollution. Nitrogen oxides are not formed from the nitrogen in the air and emission of the many of the trace elements associated with fuels are less than obtained using alternate combustion techniques.

Fluidised bed combustion technology has the ability to burn a wide range of material and can burn a combination of fuels or wastes in one unit. This versatility arises from the turbulent motion and large thermal inertia of the fluidised bed which enables the combustor to withstand the initial cooling action of fuels, particularly theoe high in ash and moisture as they enter the bed. This is typical characteristic of biomass. Agricultural wastes can be fired directly without any preparation with the exception of nutshells, which should be crushed.

6.4.1. Design of System

Properties of some biomass fuels are given in Table 6.1. The chemical composition and physical form of the biomass will influence the design of the fluidised bed combustion system. As compared to coal, the oxygen content of biomass is very high (about 40%) and contains much more volatiles and less fixed carbon. The stoichiometric air requirements will be quite different. The design of the fuel storage facilities, feeding arrangements and combustion equipment has to suit the higher volumetric feed rates necessary for biomass because of its generally low calorific value and density and to accommodate many shapes and sizes of the biomass.

Table 6.1. Combustion Data on Biomass

Type	Size (mm)	Bulk Density (kg/m^3)	Proximate Analysis				
			Moisture (%ar)	Ash (%db)	Volatile matter (%db)	Fixed carbon (%db)	Calorific Value (MJ/kg db)
1. Rice husk	*Ca* 2	150	12	11	72	17	18
2. Corn cobs	200 × 25	190	8	< 1	85	15	18
3. Baggasse	—	80	50	< 1	80	20	19
4. RDF shredded	—	130	16	15	65	20	17
Pelletised	25	500	16	15	65	20	17
5. Wood saw dust	Ca2	160	20	< 1	80	20	20
Chips	Ca10	310	20	< 1	80	20	20

Note : *ar* : as received
db : dry basis.

The high moisture content of biomass (up to 50 % as received) and volatiles (up to 80 % dry basis) give rise to some operational problems connected with controlling bed temperatures and avoiding excessive smoke production. The water in the fuel evaporates as it enters the bed and this requires heat from the bed. The volatile as released from the fuel tend to burn above the bed and donot provide heat for the bed itself. Therefore, relatively high fuel feed are needed to maintain bed temperature. However, with these high rates, incomplete combustion takes place with the limited oxygen provided by the fluidising air resulting in excessive smoke formation. This problem is aggravated with larger beds because of relatively poor lateral distribution of the fuel.

6.4.2. Types of Fluidised Beds

Various stages of fluidised beds are shown in Fig. 6.2. The pressure drop through the bed as a function of air velocity is shown in Fig. 6.3.

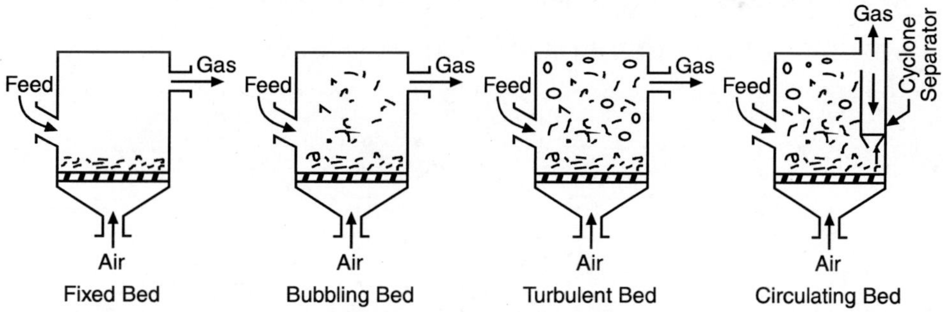

Fig. 6.2. Types of Fluidised Beds.

Feeding of fuel directly into the bed, operating the bed at high temperatures, using deep beds, using fluidising velocities designed to promote good mixing and distribution of fuel throughout the bed will help combustion of volatiles in the bed. Provision of secondary air just above the bed can help in reducing smoke emissions. Because biomass fuels have low densities, it is preferable to feed them as relatively large-sized pieces like corn, cobs wood chips, etc., into a bubbling bed.

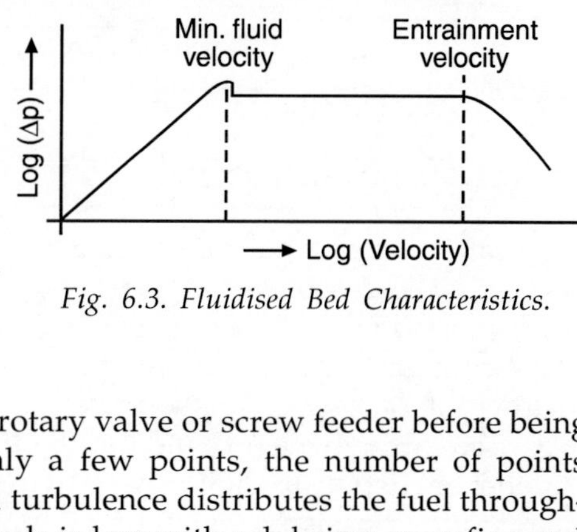

Fig. 6.3. Fluidised Bed Characteristics.

6.4.3. Feeding of Biomass

The fuel is first metered through a rotary valve or screw feeder before being dropped onto the fluidised bed at only a few points, the number of points depending on the overall bed area. Bed turbulence distributes the fuel throughout the bed. The ash content of these fuels is low with ash being very fine and easily elutriated from the fluidised bed.

6.4.4. Circulating Fluidised Bed

The finer feed stocks of biomass like rice husk and saw dust are difficult to burn satisfactorily in bubbling fluidised bed because of excessive elutriation at the fluidised velocities needed for good mixing of the fuel in bed. It is better to employ a circulating fluidised bed combustion system, for very finely divided forms of biomass. This circulating bed is characterised by high fluidising velocities in excess of 3m/s with entrainment of majority of bed particles. These are collected in a cyclone separator and returned to the combustor. The primary combustion air is fed through the main fluidising air distributor at the bottom while secondary air is introduced further up the reactor. Fine grains of sand are used as the bed material.

The fuel is screw-fed into the combustor which operates between 700°C and 1000°C depending on the fuel being burned. Because the circulating material is in a highly turbulent state and mixes rapidly, bed temperature remains very uniform. Boiler tubes are situated in the upper part of the combustor and downstream after the cyclone. It is a good system for burning biomass fuels with low densities and small size which can be difficult to burn in more conventional "bubbling" type of fluidised bed combustor.

6.5. CIRCULATING BED BOILER SYSTEM

The schematic diagram of a circulating bed boiler system is shown in Fig. 6.4.

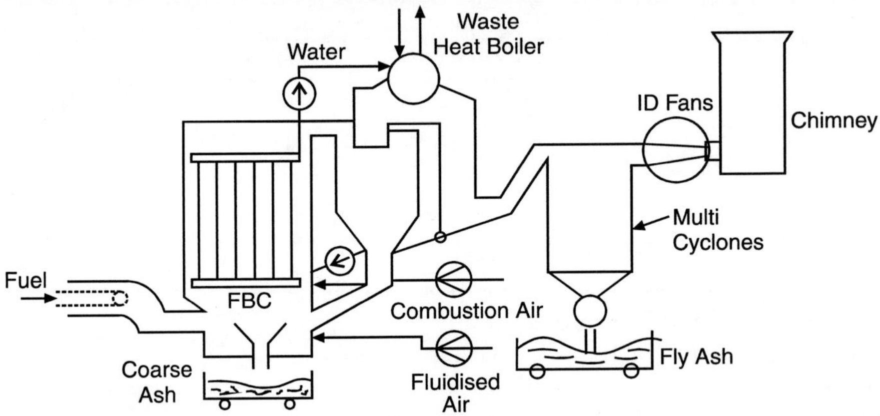

Fig. 6.4. Circulating Bed Boiler System.

The combustion of biomass in a fluidised bed requires that the bed be initially heated by means of a gas burner to a temperature of approximately 400°C. The gas burner is then shut off and biomass feed is started. The fuel ignites and from then on sustains combustion. In some cases, the bed remains hot enough to ignite the fuel directly without gas assistance even after a shut-down of 24 hours. Coarse materials burn within the bed. Finer portions are carried through the bed and burn in the hot gases over the bed. Rocks and other non-combustible foreign materials are automatically removed from the bed on a continuous basis by extracting a portion of the bed material, screening out foreign matter, and reintroducing the bed material to the combustion unit. Biogas can be easily made available for initial heating of the bed.

Analysis of the overall heat transfer process for a tube immersed in a fluidised bed shows that the main resistance to heat flow is on the fluidised bed side of the tube. Combination of fin designs and operating parameters have been evolved that lead to optimum performance. Ceramic tubes of silicon carbide in both plain and finned configuration are now commercially available. The bed material may by any clean, non-combustible material of uniform size. The system will require air clean-up equipment to satisfy an quality requirements of Air Pollution Control Board.

6.6. STEAM TURBINE CYCLE

Atmospheric fluidised bed (AFB) system has been developed for combustion of biomass. This AFB is used in conjunction with a steam turbine cycle as shown in Fig. 6.5.

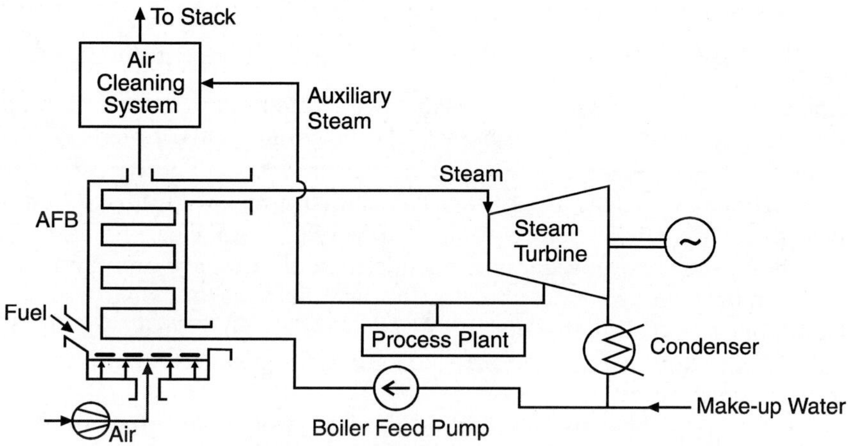

Fig. 6.5. Steam Turbine Cycle System.

Heat is transferred to water by heat transfer surfaces within the bed, in the water-cooled walls, and in the convective space above the bed. A fuel conversion efficiency of 71% has been claimed.

6.7. GAS TURBINE CYCLES

6.7.1. Indirect-fired open-cycle Gas Turbine System

An indirect-fired open-cycle system is shown in Fig. 6.6.

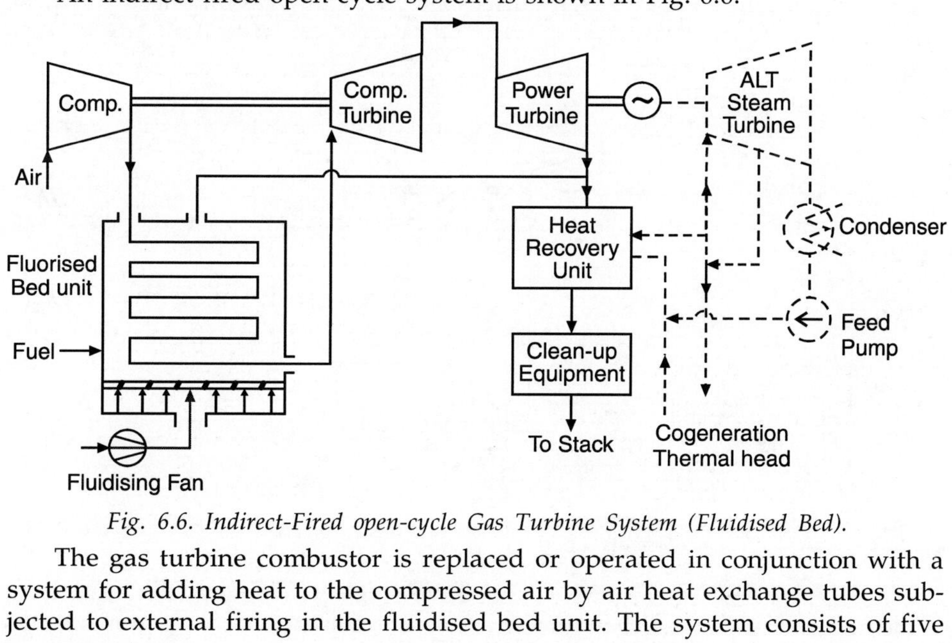

Fig. 6.6. Indirect-Fired open-cycle Gas Turbine System (Fluidised Bed).

The gas turbine combustor is replaced or operated in conjunction with a system for adding heat to the compressed air by air heat exchange tubes subjected to external firing in the fluidised bed unit. The system consists of five components:

— a fluidised-bed combustor
— a heat exchanger
— a gas turbine generator set
— cyclone separator
— emission control equipment (baghouse).

The system can be provided with a steam bottoming plant for efficient operation. The exhaust from the fluidised bed is joined to the exhaust from gas turbine. The combined stream is used to generate high pressure, high temperature steam in a heat recovery boiler. The bottoming steam plant works in the normal manner.

A primary gas clean-up system is installed after the fluidised bed to catch the larger ash particles before going to the boiler. Final clean-up can take place in the boiler by soot blowing and outside the boiler by cyclones and baghouse.

A separate air source is used for fluidisation to assure clean hot air for gas turbine operation. If some portion of compressor air is used for fluidisation, thorough clean-up is required prior to joining the main air stream for introduction into gas turbine.

Heat transfer tubes used with fluidised-bed system are both over the bed and immersed in the bed. Air from the compressor enters the over-bed tubes and proceeds down to the in-bed tubes, which operate at the highest temperature. Tube material is selected for both corrosion resistance on the outside surface and high temperature. Ceramic tubes are available with external extended surface rings.

The control of the system is achieved by control of fuel as well as air supply. The air should never be less than that required to keep the bed in fluid state.

6.7.2. Closed Cycle Gas Turbine System

In the closed system, the working fluid (gas) flows under pressure through a compressor, heat exchanger, fluidised-bed gas heater, gas turbine, again through

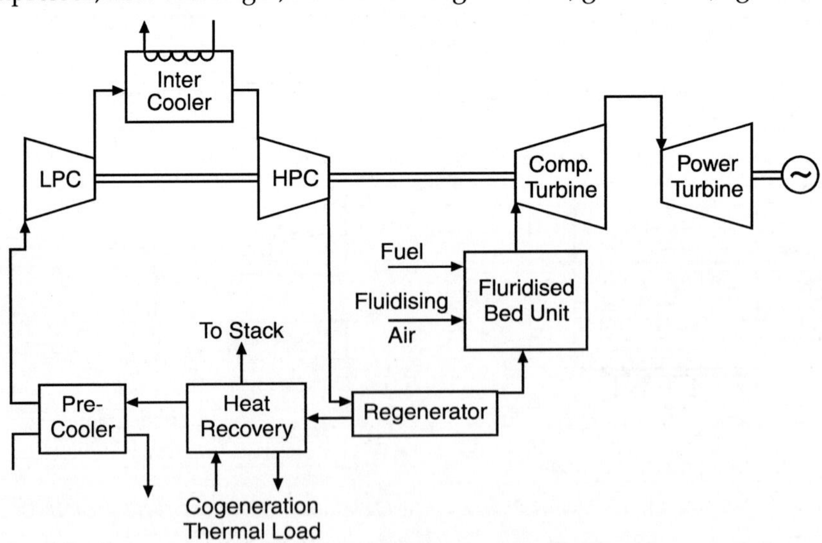

Fig. 6.7. Closed Cycle Gas Turbine System (Fluidised Bed).

the heat exchanger and then through the precooler back to the compressor inlet. The schematic diagram is shown in Fig. 6.7.

The gas may be air, nitrogen or helium, which is not in direct contact with the products of combustion. The heat rejected by intercooler and precooler can be recovered by a secondary heat transfer loop to generate low pressure steam to activate an absorption chiller for a cold storage or to produce additional power through an organic bottoming cycle. The cycle efficiency of a closed-loop gas turbine system is well above 42%. Control of air can be accomplished by passing air from the compressor discharge around the regenerator and to the turbine exhaust. While compressor power remains almost constant the turbine output decreases as a result of simultaneous decrease in turbine mass flow, expansion ratio and efficiency. The fluidised bed combustion unit will have both bed-immersed tubes and overbed tubes.

The fluidised bed unit can work in conjunction with various cycle configurations using steam turbines, gas turbines, combined cycles with and without cogeneration of thermal loads. The preferred system for burning many types of biomass is fluidised bed combustion with circulating system and bubbling beds more suitable for coarser sized biomass.

6.8 A CO-OPERATIVE RURAL POWER PLANT

All the energy needs of a village can be met locally from the available biomass using the fluidised bed technology. A project on the design of a cooperative rural power plant based on fluidised bed combustion system of biomass was undertaken at Regional Engineering college, Kurukshetra. The salient features of the project are described below.

6.8.1. Survey Data of Village 'BIROHAR' in District Kurukshetra

Total population	:	3000
No. of houses	:	400
Area under cultivation	:	1200 acres
Wheat and rice	:	500
Sugarcane	:	400
Mustard	:	200
Miscellaneous	:	100
Method of irrigation	:	100 pumps of 5 kW each with sprinkler system.
Other electrical loads :		
Flour mills	:	2 × 7.5 kW
Domestic flour mills	:	100 × 1.0 kW
(Also used for shredding of cattle feed in addition to milling)		
Thrasher	:	20 × 7.5 kW

6.8.2. Biomass Availability

Saw dust, wood chips from saw mill : 3 tonne per day.

Agricultural residues from wheat and mustard crops

@1.6t/acre/year	:	320tonne per year
Rice husk	:	800 tonne per year
Baggage (dried 2tonne/acre)	:	800 tonne per year
Total biomass available	:	7.5 tonne per day

Total biomass required for the proposed plant of $3MW_t$ capacity based on 35% overall efficiency and biomass heating value of 15 MJ/kg is 25t per day. Therefore, it is required to collect the additional requirement from the surrounding villages or by undertaking special fuel crop plantation.

Agricultural residues are seasonal depending upon the crop. However, some overlay and with properly managed storage areas, a fairly uniform feed supply can be ensured. The biomass can be stored in an enclosed area of 15m × 15m with a heap height of 6m. This is worked out on the basis of 60 day's inventory and bulk density of biomass of 150 kg/m^3. A shredding machine is required to keep the feed size of 6 cm.

The biomass, sized and dried is fed to the combustion chamber through belt conveyors and screw feeder 25 cm diameter × 16 m long driven by 1 kW motor. The feed control is automatically affected by regulating the speed of the motor. A 600 mm rotary valve is used to feed biomass to the combustor under pressure.

6.8.3. Proposed Focal Village

1. Industrial Estate

Saw mill	:	7.5 kW
Toy making machine	:	7.5 kW
Milk chilling plant	:	15 kW
Rice mill	:	15 kW
Repair workshop	:	35 kW
Flour mill	:	15 kW

2. Agricultural Load

Irrigation pump sets (100 × 5 kW)	:	500 kW
Thrashers	:	15 kW

3. Domestic Load

Fans, lighting, radio and TV sets 400 × 500 W	:	200kW
Domestic flour mills 100 × 1.5 kW	:	150 kW

4. Community Load

Street lights 50 × 40W	:	2 kW
Hospital, school and community centre	:	16 kW

5. Total connected load : 1000 kW

Required installed capacity with diversity factor 1.5	:	660 kW

6.8.4. Plant Specifications

Capacity	:	3 MWt
Plant	:	Circulation type

Biomass (mixture of saw dust, baggage, rice husk, crop, residues	:	25t/day
Reactor size	:	1.4m dia × 1.4m bed height.
Bed material (lined with fire bricks)	:	2.8t of 16 mesh sand.
Fluid air	:	3.65 kg/s at a velocity of 2m/s.
Fluid air system	:	Roots blower 2.86 m^3/s discharge and 1.9 bar pressure with 375 kW motor.
Parasite power requirement of power plant	:	340 kW
Total plant capacity	:	1000 kW.

6.8.5. Component Design

1. Distributor Plate

Three compartments for differential velocities to ensure circulation of sand and biomass inside the furnace. The minimum velocity through sieve is 2m/s and velocities in the three chambers are 2.4m/s, 2.8m/s and 3.2m/s.

2. Dust Collection System

A cyclone separator is provided after the heat recovery unit and before the stack to remove dust particles. The cyclone separator with 98% efficiency will be 1150 mm diameter and 2300 mm height.

3. Steam Tubes

Ceramic tubes of silicon carbide with 2 fins per centre (disk type), 80 mm diameter and 1.6 mm thick 234 tubes of 1500 mm length will be required.

4. Steam Parameters

Super heated steam at 20 bar and 275°C.

Flow rate : 2.7 t/hour

5. Automatic control and Interlocking

The aim of control system is to meet the variable load demand effectively. The main requirements are :

(*i*) to maintain highest combustion efficiency

(*ii*) to regulate air supply in fluidised bed.

(*iii*) to control fuel feed

(*iv*) to control steam flow rate

The controlling parameter used is steam pressure which is used to control feedwater rate, fuel supply rate and air quantity. Electrical system is used for transmitting signals to water control valve, fuel feed motor and blower damper.

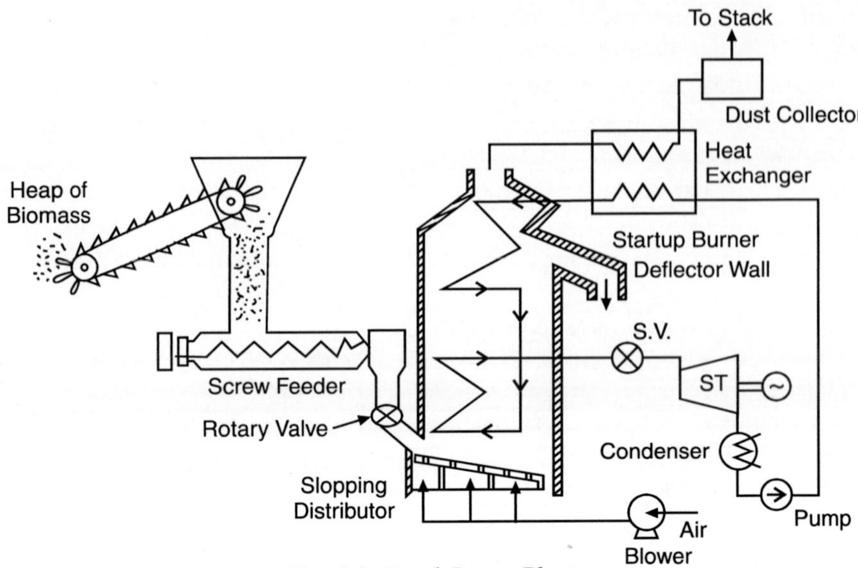

Fig. 6.8. Rural Power Plant.

6.9. BIOMASS GASIFICATION

Gasifier is essentially a thermochemical reactor in which the biomass undergoes partial oxidation and producer gas is obtained. The agricultural and forest residues can first be dried in solar driers to restrict the moisture content below 15 per cent. These are pressed in compaction briquetting machines to prepare feed size of 50 – 75mm for the gasifier. The biomass briquets are ignited in the gasifier and air flow is controlled to ensure partial combustion. The producer gas is reduced to CO and H_2 by carbon present in the burnt lower layers. The main constituents of fuel gas are CO, H_2 and CH_4.

The gas produced by the gasifier is hot and contains tar, vapours and soot particles. To make it engine worthy, it is cooled to near ambient conditions and cleaned to remove tar and dust by cross current water scrubbers. Air and clean gas are supplied to the engine in pre-specified proportions. On an average 1 kg of biomass produces 2.5 nm^3 of fuel gas and 1 kWh of shaft energy to drive the irrigation pump.

The diesel engine utilizes the gas as a supplementary fuel operating on dual-fuel mode. The solar driers, briquetting machines and gasifiers are now made in India. The total cost of a biomass gasifier with gas cooler and scrubber of $15m^3$/hr capacity, 5 kW/1500 rpm diesel engine and 100 × 100 mm centrifugal pump with necessary controls and accessories is about Rs. 37000.

There are a variety of types of biomass gasifiers.

1. Fixed bed gasifiers. The three main designs of fixed bed gasifiers are up-draught, down-draught and cross draught gasifiers depending upon the direction of flow of air.
2. Fluidised bed gasifiers which have many advantages over fixed bed gasifiers.
   (*i*) Quick start up
   (*ii*) High combustion efficiency

(*iii*) High output rate
(*iv*) Consistent rate of combustion.
(*v*) Usage of fuel with high moisture content.
(*vi*) Rapid response to fuel input changes
(*vii*) Fuel flexibility
(*viii*) Good heat storage capacity
(*ix*) Compact size
(*x*) Reduced emission of harmful nitrogen oxides.
(*xi*) Simple operation

Fig. 6.9. gives the flow diagram of oxygen donor gasifier commercial plant along with various applications of fuel gas.

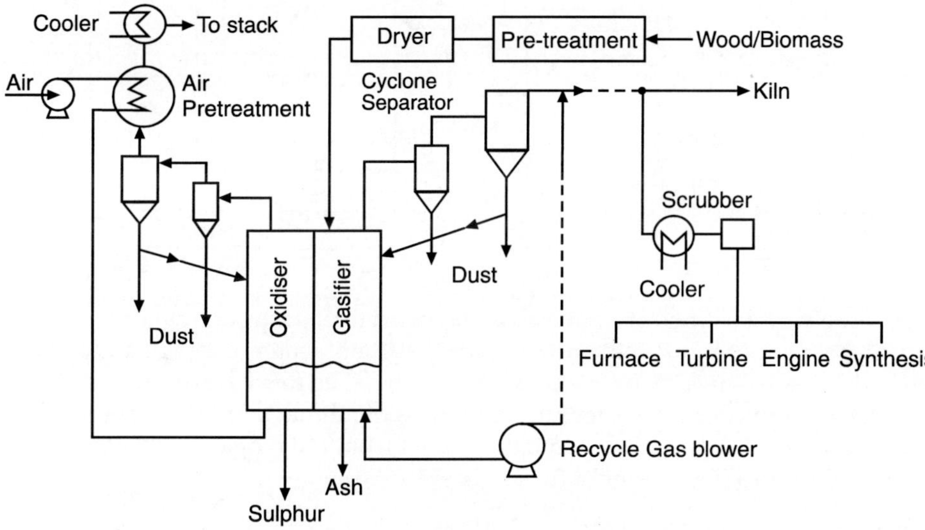

Fig. 6.9. Biomass Gasifier.

6.9.1. Chemistry of Gasification Process

The reactions take place in three zones of the gasifier bed : oxidation, reduction and distillation.

$$\text{Biomass (C)} + O_2 + 3.79N_2 = 3.79N_2 + CO_2 + \text{Heat}$$

The oxidation reaction is exothermic and 395,000 kJ/kg atom of carbon in biomass heat is produced.

The carbon dioxide formed is reduced in the presence of carbon. Over 90% of CO_2 is reduced to CO at 900° C.

$$C + CO_2 + 3.79N_2 + \text{Heat} = 3.79N_2 + 2CO$$

It is an endothermic reaction and 172,000 kJ/kg atom of carbon heat is absorbed in the reaction.

The oxygen requirement is 0.25 – 0.3 kg/kg of biomass. The dry gas has the following composition by volume.

H_2	: 36.2%
CO	: 44.4%
CO_2	: 19.1%
CH_4	: 0.3%

6.10. BIOGAS GENERATION

Most organic materials undergo a natural anaerobic digestion in the presence of moisture and absence of oxygen. 60 – 80% of carbon of biomass is converted to a gas called biogas. This contains CH_4, CO_2, N_2 and traces of H_2S. The biogas has a heating value of 18.8 to 26.4 MJ/m^3. The biogas can be upgraded to a synthetic natural gas (SNG) by removing CO_2 and H_2S. The heating value of SNG is 37 MJ/m^3. One kilogram of dry organic material produces 0.12 to 0.18m^3 of methane with an energy content of 4.5 to 7 MJ/kg raw material.

A biogas plant is schematically shown in Fig. 6.10.

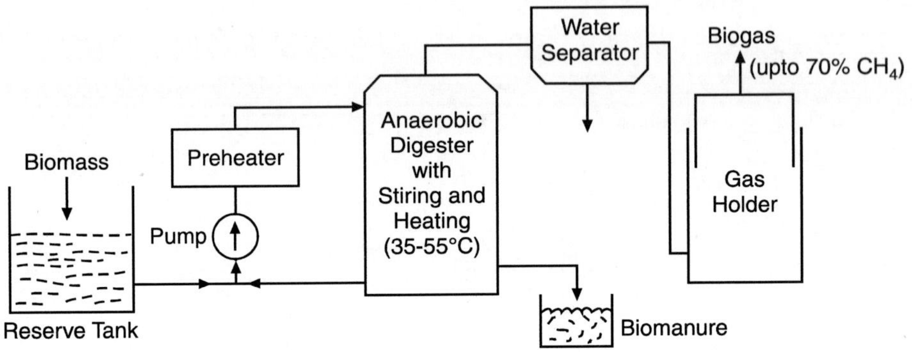

Fig. 6.10. Biogas Plant.

The various types of biomass can be vegetable waste, animal dung, silt from sewage plant of municipal and industrial effluent water, garbage and vegetable residues, waste from food processing industry. The rate of production of biogas is 0.3 to 0.7 m^3 per kg of organic dried mass of animal and vegetable waste. The water vapour and sulphur can be removed from the biogas and cleaned biogas can be used to drive engine for cogeneration plant.

6.11. GOBAR GAS PLANT

Cowdung is a wonderful waste to meet the fuel and fertilizer requirements in rural areas. With the help of modern technology of anaerobic digestion, the biogas obtained form cowdung can be effectively used to meet the energy requirements for cooking and heating by direct burning of biogas. This gas can be fed to engines to generate power for lighting, pumping of water, agriculture and small village industries. Of course, spent slurry is a useful by-product with nitrogen content of 1.5 to 2% to be used as a good manure.

The cattle population in India is 235 million heads with a dry dung yield of about 170 × 10^6 tonnes/year.

Livestock and indeed human waste have much potential as an energy source via anaerobic digestion. The wastes are treated with the simultaneous generation of methane fuel and retention of nitrogen and other vital plant nutrients in the remaining sludge which can be recycled as a fertilizer.

Biogas generation is a versatile process ranging from small domestic Gobar gas plants catered by a few cattles to a much bigger community type plant where relatively large volume of cowdung is continuously available. These depend greatly on a hot climate and manual labour for its functioning.

6.11.1. Rural Energy Balance

In India, more than one lakh biogas plants developed by Khadi and Village Industries Commission are already operating. There is a potential of producing about 22,000 × 10^6 cubic metres of biogas. The biogas has approximately 65% CH_4 and has a heating value of 21 MJ/m^3. For a typical village of population 500, the number of cattle would be around 250. The amount of dry dung collected from them is 180 tonnes/year. This would be sufficient to meet the complete energy requirements of the village population except for two or three months of winter when there will be a shortfall of 20– 40%. However, with further optimization of the anaerobic digestion process this shortage could conceivably be reduced to zero. Given the appropriate innovative technology, and equally important, the willingness to use it, the present quantity of cowdung can be used to fully meet the energy requirements of rural India.

6.11.2. Biogas Engines

The requirement of Indian farmers for engines varies from 3.7 kW to 7.5 kW in vertical high speed and vertical slow speed types all over the country. Normally, the engine runs on an average for 5 hours a day. The biogas engines are essentially diesel engines where air intake is connected to biogas supply. The diesel is required for starting the engine and about 20 percent diesel and 80 per cent biogas are mixed automatically to produce power for running the pumpset. The biogas consumption is about 0.46m^3/kWh which can be obtained from 13.5 kg of cattle dung, *i.e*, daily produce of one cattle. The price of a typical 3.7 kW, 1500 rpm biogas engine with centrifugal pump of 100 × 100 mm including accessories and trolley is about Rs, 6000. It is subsidised by the Government and can be financed by the State Agro Industries, Land Development Banks and all nationalized banks. The cost of a typical gobar gas plant of 85m^3 capacity is about one lakh rupees.

6.11.3. Biogas Digesters

Biogas is produced by digestion in the presence of anaerobic organisms in the absence of oxygen at ambient pressure and temperatures of 35 – 70°C. The container in which digestion takes place is known as the digester. The anaerobic digestion consists of enzymatic hydrolysis to breakdown cellulosic biomass into simple compounds, these organic compounds are then broken down into organic acids, these organic acids are then converted into methane (CH_4) and carbon dioxide (CO_2).

The efficiency of biogas (CH_4 + CO_2) generation depends upon the following factors:

1. Acid formers and methane fermentors must remain in a state of dynamic equilibrium which can be achieved by proper design of digester.
2. A *pH* value between 6.5 to 8 must be maintained for best fermentation and normal gas production.
3. For anaerobic digestion, temperature variation should not be more than 2 to 3°C. Methane bacteria work best at 35 – 38°C.
4. A specific ratio of carbon to nitrogen (C/N ratio) must be maintained between 25:1 and 30:1 depending upon the raw material used.
5. The water content should be around 90% of the weight of the total contents.

6. The slury should be agitated to improve the gas yield.
7. Loading rate should be optimum. If digester is loaded with too much raw material, acids will accumulate and fermentation will be affected.

Digester Design

1. There are many shapes of digester tanks used in practice.
2. Two types of covers are used. Fixed cover has low cost and more resistance to corrosion. Floating type cover has the following advantages:
   (*i*) The danger of mixing oxygen with the gas is less. Therefore, danger of explosive mixture formation is reduced.
   (*ii*) Fresh waste can be added to the tank and slurry can be withdrawn without difficulty while digester is in operation. No special pressure equalising device is required.
   (*iii*) The solids are continuously submerged and there is less problem of floating matter.

Capacity of the Digester

The capacity of the digester tank can be estimated as follows

$$\text{Volume of digester} = \left(\frac{Q_1 + Q_2}{2}\right)t$$

where Q_1 = Daily rate of raw waste added
Q_2 = Daily volume of waste after digestion
t = Digestion time in days.

Digestion time to produce good sludge depends upon the temperature in the tank. In winters, the sewage sludge digestion tanks are heated to 35°C.

The following guidelines may be used to fix optimum size of a biogas plant:

1. Daily rate of waste to be digested
2. Type of waste
3. Period of digestion
4. Method of stirring, if any
5. Arrangement for raw waste feeding and discharge of digested slurry
6. Climatic conditions
7. Mix of raw waste
8. Water table and sub-soil conditions
9. Type of dome.

No separate heating and stirring may be used for agriculture waste.

6.11.4. Types of Digesters

There are numerous types and designs of biogas digesters. Mainly two types of digesters are used for community biogas generation.

1. KVIC Community-type Gobar Gas Plant. The Khadi and Village Industries Commission of India has designed and patented a "Grah Luxmi" Gobar Gas plant. It has been adopted in Indian villages on a large scale.

The digester is pit made of masonary. It has a diameter of 1.2 to 6m and well depth of 3.5 to 6 m. A vertical wall divides the well into two semi-cylindrical compartments. The partition wall is submerged in the slurry. Cement pipes of 10 cm diameter serve as inlet and outlet pipes. The dung and water are mixed in the ratio of 4:5 in the feed tank and fed to the well.

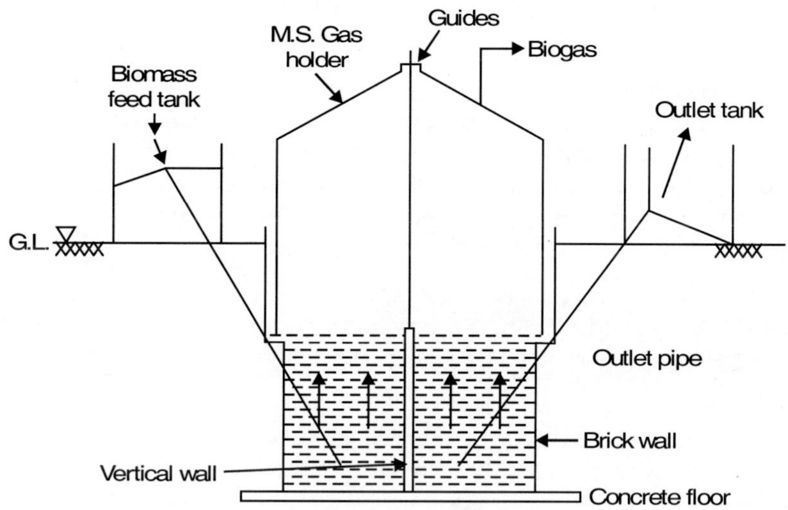

Fig. 6.10(a) KVIC Gobar Gas Plant

The digester can hold the raw material for 60 days. When gobar slurry is added, fermented slurry flows out in the outlet tank and collected in a composed pit.

A floating type mild steel drum fits the digester at the top and sinks into the slurry. When gas is generated, the float cover rises and floats freely on the surface of the slurry. A central guide pipe prevents its tilting. The floating drum also acts as gas seal and ensures a gas pressure of 10–15 cm of water gauge. The biogas is tapped from the drum top.

The cost of steel drum is 40% of total plant cost and corrosion of drum demands regular maintenance.

2. Chinese Digester

In KVIC gobar gas plant, biogas is the main product and manure is a by-product. Chinese use the digester for manure and biogas as by-product.

It is a fixed dome design and gas is available at variable pressure. It is totally made of masonary. The construction cost may be 40% of KVIC design and it has very little maintenance requirements.

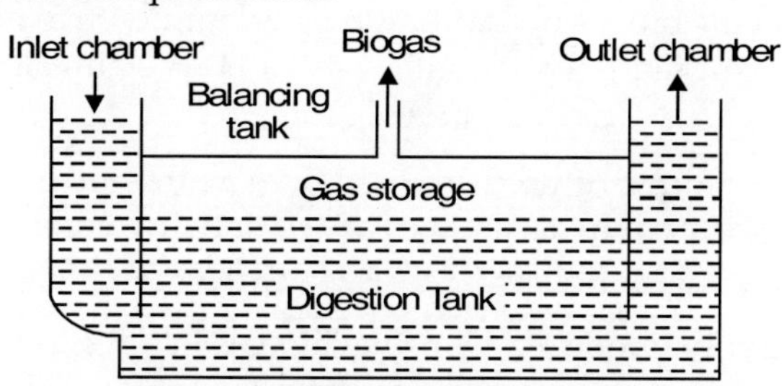

Fig. 6.10.(b) Chinease Digester.

6.11.5. Raw Materials for Biogas Generation

Forestary and agricultural residues can be best utilized to generate electricity or process heat by fluidised bed combustion. These can also be dried and used for gasification. Cowdung is a wonderful waste for generation of biogas by anaerobic digestion. Other raw materials are :

1. Water Hyacinth.

It grows as floating water plant in the rivers and canals. It contains 95 percent water and 5 percent cellulose, lignin, etc. It grows in waste waters thereby cleaning the ponds. It gives about 350–420 litres of biogas per kg of dry weight. It is one of the important sources of biogas production next to animal wastes.

2. Algae

Algae plants grow in lakes, tanks, etc. The best place for their cultivation is in shallow land ponds. It can be anaerobically fermented to produce methane with a lower calorific value of 15000 kJ/kg of dry algae.

3. Ocean kelp

It is a kind of sea weed which grows in the coastal areas and also in the high seas. Yield rates range from 300–500 wet tonnes/acre/year.

4. Grasses

Some fast growing grasses can also be used for biogas generation. Napier grass, Sudan grass and Bangola grass give the best yield. The cut grass is mixed with water and fed to bio-gas digesters.

6.11.6. Bio-Gas Applications

Biogas is a flammable fuel gas with 60 percent methane and rest CO_2. Its heating value is about 18 kJ/m^3. The gas can be upgraded by removal of CO_2 with water scrubbing and the gas with high heating value can be used in I.C. engine. The main applications of biogas are:

1. Cooking

There are thousands of homes in rural areas using biogas as fuel for cooking. Low cost burners have been designed for the hot plate.

2. Domestic lighting and heating

The biogas supply from the digester is sent through a rubber hose where a nozzle is used for the lamp and stove. The gas sprays out from the nozzle at a very high velocity and mixes with air drawn into the mixing chamber due to low pressure after the nozzle. The brightness and force of combustion of stove and lamp depend on the biogas pressure. A biogas lamp with a mantle can run for six hours per m^3 biogas and give a luminosity of 60 W equivalent of electrical light.

3. I.C. Engines

Biogas can be directly used in SI engines with suitable modification in the carburettor. However, bio-gas engines are essentially diesel engines (Para 6.11.2).

4. Fuel Cell

Electricity can be produced by using bio-gas in a fuel cell with air as oxidant. The electrolyte is usually potassium hydroxide (Table 11.1)

6.11.7. Advantages

1. The initial investment is low due to simple plant.
2. The raw materials for digester can be grown and cultivated like conventional agriculture.
3. The technology is very suitable for rural areas.
4. Biogas is locally generated and can be easily distributed for domestic use.
5. The raw material utilization also helps to keep the villages clean.
6. The by-products like nitrogen rich manure can be used with advantage.

The main disadvantages are requirement of large land areas for cultivation of biomass and difficulties in the distribution of bio-gas as the same cannot be liquefied.

6.12. BIO FUEL PRODUCTION

Non-petroleum liquid fuels find use when petroleum fuels are scarce or costly. Methanol (CH_3OH) and ethanol (C_2H_5OH) are used as fuels in S.I. engines. Ethanol due to its very high heat of vaporization is used in racing cars.

Alcohals are mainly used for blending with gasoline Ethanol or ethyl alcohol can be produced by fermentation of carbohydrates which occur naturally and abundantly in some plants like sugarcane and from starchy materials like corn and potatoes. Methanol can be produced from municipal solid wastes and specially grown biomass.

6.12.1. Manufacture of Ethanol

The starch in grain is converted into sugar by means of enzymes and the sugar is then fermented with yeast to produce a dilute alcohol solution. Distillation is used to seperate and purify the alcohol solution.

(a) Ethanol from Grains

The grains (corn, wheat etc.) are ground and cooked with water to covert starch to sugar with the enzymes. The sugar is further fermented with yeast to produce raw ethanol. It is distilled to remove impurities and water. The extractive distillation with benzene yields anhyodrous ethanol. The distillers dried grain (DDG) is dried and removed for sale as cattle feed. The capacity is about 1.4 million barallel per stream day.

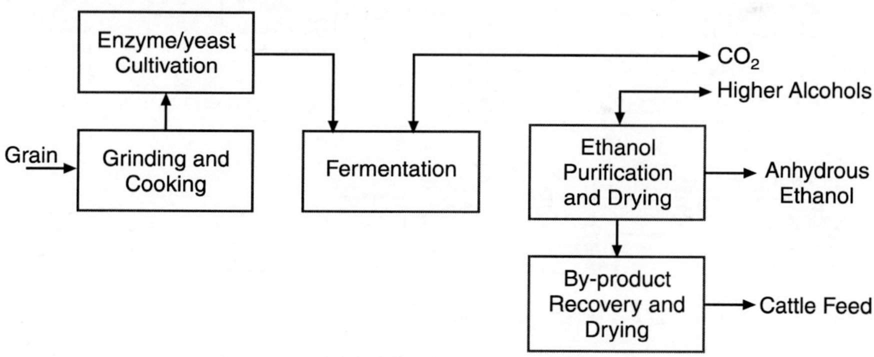

Fig. 6.11. Ethanol Production from Grain.

(b) Ethanol from sugarcane

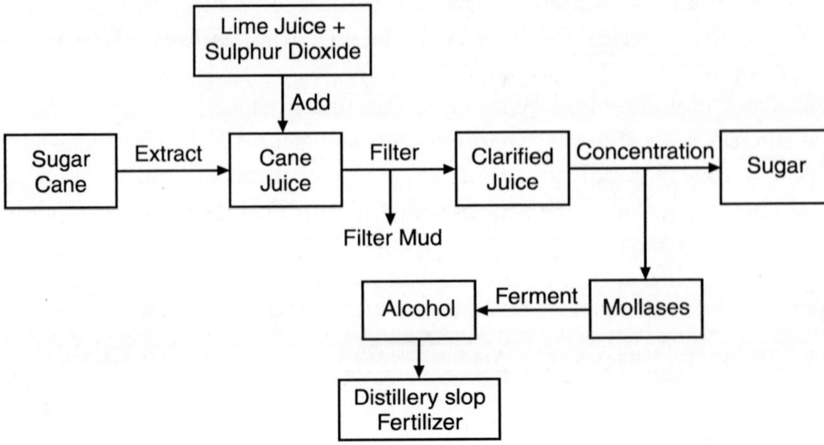

Fig. 6.12. Ethanol Production from Sugarcane

The cane is cut and ground and cane juice is extracted by maceration. After classification and concentration through evaporation , the juice is fermented with yeast to produce raw ethanol. A series of distillation steps including a final extractive distillation with benzene are used to obtain anhydrous ethanol.

The normal yield of ethanol is about 8.73 litres of alcohol per tonne of cane processed in a sugar factory. The potential of ethanol production in India is about 475 million litres per year.

6.12.2. Manufacture of Methanol

The wastes are shredded and passed through magnets to remove iron. The iron-free wastes are gasified with O_2. The product synthesis gas is scrubbed by water to remove particulates, entrained oil, H_2S and CO_2. Co-shift conversion for H_2/Co/Co ratio adjustment, methanol synthesis and methanol purification are accomplished to produce methanol.

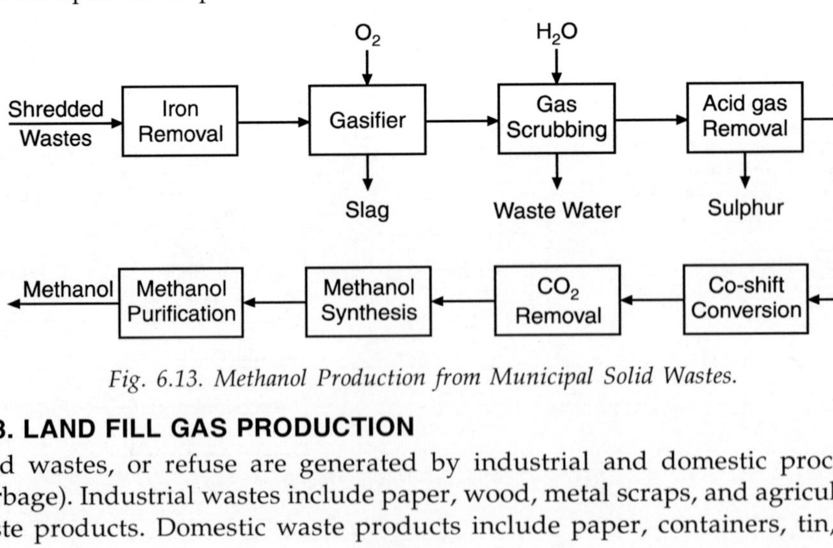

Fig. 6.13. Methanol Production from Municipal Solid Wastes.

6.13. LAND FILL GAS PRODUCTION

Solid wastes, or refuse are generated by industrial and domestic processes (garbage). Industrial wastes include paper, wood, metal scraps, and agricultural waste products. Domestic waste products include paper, containers, tin, aluminium, food scraps and sewage. In developed countries the solid waste production can be as high as 1 metric ton per person per year. A typical sample of municipal waste can be as given in table. 6.2. Some wastes like paper, metals and woods can be recycled for reuse. Much of it can be burned for heating and generating steam for electric power plants. Refuse burning in incinerators has been a wide practice in many parts of the world. The energy potential of refuse is not too great by burning. But refuse has low Sulphur content.

Table 6.2. Typical Composition of Municipal Waste

Material	% age	Material	% age
Paper	50.7	Leather, rubber, plastic	3.3
Food	19.1	Wood	2.9
Metal	10.0	Textiles	2.1
Glass	9.7	Miscellaneous	1.7

The main problems are the wide assortment of constituents, a high moisture content, dangers of explosions due to careless volatile fuel dumping and metal sparks during processing and wide variations in heat content. Therefore, a mixture of solid waste and fossil fuel, with refuse supplying 10 to 20 percent of the heat input to the boilers is being used for power plants. The costs of collection, transportation and processing (removal of ferrous metals and non-combustibles, shredding and hammer milling) have to be considered against its use for land filling.

The landfilling yields gas @ $8m^3$/annum. The approximate composition of landfill gas is :

CH_4 : 59%
CO_2 : 40%
N_2 : 1%
$HC + H_2S$: Traces

The lower heating value is only $5MJ/m^3$. In case of large-scale land-fill gas, it can be processed to be used as a substitute for natural gas. In an absorption or membrane separation process, CO_2 is removed and the methane content can reach 98%. A gas turbine power plant working on landfill gas is operating on the suburbs of India's capital. Fig 6.13. shows the layout of a landfill gas plant.

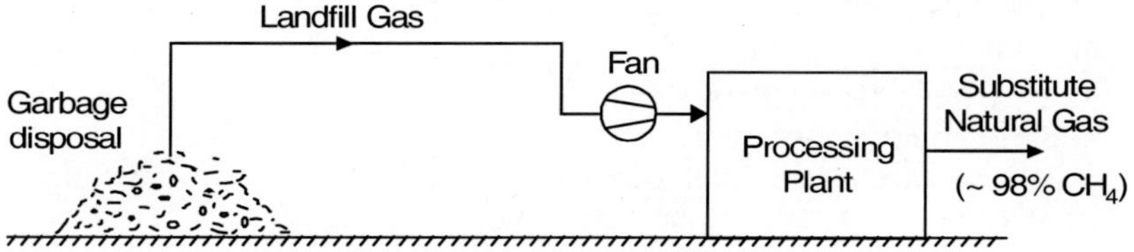

Fig. 6.14. Landfill Gas Plant.

6.14. ENERGY PLANTATION

Biomass energy concepts under study are resulting in the cultivation of large forests in areas not suitable for food production. Energy plantation may yield

10 to 20 tons/acre per year. The energy plantation would be perhaps 125 to 500 km^2 in land area. The trees are to be harvested by automated means, then chipped and pulverized for burning in a power plant that would be located in the middle of the energy plantation.

The choice of plants to be cultivated in India are Casuarina, Eucalyptus and Sorghums, etc. The properties of common species recommended for energy plantations are given in Table 6.3.

Table 6.3. Common Species for Energy Plantation

Name of Species	Yield m^3/ha/yr	Calorific Value MJ/kg
1. Casuarina equisilifolia	7–10	20
2. Eucalyptus camaldulensin	7–10	19
3. Lencaena leucocephala	30–40	17—19
4. Acavia auriculiformis	10–12	19–20
5. Albizia lebbek	5	22

Other schemes envision acquatic farms growing algae, tropical grasses, floating kelp, water hyacinth and others. In controlled environments, they could yield several hundred tons/acre year. One interesting idea is to use hot condenser cooling water from a power plant to grow algae in large quantities or increase the yield of other crops. Fast growing trees, sugar, starch and oil containing plants can also be cultivated for bio-energy. The sugar beat and cane have about 5% efficiency of the solar energy utilization. Special studies are required to increase the efficiency of solar energy use in the growth of crops to 10—11%.

TUTORIAL SHEET NO. 6.1.

1. What is biomass ? How is it useful for bio-energy production ?
2. Describe the availability of biomass in India and the world.
3. What are biomass conversion technologies ? Draw a schematic diagram to explain various conversion technologies and products.
4. Explain the following in relation to biomass
   (*a*) Combustion
   (*b*) Thermochemical conversion
   (*c*) Bio-chemical conversion
5. What is fluidised bed combustion of biomass ? Which type of fluidised bed is suitable for fine biomass and coarse biomass ?
6. Describe a circulating bed boiler system.
7. Describe the following as applicable to biomass combustion
   (*a*) Steam turbine cycle
   (*b*) Gas turbine cycle
8. Explain the concept of a Cooperative Rural Power Plant.

9. What is biomass gasification technology ? What are the advantages of fluidised bed combustion for biomass gasification ?
10. What is the principle of biogas generation by anaerobic digestion process ? Describe a Gobar Gas Plant.
11. What are bio-fuels ? How are these manufactured from biomass ?
12. Write notes on :
    (*a*) Landfill Gas Production.
    (*b*) Energy Plantations.
13. Classify various types of biogas generators. Describe a community type Gobar Gas Plant developed by Khadi and village Industries Board of India. Also list the requirements for its operations.
14. Discuss the properties of biogas and various materials used for its production.
15. Discuss how the bio-mass can be converted into electrical energy.

❑❑❑

Chapter 7

GEOTHERMAL ENERGY

7.1. INTRODUCTION

Geothermal energy is primarily energy from the earth's own interior. It is classified as renewable because the earth's interior is and will continue in the process of cooling for the indefinite future. Hence, geothermal energy from the earth's interior is almost as inexhaustible as solar or wind energy, so long as its sources are actively sought and economically tapped. The centre of the earth is estimated at temperatures upto 10,000 K due to decay process of radioactive isotopes. The total steady geothermal energy flow towards earth's surface is 4.2×10^{10} kW. But the average energy flow density is only 0.063 W/m^2 and average temperature gradient is 30 K per kilometer of depth. Locations with very high temperature gradients are suitable for geothermal energy utilization. Geothermal energy can be used for heat and power generation.

7.2. TYPES OF GEOTHERMAL ENERGY

The molten mass of the earth is called magma. The thickness of crust above magma is about 32 km in average. At certain places due to earth's tremours, magma came close to earth's surface. The hot magma near the surface causes active volcanoes, hot springs and geysers where water exits. It also causes steam to vent through the fissures in the form of geothermal energy.

Geothermal energy reservoirs are liquid-dominated and steam dominated. Some are only hot rocks where there is no ground water. Geothermal sources are :

1. Hydrothermal systems
2. Geopressured systems
3. Petrothermal systems

7.2.1. Hydrothermal Systems

Water is heated by the hot rocks. These can be vapour-dominated and liquid-dominated systems.

1. **Vapour-dominated systems** : In these systems, water is vaporized into steam which reaches the earth's surface at 8 bar and 205°C in dry condition. The steam can be used to produce power by Rankine Cycle with minimum costs.

However, the steam is associated with corrosive and erosive materials. Such systems are very rare in the world.

2. Liquid-dominated systems : Hot water is trapped underground at a temperature of 174 to 315°C. This water is brought up either by drilling wells or pumped up. Due to pressure drop, hot water flashes into two phase mixture of low quality. The water contains high concentration of dissolved solids and power production becomes difficult due to scaling in pipes and heat exchangers. The liquid-dominated systems are much more plentiful.

7.2.2. Geopressured Systems

Geopressured system is hot water or brine at about 160°C trapped underground at about 2400 to 9100m depth. The pressure is more than 1000 bar. Although it has great thermal and mechanical potential for power generation, but due to low temperatures and great depth, it may not be economical to drill for this brine. But this brine has recoverable methane and there may be economic feasibility to generate electricity involving combustion of methane as well as heat from the thermal content of water.

7.2.3. Petrothermal Systems

When there is no underground water, there are hot dry rocks (HDR) at 150 to 298°C near the earth's surface. This energy is called petrothermal energy and accounts for 85% of geothermal resources. Water will have to be pumped into and back out to the surface. Fracturing methods for drilling wells into the rocks can be by (*i*) high pressure water or (*ii*) nuclear explosions.

7.3. ENVIRONMENTAL CONSIDERATIONS

Steam and water from hydrothermal system contain the following :

1. Dissolved solids in water.
2. Entrained solid particles
3. Non condensable gases (CO_2, CH_4, H_2, N_2, and NH_3, H_2S)
4. Sand

The above solid and non-condensable gases have to be removed before using steam or water for power generation. These create the following environmental problems.

1. The non-condensable gases enter the equipment with the fluid and partly escape to the atmosphere via particle centrifugal separators, condenser ejectors and cooling towers.
2. The gas ejectors have to be specially designed to handle large quantities of non-condensable gases as compared to conventional steam power plants.
3. The corrosive action of these gases in wet conditions requires the use of stainless steel for handling wet steam and condensate.
4. Special materials of construction are required for steam turbines, nozzles, condensers, cooling towers, pipes and valves and open structures due to presence of corrosive atmosphere.
5. All electrical equipment need protected copper to withstand the corrosive action of H_2S on bare copper.
6. H_2S and NH_3 are corrosive and toxic and therefore objectionable for escape into atmosphere.

7. Large extraction of geothermal fluids and reinjection into the ground pose the possibility of seismic disturbances.
8. Exhausts, blowdowns and centrifugal separation are sources of noise pollution and suitable silencers have to be installed.
9. Geopressed water carriers large quantities of sand resulting in increased erosion and scaling problems.

7.4. VAPOUR DOMINATED POWER PLANT

The energy conversion in geothermal power plants is based on the Rankine cycle. The basic layout and T-s diagram for a vapour-dominated geothermal power plant are shown in Fig. 7.1 and 7.2 respectively. These systems are most suitable for electricity generation with lowest cost and the least number of serious problems. But these are the rarest form of geothermal systems. Examples are Geysers in USA, Larderello in Italy and Matsukawa in Japan.

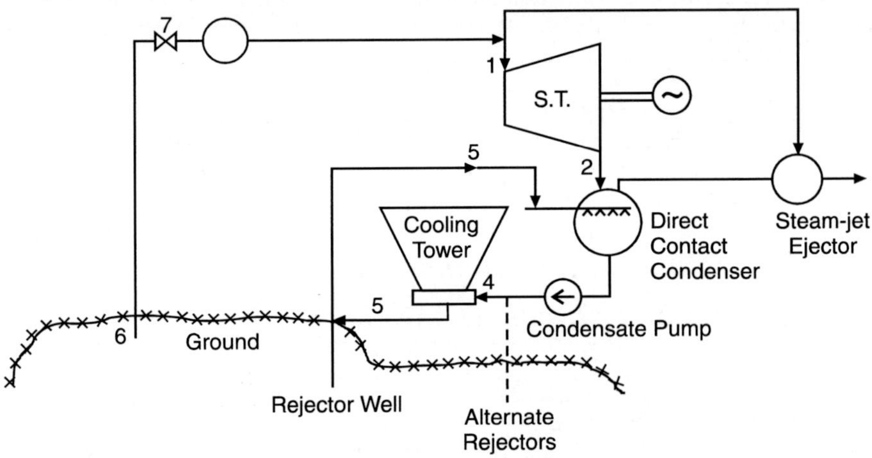

Fig. 7.1. Vapour-dominated Power Plant

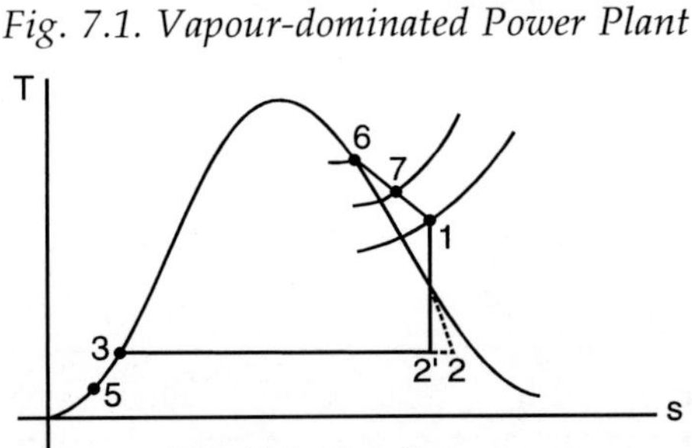

Fig. 7.2. T-s diagram

Dry steam at 35 bar and 200°C is available at the bottom of the well gets superheated due to pressure drop to 7 bar. The particulate matter is separated in a centrifugal separator. The process 6-7 and 7-1 are throttling processes so that enthalpy of steam remains constant. After passing through steam turbine and condensor, it is reinjected back to earth.

H_2S Removal

The recent environmental regulation, the toxic and noxious H_2S is removed in the shell-and-tube condenser by a process called Stretford process. 90% H_2S is removed and commercial-grade sulphur is produced as a by-product.

Example 7.1. *A 100 MW vapour-dominated system uses saturated steam from a well with a shut-off pressure of 28 bar. Steam enters the turbine at 5.5 bar and condenses at 0.15 bar. The turbine polytropic efficiency is 0.82 and the turbine-generator combined mechanical efficiency is 0.9. The cooling tower exist is at 20°C. Calculate the necessary steam flow, the cooling water flow and the plant efficiency and heat rate if reinjection occurs prior to cooling tower.*

Solution : Refer Fig. 7.1 and 7.2 and the steam tables

$$h_6 = h_g \text{ at 28 bar} = 2801.7 \text{ kJ/kg}$$

$$h_1 \text{ at 5.5 bar} = 2801.7 \text{ kJ/kg}$$

$$t_1 = 176°\text{C (20°C superheat)}$$

$$s_1 = 6.897 \text{ kJ/kg-K}$$

$$v_1 = 0.356 \text{ m}^3\text{/kg}$$

$$s_{2s} \text{ at 0.15 bar} = s_1 = 6.897 = 0.7327 + x_{4s} \text{ (7.306)}$$

$$\therefore \quad x_{2s} = 0.8437$$

$$h_{2s} = 218.7 + 0.8437 \text{ (2377.24)}$$

$$= 2224.38 \text{ kJ/kg}$$

$$\text{Isentropic work} = h_1 - h_{2,s}$$

$$= 2801.7 - 2224.38 = 577.32 \text{ kJ/kg}$$

$$\text{Actual turbine work} = 0.82 \times 577.32 = 473.4 \text{ kJ/kg.}$$

$$h_2 = h_1 - W_T$$

$$= 2801.7 - 473.4 = 2328.3 \text{ kJ/kg}$$

Ignoring pump work

$$h_3 = h_4 = 218.7 \text{ kJ/kg}$$

$$h_5 = h_f \text{ at 20°C} = 88.5 \text{ kJ/kg}$$

$$\text{Turbine steam flow} = \frac{100 \times 10^3}{473.4 \times 0.9} = 234.7 \text{ kg/s}$$

$$= 0.845 \times 10^6 \text{ kg/hr}$$

$$\text{Turbine volume flow} = \frac{0.845 \times 10^6 \times v_1}{60}$$

$$= \frac{0.845 \times 10^6 \times 0.356}{60}$$

$$= 5013.4 \text{ m}^3\text{/min}$$

Cooling water flow to condenser, $\dot{m}_5$

$$\dot{m}_5(h_3 - h_5) = \dot{m}_2(h_2 - h_3)$$

$$\therefore \quad \dot{m}_5 = \frac{(2328.3 - 218.7)}{(218.7 - 88.5)} \dot{m}_2$$

$$= \frac{2109.6}{130.2} \times 0.845 \times 10^6$$

$$= 13.7 \times 10^6 \text{ kg/hr}$$

$$\text{Heat added} = h_6 - h_4$$

$$= 2801.7 - 218.7 = 2583 \text{ kJ/kg}$$

$$\text{Plant efficiency} = \frac{W_T \times 0.9}{(h_6 - h_4)} = \frac{473.4 \times 0.9}{2583} = 16.49\%$$

$$\text{Plant heat rate} = \frac{3600}{0.1649} = 21831.4 \text{ kJ/kWh.}$$

7.5. LIQUID-DOMINATED SYSTEMS

Liquid-dominated systems are abundantly available and donot require development of special technologies. The following types of power systems are discussed.

1. The flashed-steam system
2. Binary cycle
3. Total-flow system

7.5.1. The Flashed-Steam System

Hot water is available above 150°C to 315°C underground. When tapped, the water can flow naturally under its own pressure or be pumped to the surface of the earth. The drop in pressure causes it to partially flash into steam and a liquid-dominated, low-quality, two-phase mixture of water and stream is available at the well head. The water contains dissolved solids. The flow diagram and T-s diagram of a flash-steam system are shown in Fig. 7.3 and 7.4 respectively.

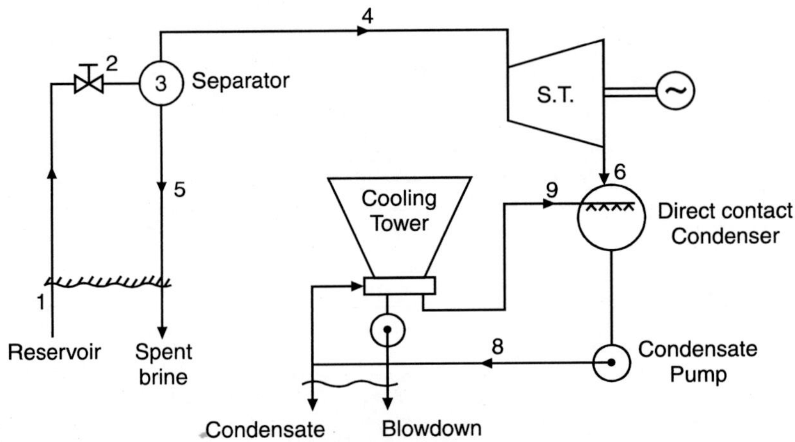

Fig. 7.3. Single-flash steam power plant

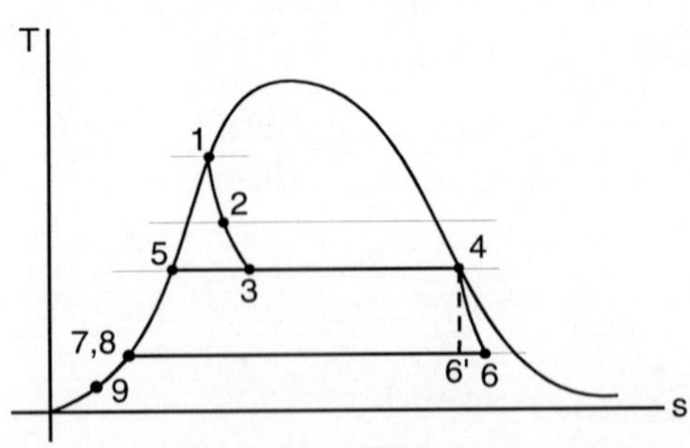

Fig. 7.4. T-s diagram of single-flash steam cycle

Hot water from reservoir (1) reaches the well head (2). Pressure p_2 is lower than p_1 and process 1-2 is a constant enthalpy throttling process. The two-phase mixture of low quality (2) is passed through a flash separator (3). The quality of steam is higher at point 3. The dry saturated steam (4) at pressure of about 8 bar is expanded in the steam turbine. The separated brine (5) is reinjected into the ground.

The exhaust steam from turbine is mixed with cooling water in a direct-contact condenser. The mixture is cooled in a cooling tower.

7.5.2. Improved Flashed-Steam System

In order to recover large quantity of heat energy from brine at point (5), some improvements in the cycle are carried out.

1. Double Flash System

Depending upon the original water conditions, the brine at point 5 is admitted to a second, lower-pressure separator, where it flashes to a lower pressure steam. This steam is admitted to a a low-pressure stage in the turbine as shown in Fig. 7.5. The new low-pressure brine contains less energy and therefore cycle energy loss is reduced. The corresponding *T-s* diagram is shown in Fig. 7.6. It uses an innovative steam condenser and gas extraction system and a dual - admission steam turbine. An example of the double-flash system is the 50 MW Hatchobane plant, Kyushu in Japan.

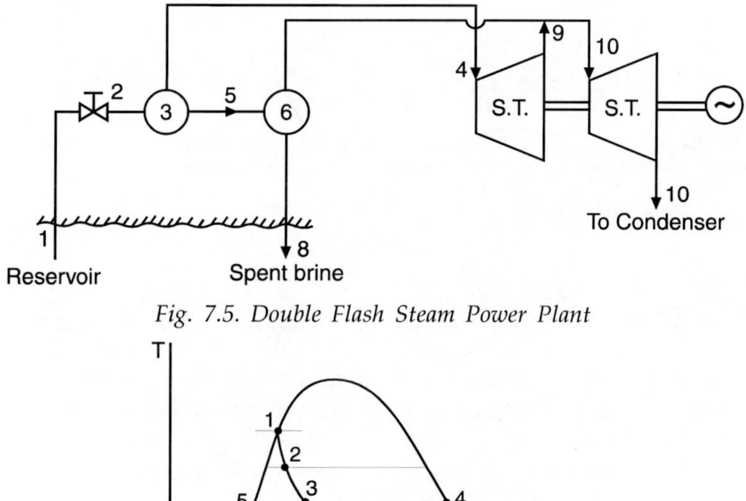

Fig. 7.5. Double Flash Steam Power Plant

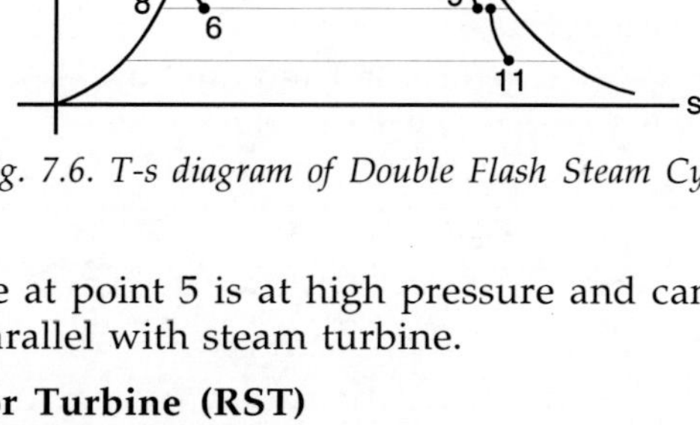

Fig. 7.6. T-s diagram of Double Flash Steam Cycle

2. Water Turbine

The spent brine at point 5 is at high pressure and can be used to drive a water turbine in parallel with steam turbine.

3. Rotary Separator Turbine (RST)

The fluid leaving the well head (point 2) is a two-phase mixture of steam and water. It is partially expanded in a nozzle. The quality and kinetic energy of fluid increase. This also helps in the separation of two phases. The separation takes place in a rotating drum by centrifugal action.

The steam is admitted to the steam turbine and high velocity water is used in a special liquid turbine. The discharge from the turbine is reinjected into the ground.

7.5.3. Binary Cycle

Water in lower temperature ranges is unsuitable for power production. It can be used directly for domestic and industrial process heating. Hot water can also be used in heating an organic fluid with low boiling point and can be used to run a Rankine Cycle. The working fluids can be isobutane, Freon-12, ammonia or propane.

The flow diagram of a binary cycle is shown in Fig. 7.7.

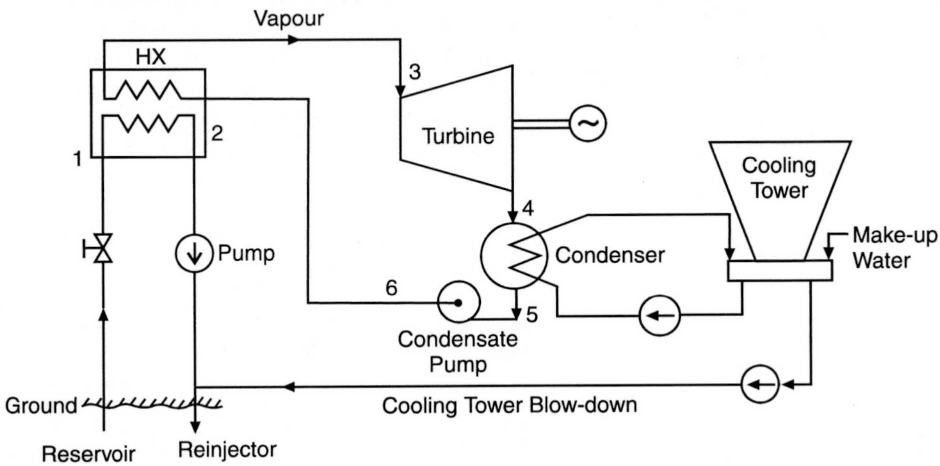

Fig. 7.7. Binary Cycle System

Hot brine from underground reservoir circulates through a heat exchanger (HX) and is pumped back to ground. The organic fluid is heated to superheated vapour and is used in a standard closed Rankine cycle. The vapour drives the turbine and is condensed in a surface condenser. The condenser is cooled by water from a natural source or a cooling tower circulation system. There is no problem of corrosion and scaling in the working cycle components.

Kamchatka binary cycle plant in Russia is 680 kW plant using hot water at 80°C. The working fluid is Freon-12.

7.5.4. Total-Flow Power Plant

Hot brine from the well is throttled and full flow is expanded in a two-phase *expander as shown in Fig. 7.8. A barometric condenser is used. As seen from T-*s diagram, thermodynamically maximum available energy can be converted into mechanical work.

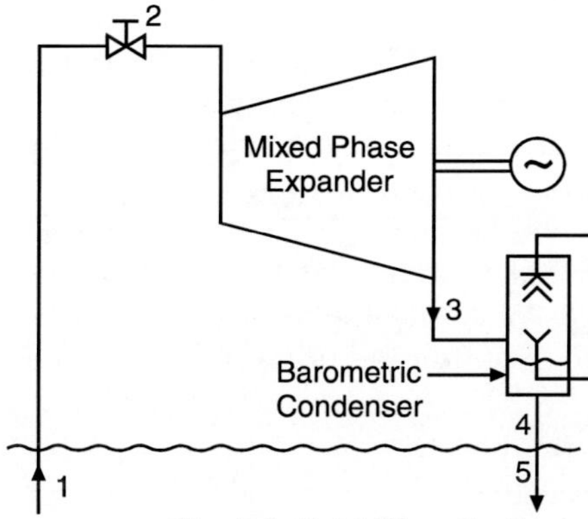

Fig. 7.8. Total-Flow Geothermal Power Plant

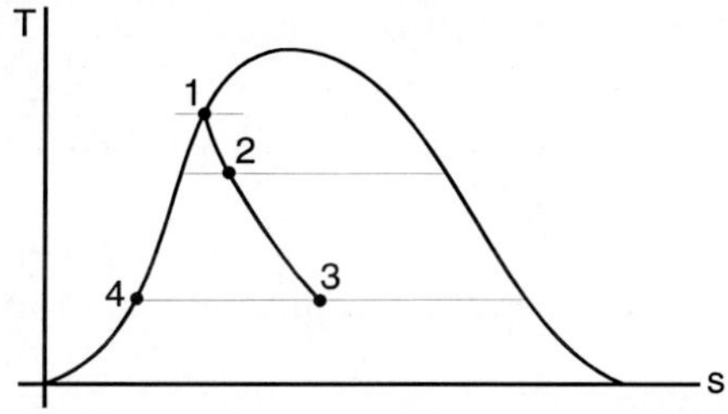

Fig. 7.9. T-s diagram of Total-Flow Power Plant

Comparison

The analytical comparison of various liquid-dominated geothermal system is given in Fig. 7.10. The total-flow system produces the highest specific power per unit mass-flow rate at the well head. These calculations are based on the following assumptions :

Condenser temperature = 45°C

Turbine efficiency = 85%

In-plant power requirements = 30% of gross.

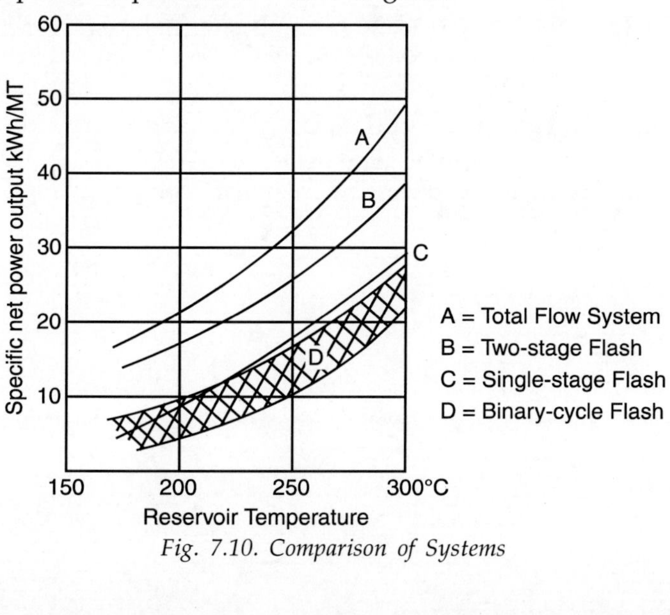

Fig. 7.10. Comparison of Systems

7.6. PETROTHERMAL SYSTEMS

The thermal energy of hot dry rock (HDR) is extracted by pumping water (or other fluid) through a well drilled to the lower part of the fractured rock. The water flows through the fractures, gets heated and travels up through the second drilled well. The hot water is used to produce electricity as in case of liquid-dominated geothermal system.

It offers more flexibility in design and operation. The water flow rate and temperatures can be controlled for a site by drilling to various depths. The operator of the plant can change the pumping pressure and therefore, flow rates to suit the load conditions.

The limitations of petrothermal systems are :

1. Leakage of water underground.
2. Necessity of make-up water from resources above ground.
3. Effect of water on rock composition.
4. Material carryover with water.
5. High cost as two wells have to be drilled instead of one as in hydrothermal systems.
6. Very detailed studies on the mechanical thermodynamic and economic aspects are necessary before commercial exploitation becomes feasible.

7.7. HYBRID SYSTEMS

The high-temperature heat of fossil-fuel combustion systems can be used in the high-temperature end of a Rankine Cycle and low-temperature heat of geothermal heat can be used in the low-temperature end of a conventional power plant. Such hybrid systems thus combines the high efficiency of a high-temperature cycle with a natural source of heat for part of heat addition. The consumption of expensive and nonrenewable fossil fuel can be reduced substantially. There are two possible arrangements for hybrid plants.

1. Geothermal Preheat Hybrid System.

The low temperature geothermal heat is used for feedwater heating in a conventional steam plant. No steam is required to be bled from the low-pressure stages of steam turbine and feed water heating is carried out using geothermal brine.

2. Fossil-Superheat Hybrid System

The geothermal fluid can be superheated in a fossil-fired superheater. A flow diagram and *T-s* diagram of such a plant are shown in Fig. 7.11 and 7.12 respectively.

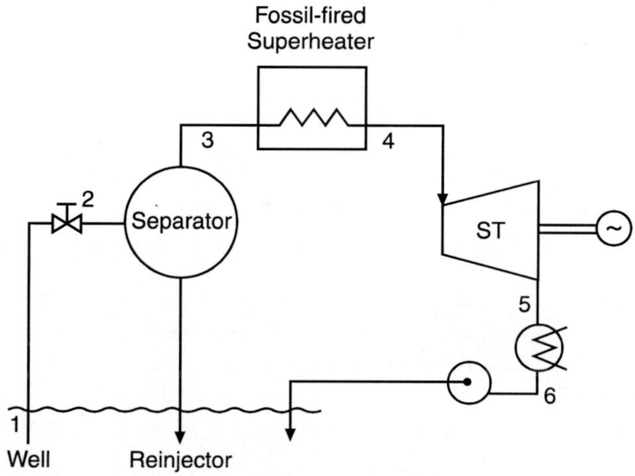

Fig. 7.11. Fossil-Superheat Hybrid System

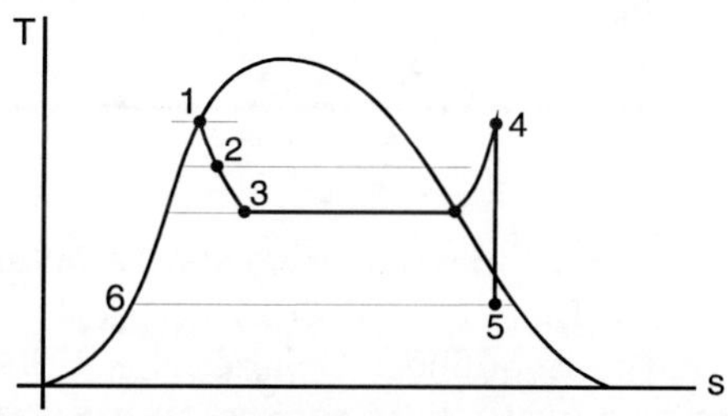

Fig. 7.12. T-s diagram of Hybrid System.

TUTORIAL SHEET No. 7.1.

1. What is geothermal energy ? Classify various types of geothermal energy systems as available.
2. Compare the following systems
   (*i*) Hydrothermal system
   (*ii*) Geopressured system
   (*iii*) Petrothermal system
3. What are the operational and environmental constraints in the design of geothermal power plants ?
4. With the help of a line diagram and *T-s* diagram, explain a vapour-dominted geothermal power plant.
5. Discuss the various types of liquid-dominated geothermal power plants. Compare their performances.
6. Write notes on :
   (*i*) Binary cycle geothermal power plant.
   (*ii*) Total-flow power plant.
7. Explain the following :
   (*i*) Petrothermal systems.
   (*ii*) Hybrid systems.
8. Discuss the advantages, disadvantages and applications of geothermal energy.

Chapter **8**

SOLAR THERMAL ENERGY

Introduction ; Extraterrestrial Solar Radiation, Terrestrial Solar Radiation ; Total Solar Radiation Incident on an Inclined Solar Collector ; Solar Collectors ; Performance evaluation of Flat Plate Collector ; Performance Evaluation of Concentrating Collector ; Performance Comparison of Different Collectors ; Solar Water Heaters ; Solar Thermal Power Plants ; Solar Ponds ; Solar Energy Storage ; Miscellaneous Applications ; Advantages and Limitations of Solar Energy.

8.1. INTRODUCTION

Solar energy is an essentially inexhaustible source potentially capable of meeting significant portion of the world's future energy needs with a minimum of adverse environmental consequences. Solar energy is the most promising of the unconventional energy sources. In sheer size, it has the potential to supply all energy needs : electric, thermal process, and chemical, and even transportation fuels.

The sun is a sphere of intensely hot gaseous matter, continuously generating heat by thermonuclear fusion reactions converting hydrogen atoms into helium atoms. This energy is radiated from the sun in all directions and a very small fraction of it reaches the earth. The earth and its atmosphere receive continuously 1.7×10^{17} W of radiations from the sun. A world population of 10 billion with a total power need per person of 10 kW would require about 10^{11} kW of energy. It is thus apparent that if irradiation on only one percent of the earth's surface could be converted into useful energy with 10% efficiency, solar energy could fulfill the energy needs of the entire population.

The immense magnitude of direct solar energy can be illustrated by the following estimates.

The avarage solar constant at sea level = 1 kW/m^2

The normal surface of the earth presented to the sun

$$(\pi R)^2 = \pi(3960 \text{ miles})^2 \times (1610 \text{ m/mile})^2$$

$$\text{Total solar energy at sea level} = 1 \times \pi \times (3960)^2 \times (1610)^2$$

$$= 1.28 + 10^{14} \text{ kW.}$$

If 1% of earth's surface is used for converting solar energy into useful energy at an efficiency of 10% (a figure well below that of some available solar cells), the total energy generated per year

$$= 1.28 \times 10^{14} \text{ kW} \times 0.01 \times 0.1 \times 24 \times 365$$

$$= 11.2 \times 10^{14} \text{ kWh}$$

This is the present annual rate of consumption of the whole world. Another interesting comparison is that the total fossil energy on earth is less than the solar energy incident upon the earth in one year.

Solar energy is, however, very diffuse, cyclic and often undependable. It, therefore, needs systems and components that can gather and concentrate it efficiently for conversion to any of the uses and that can do the conversion as efficiently as possible.

8.2. EXTRATERRESTRIAL SOLAR RADIATION

The mean extraterrestrial radiation normal to the solar beam on the outer fringes of the earth's surface is called solar constant. Its value is approximately 1.353 kW/m^2. This varies by ±3.4 percent during the year due to elliptical orbit of the earth. The radiant energy from the sun is distributed over a range of wave-lengths and can be approximated by the spectrum of a black body at 5762 K.

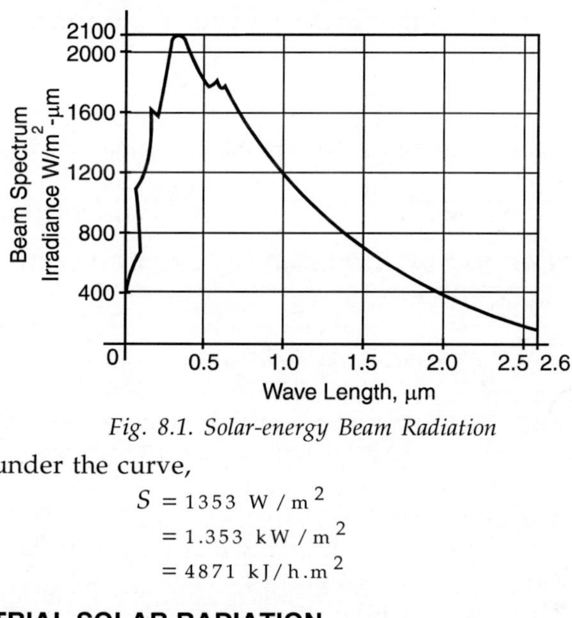

Fig. 8.1. Solar-energy Beam Radiation

The area under the curve,

$$S = 1353 \text{ W/m}^2$$
$$= 1.353 \text{ kW/m}^2$$
$$= 4871 \text{ kJ/h.m}^2$$

8.3. TERRESTRIAL SOLAR RADIATION

Extraterrestrial solar radiation is all of the beam-radiation type, also called direct radiation. This is received from the sun in essentially straight rays or beams that are unscattered by the atmosphere. The solar energy falling on the earth's surface is called terrestrial radiation. The rate of terrestrial energy falling on a unit surface area in W/m^2 is variably referred to as radiation, irradiation, irradiance, insolation, or energy flux. Terrestrial radiation is not constant and varies significantly.

1. It varies daily because of earth's rotation around the sun.
2. It changes seasonally because of sun's declination angle, *i.e.*, the angle between the sun's rays and the earth's equatorial plane normal to the polar axis.
3. It changes due to spatial changes between the sun and the earth due to elliptical motion of earth. During summer, the earth's axis is tilted

towards the sun and northern hemisphere receives more radiation than the southern hemisphere and vice-versa during winter.

The extraterrestrial radiations are attenuated by the following :

1. **Scattering :** A part of a radiation beam is scattered laterally and attenuated by the air molecules, water vapour and dust in the atmosphere. The scattered and diffuse radiation is mostly of shorter wavelengths.
2. **Absorption :** The solar radiation is absorbed by Ozone (O_3), water vapour (H_2O) and Carbon dioxide (CO_2)

About 25-50 percent of solar energy is lost by scattering and absorption. The mean annual horizontal surface irradiance for India is about 200 W/m^2 and for Thar desert, it exceeds 250 W/m^2.

The global solar radiation incident on the earth surface comprises both beam (direct) and diffuse solar radiation. The intensity of global radiation on one square meter of the earth surface in a unit time depends on the geographical latitude, season of the year, time of the day and on weather conditions especially cloud cover of the sky.

The annual global solar radiation varies widely from 800 kWh/m^2 to 2400 kWh/m^2 depending upon the location. Another parameter of solar radiation at a particular location is the daily and annual sunshine hours. The value can be anything from 1400 to 3500 hours per year. Solar radiation data for various selected cities of the world have been measured & recorded.

8.4. TOTAL SOLAR RADIATION INCIDENT ON AN INCLINED SOLAR COLLECTOR

Solar collectors are installed either in a fixed position or track the sun's motion in the sky. The total solar radiation incident on an inclined flat plate collector consists of (1) beam radiantion ; (2) diffuse sky radiation and (3) diffuse radiation reflected from the earth surface. The flat plate collectors are installed in fixed position inclined towards the equator, *i.e.*, to the north in southern hemisphere and to the south in the northern hemisphere. Typically, the collector tilt angle β is selected as follows :

$\beta = \phi \pm 15°$. where ϕ is the latitude in degrees of the location. The hourly beam solar radiation is given by (Northern hemisphere)

$$I_{bc} = I_b \cos \theta_c / \cos\theta$$

$$= \frac{I_b[\cos(\phi - \beta)\cos\delta\cos\omega + \sin(\phi - \beta)\sin\delta}{\cos\phi\cos\delta\cos\omega + \sin\phi\cos\delta}$$

where I_{bc} = Hourly beam solar radiation on inclined collector surface [kWh/m^2-hr]

I_b = Hourly beam solar radiation incident on a horizontal surface [kWh/m^2-hr]

θ = Angle of incidence of beam solar radiation on a horizontal plane [°]

θ_c = Angle of incidence of beam solar radiation on tited solar collector towards equator [°]

β = Angle of tilt of solar collector [°]

ϕ = Geographical latitude [°]

δ = sun's declination [°]
ω = hour angle [°]

The geographical latitude ϕ varies from 0° (equator) to +90° (north pole) in northern hemisphere and 0° to –90° (south pole) in the southern hemisphere. ϕ for Delhi is 28.6°N.

The sun's declination δ varies between –23.45° and +23.45° over the year as shown by a graph in Fig. 8.2. The seasonal variation of the terrestrial radiation on a horizontal surface at any one location on the earth's surface is accounted by the variation of δ.

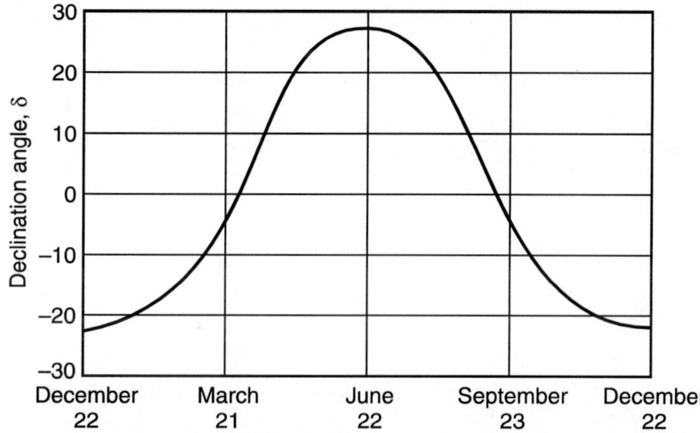

Fig. 8.2. Seasonal variation in the sun's declination angle, δ

The hour angle ω is equal to zero at noon (the sun is at zenith). One hour is equivalent to 15°. The hour angle is positive in the morning, 0° at noon and negative in the afternoon.

Example : At 10 a.m., ω = +30°, and at 2pm ω = –30°.

The daily beam radiation incident on an inclined collector surface tilted towards the equator is given by the following relation. The collector surface faces true south in the northern hemisphere and true north in the southern hemisphere.

$$E_{bc} = \frac{E_s\left[\cos(\phi-\beta)\cos\delta\cos\omega_{sc} + \left(\frac{\pi}{180}\omega_{sc}\right)\sin(\phi-\beta)\sin\delta\right]}{\left[\cos\phi\cos\delta\cos\omega_s + \left(\frac{\pi}{180}\omega_s\right)\sin\phi\sin\delta\right]}$$

where E_{bc} = Daily beam radiation [kWh/m^2 per day]

E_b = Daily beam radiation incident on a horizontal surface [kWh/m^2 per day]

ω_{sc} = sunset hour angle for the inclined collector surface [°]

ω_s = sunset hour angle for the horizontal surface [°]

The sunset hour angles can be calculated as :

$$\omega_s = \cos^{-1}(-\tan\phi\tan\delta)$$

$$\omega_{sc} = \cos^{-1}[-\tan(\phi-\beta)\tan\delta]$$

The hourly total solar radiation incident on inclined collector surface can be calculated as :

$$I_c = I_b R_b + I_d \frac{(1+\cos\beta)}{2} + I_{\rho G} \frac{(1-\cos\beta)}{2}$$

where I_C = Total hourly solar radiation incident on unit surface area of inclined collector surface [kWh/m^2 per hour]

I = Hourly global solar radiation on a horizontal surface [kWh/m^2-hr]

I_b = Hourly beam radiation on a horizontal surface [kWh/m^2-hr]

I_d = Hourly diffuse radiation on a horizontal surface [kWh/m^2-hr]

ρ_G = Reflectivity of earth's ground surface

The ratio of total solar radiation per hour on the collector surface to the global solar radiation per hour on a horizontal surface,

$$R = \frac{I_C}{I} = \frac{I_{bC}}{I_b + I_d}$$

$$= \frac{I_{bC}}{\dfrac{I(1+\cos\beta)}{2}} + E_{\rho G} \frac{(1-\cos\beta)}{2}$$

The total solar radiation per day on a unit area of inclined collector surface is :

$$E_C = E_R$$

$$= E_{bc} + E_d \frac{(1+\cos\beta)}{2} + E_{\rho G} \frac{(1-\cos\beta)}{2}$$

where, E_c = Total solar radiation per day per unit area of inclined collector surface [kWh/m^2 per day]

E = Total global solar radiation per day per unit area on horizontal surface [kWh/m^2 per day]

E_b = Total beam solar radiation per unit area per day on horizontal surface [kWh/m^2 per day]

E_d = Total diffuse solar radiation per unit area per day on horizontal surface [kWh/m^2 per day]

$$R = \frac{\text{Total solar radiation on tilted collector surface per day}}{\text{Global solar radiation on a horizontal surface per day}}$$

$$= \frac{E_{bc}}{E} = \frac{E_{bc}}{E_b + E_d}$$

$$= E\frac{1+\cos\beta}{2} + E_{\rho G} \frac{(1-\cos\beta)}{2}$$

The above relations have been derived by the use of spherical trigonometry.

Example 8.1. *Calculate total insolation per hour on a flat-plate collector installed in New Delhi. The site latitude is 28.6°, the date is June 15 and time is 1 p.m. (solar time). The collector surface is tilted at an angle 20° with the horizontal facing south.*

The total radiation per hour on a horizontal plane is 2.5 MJ/m²-hr and ratio of diffuse radiation is 0.37. The ground reflectivity is 0.2.

Solution : June 15 is 166th day of the year.

The angle of declination,

$$\delta = 23.45° \sin\left[\frac{360\,(284+n)}{365}\right]$$

$$= 23.45° \sin\left[\frac{360\,(284+166)}{365}\right] = 23.3°$$

$$R_b = \frac{I_{bc}}{I_b}$$

$$= \frac{[\cos(\phi-\beta)\cos\delta\cos\omega + \sin(\phi-\beta)\sin\delta]}{[\cos\phi\cos\delta\cos\omega + \sin\phi\sin\delta]}$$

$$\phi = 28.6°$$

$$\beta = 20°$$

$$\omega \text{ for 1 p.m.} = -15°$$

$$\therefore \quad R_b = \frac{\cos(28.6°-20°)\cos 23.3°\cos(-15°) + \sin(28.6°-20°)\sin 23.3°}{\cos 28.6°\cos 23.3°\cos(-15°) + \sin 28.6°\sin 23.3°}$$

$$= 0.96703$$

Total radiation per hour per unit area on collector surface,

$$I_c = I\left[R_b\left(1-\frac{I_d}{I}\right) + \frac{I_d}{I}\left(\frac{1+\cos\beta}{2}\right) + E_{\rho_G}\frac{(1-\cos\beta)}{2}\right]$$

$$= 2.5\left[0.96703\,(1-0.37) + 0.37\frac{(1+\cos 20°)}{2} + 0.2\frac{(1-\cos 20°)}{2}\right]$$

$$= 2.43526 \text{ MJ/m}^2\text{-hr}$$

$$R = \frac{I_C}{I} = \frac{2.43526}{2.5} = 0.9741$$

8.5. SOLAR COLLECTORS

A solar collector is a receiving device which absorbs the incident solar radiation and heat a fluid like water or air. The solar radiation is converted into useful heat which can be used as such or can be converted into electrical power. The collectors can be classified into low temperature, medium temperature and high temperature collectors. Mainly there are two types of collectors.

1. Flat plate collector
2. Concentrating collector

8.5.1. Flat Plate Collectors

These are used for temperatures below 100°C. These can use both beam and diffuse solar radiation. They are installed in a fixed tilted position optimally oriented towards the equator.

The main components of a flat plate collector are shown in Fig. 8.3.

1. Transparent cover
2. Absorber plate
3. Heat transfer fluid tube
4. Thermal insulation
5. Casing

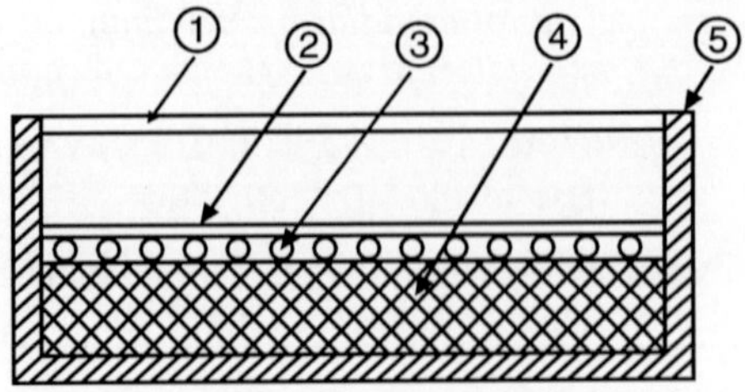

Fig. 8.3. Flat Plate Collector

Flat plate collectors used for water heating and space heating can produce 250 to 400 kWh per year and per square meter of useful heat. Evacuated tube collectors can produce useful heat at a temperature upto 250°C. The efficiency of the collector reduces with the increase of fluid temperature.

Evacuated Tube Collector

The efficiency of a collector is calculated from intensity of solar radiation, optical and heat losses. Evacuated tube collectors are used for reduction of losses. The main applications are water heating, space heating, refrigeration and airconditioning and process heat production. Energy in the range of 300 to 600 $kW\ h/m^2$ per year can be attained.

8.5.2. Concentrating Collectors

These are used for medium and high temperature applications. Various types of concentrating solar collectors are shown in Fig. 8.4. It consists of a concentrating device which may be a reflecting mirror or Fresnel lense. The beam radiations are concentrated and focussed on to the absorber. Different types of collectors have different values of concentration ratio.

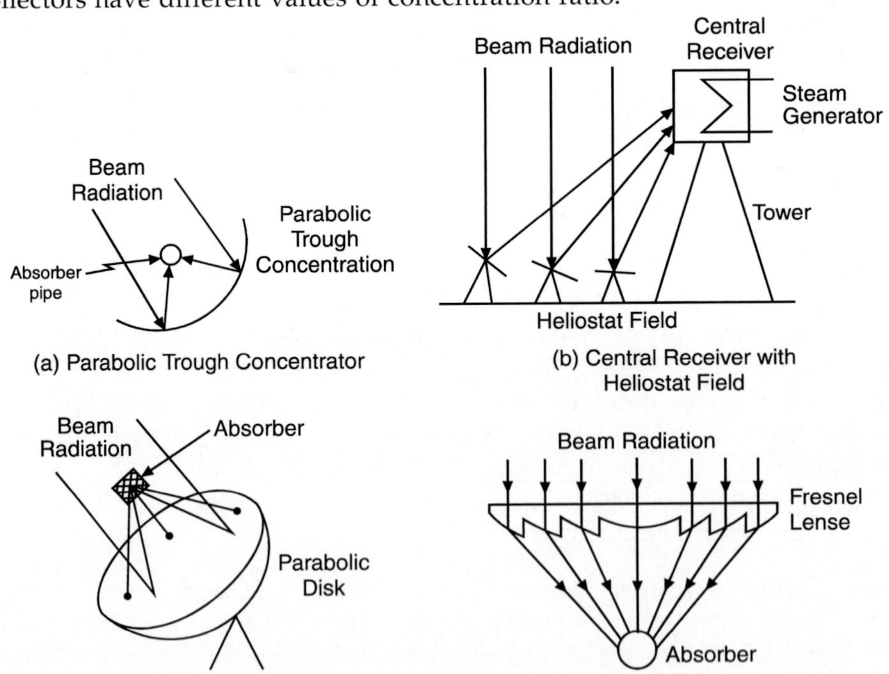

(a) Parabolic Trough Concentrator

(b) Central Receiver with Heliostat Field

(c) Parabolic Disk Concentrator

(d) Fresnel Lense Concentrator

Fig. 8.4. Concentrating Solar Collectors

The concentration ratio,

$$C = \frac{A_a}{A_r}$$

where A_a = Aperture area of the concentrator
A = area of absorber

Parabolic trough concentrator is a linear concentrator. The value of C may be 20 to 100.

Parabolic disk concentrator is a point focuss concentrator. The value of C may be 100 to 4000 for parabolic disk and heliostat field concentrator.

The mirrors and Fresnel lenses can focus only beam radiation. These concentrators are rotated through tracking mechanism to follow the motion of the sun in the sky. Temperatures as high as 1000°C can be achieved. These are mainly used in solar plants for power generation and process heat supply.

8.6. PERFORMANCE EVALUATION OF FLAT PLATE COLLECTOR

The useful heat output of a flat-plate solar collector is given by :

$$Q_c = A\left[I_c F_R (\tau\alpha)_e - F_R U_c (T_{in} - T_a)\right] \quad [W]$$

where A = surface area of collector or absorber [m^2]
I_c = Intensity of solar radiation incident on the collector [W/m^2]
F_R = Heat removal factor of the collector
$(\tau\alpha)_e$ = Effective product of transmissibility τ of the transparent cover and absorptivity α of the absorber.
U_c = Overall heat loss coefficient of collector [W/m^2]
T_{in} = Fluid inlet temperature [°C]
T_a = Ambient air temperature [°C]

The efficiency of the solar collector is defined as the ratio of the useful heat output of the collector and the solar energy flux incident on the collector.

$$\eta_c = \frac{Q_c}{A I_c}$$

$$= \frac{F_R (\tau\alpha)_e - F_R U_c (T_{in} - T_a)}{I_c}$$

Let $$\frac{(T_{in} - T_a)}{I_c} = X$$

The performance curve of a collector can be plotted between η_c and X.

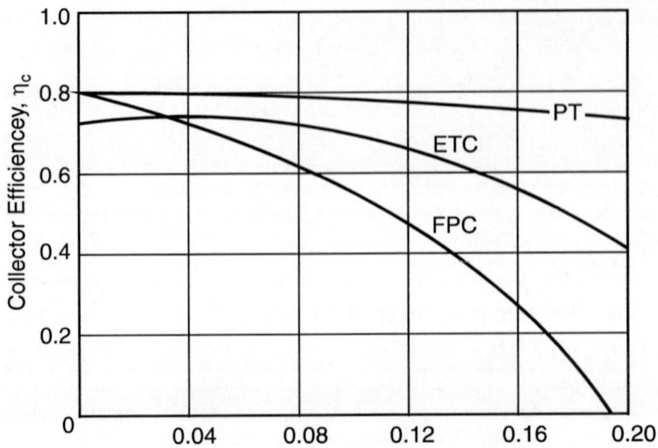

$$X = \left(\frac{T_i - T_o}{I_c} \right) (m^2 k / W)$$

PT = Parabolic Trough Concentrator
ETC = Evacuated Tube Collector
FP = Flate Plate Collector

Fig. 8.5. Performance curves of collectors.

The following parameters can be found out from the performance curves.

1. At $X = 0$, $\eta_c = F_R(\tau\alpha)_e$ = Effective optical efficiency.
2. The value of effective overall heat loss coefficient $F_r U_c$.

The outlet temperature of the fluid from the collector can be found out as

$$T_{out} = T_{in} + \frac{Q_c}{\dot{m} C_p} \quad [\,°C]$$

Where T_{in} = Inlet temperature of fluid [°C]
Q_c = Useful heat output of collector [W]
$\dot{m}$ = mass flow rate of fluid [kg/s]
C_p = Specific heat of fluid [J/kg-K]

The stagnation temperature is the temperature of the absorber when there is no fluid flow.

$\therefore \quad Q_c = 0$ and $\eta_c = 0$ and if $T_{in} = T_{stag}$

$$\therefore \quad T_{stag} = T_a + \frac{I_C F_R (\tau\alpha)_e}{(F_R U_c) R}$$

T_{stag} is high if I_c is high, $(\alpha\tau)_e$ is high and $(F_r U_c)$ is low.

Example 8.2. *A flat plate collector is working under the following conditions :*

1. *The intensity of solar radiation on the collector surface* $= 760\ W/m^2$
2. *Inlet temp. of fluid* = 43°C
3. *Ambient temperature* = 26°C
4. *Effective optical efficiency* = 0.82
5. *Effective heat loss coefficient* = 4.1 W/m^2-K
6. *Mass flow rate of water* $= 0.017\ kg/s/m^2$
7. C_p *for water* = 4187 J/kg-K

Calculate outlet temperature of water, stagnation temperature and useful heat output.

Solution : 1. Useful heat output per m^2 of collector area

$$q_c = I_c F_R (\tau\alpha)_e - F_R U_c (T_{in} - T_a)$$
$$= 760 \times 0.82 - 4.1\ (43 - 26)$$
$$= 553.5\ \text{W/m}^2$$

2. $$T_{out} = T_{in} + \frac{q_c}{m C_p}$$

$$= 43 + \frac{553.5}{0.017 \times 4187} = 50.78°\text{C}$$

3. Collector stagnation temperature,

$$T_{stag} = T_a + \frac{I_c F_R (\tau\alpha)_e}{F_R U_c}$$

$$= 26 + \frac{760 \times 0.82}{4.1} = 178°\text{C}$$

8.7. PERFORMANCE EVALUATION OF CONCENTRATING COLLECTOR

The useful heat output,

$$Q_c = F_R A_a \left[I_{bc}\, \eta_{opt} - \left(\frac{U_c}{C}\right)\right](T_{in} - T_a) \quad [\text{W}]$$

where F_R = Heat removal factor of the collector.
A_a = Unshaded aparture area [m^2]
I_{bc} = Intensity of beam radiation on aparture of the collector (W/m^2)
η_{opt} = Optical efficiency of the collector.
U_c = Total heat loss coefficient of collector [W/m^2-K]
C = Concentration ratio of the collector
T_{in} = Inlet temperature of fluid [°C]
T_a = Ambient temperature [°C]

Optical Efficiency

It is the ratio of solar radiation absorbed and the beam radiation on the concentrator

$$\eta_{opt} = \eta_{opt\,(0°)}\, C_{opt}$$
$$= \rho\, \gamma\, \tau\alpha_a\, C_{opt}.$$

where $\eta_{opt(0°)}$ = Optical efficiency at 0° incidence angle of beam
C_{opt} = Correction factor for deviation from 0°
ρ = Reflectivity of mirror
γ = Intercept factor
τ = Transmissivity of the cover (if available)
α_a = Absorptivity of absorber

Efficiency of the collector,

$$\eta_c = \frac{Q_c}{A_a I_{bc}}$$

$$= F_R \eta_{opt} - \left(\frac{F_R U_c}{C I_{bc}}\right)(T_{in} - T_a)$$

Figure 8.6. shows the efficiency of a collector as a function of absorber temperature for different concentration ratios.

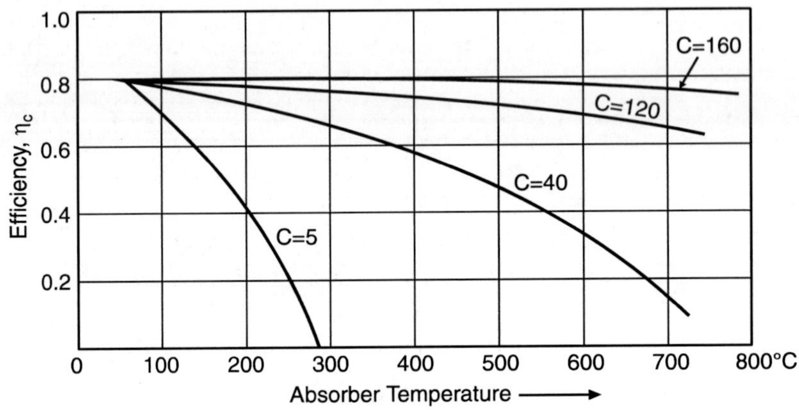

Fig. 8.6. Performance Curve of Concentrating Collector.

There can be phase change of fluid in a concentrating collector. The outlet enthalpy

$$h_{out} = h_{in} + \frac{Q_c}{m} \quad [J/kg]$$

where h_{in} = enthalpy of inlet fluid [J/kg]

Q_c = Useful heat output [W]

$\dot{m}$ = mass flow rate of fluid [kg/s]

8.8. COMPARISON OF PERFORMANCE OF DIFFERENT COLLECTORS

The technical and economic parameters of different collectors are tabulated in Table 8.1 for comparison

Table 8.1.

S. No.	Type of collector	Effective optical efficiency $F_R(\tau\alpha)$	Effective heat loss coeff. $F_R U_c$ (W/m²-k)	Operating temp. (0°C)	Cost $/m²
1.	Flat plate collector with no cover	0.9	15-20	30	120-170
2.	Flat plate collector with single cover	0.8	35-40	< 100	300-500
3.	Evacuated tube collector	0.7	1.5-1.8	< 250	700-1000
4.	Parabolic trough collector	0.8	0.2-0.7 (C 20 to 100)	400	~700

8.9. SOLAR WATER HEATERS

Solar plants for water heating can be used for the following :

1. Swimming Pool. Solar water heaters can be used for heating water to 23 to 28°C for swimming pools.
2. Domestic hot water at 45-60°C
3. Space heating : Water is heated to 30-90°C.
4. Refrigeration and space cooling. Heat is supplied to refrigerant vapour generator at 95 to 150°C.

Domestic water heating is most widely used. There are two types of water heating plants.

1. Single loop, thermosyphon circulation solar water heaters : These can be used with ambient temperature above 0°C. It consists of a solar collector, hot water storage tank and return tubes. The water is returned due to gravity and rises due to thermosyphon effect.

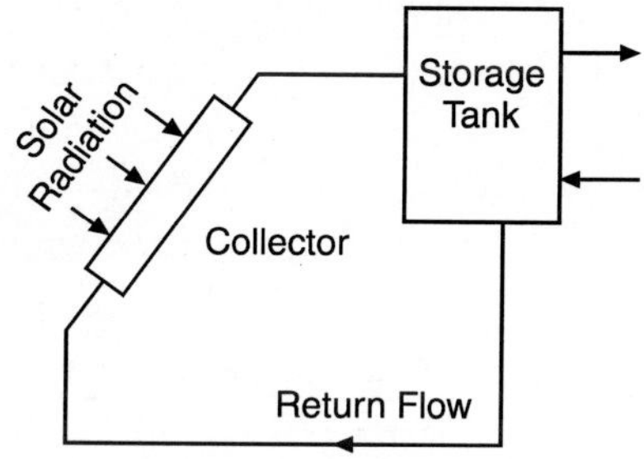

Fig. 8.7. Simple Solar Water Heater

2. Forced circulation water heater. It is used for supply of hot water in summer and winter even with ambient temperature below 0°C. It also includes a circulation pump with automatic control and a heat exchanger. An antifreeze solution of polypropylene or polyethylene is used as collector fluid. Water is used for storage in the hot water tank. A back-up heating facility is also used in addition to solar heating.

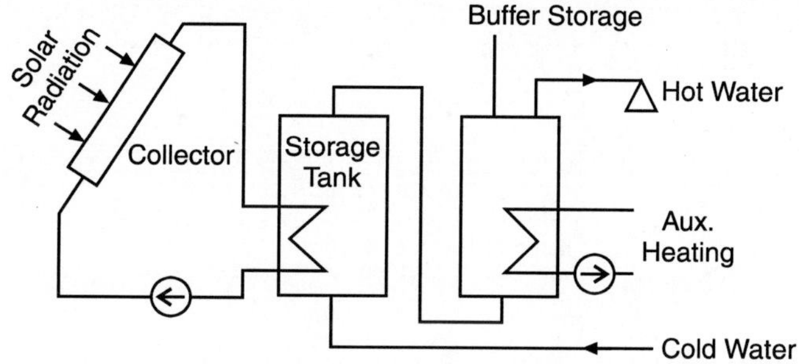

Fig. 8.8. Forced Circulation Water Heater

The heat quantity for hot water per month is

$$Q_{HW} = d_w \, N_{per} \, C_{p_w} (T_{HW} - T_{CW}) N_D \quad [J]$$

where
d_w = daily hot water demand in kg/day/person.
N_{per} = No. of hot water consumers
C_{p_w} = Specific heat of water [4187 J/kg-K]
T_{HW} = Hot water temperature [°C]
T_{CW} = Cold water temperature [°C]
N_D = No. of days in the month

Heat required per month for space heating

$$Q_{heat} = U_B \, A_B \, DD \; [\text{J/month}]$$

where U_B = Average heat loss Coeff. of building [W/m²-K]
A_B = Total external surface area of heated building [m²]
DD = No. of degree-days per month [K-days]

The surface area of collector required,

$$A = \frac{Q_{HW}}{q_{sol.}}$$

$$= \frac{m_{HM} C_{pw} (T_{HM} - T_{CW})}{E_c \, \eta_{sol.}} \text{ [m}^2\text{]}$$

where Q_{HW} = Amount of heat required for hot water supply [J]
$q_{sol.}$ = Heat output of collector [J/m²]
m_{HW} = Hot water required [kg]
E_C = Solar radiations on collector [J/m²]
η_{sol} = Efficiency of solar plant.

Example 8.3. *The following data may be used for the design of a solar water heater*

1. *Solar radiation = 5 kWh/m²/day*
2. *Hot water required = 1000 kg/day*
3. *Hot water temperature = 45°C*
4. *Cold water temperature = 14°C*
5. C_{p_w} *= 1.163 Wh/kg-K*
6. *Mean efficiency of water heater = 48%*

Piping and storage heat losses may be neglected. If a single plant has an area of 2.2 m², find out the total area required and number of solar collector modules.

Solution : 1.

$$Q_{HW} = m_{HW} C_{p_w} (T_{HW} - T_{CW})$$
$$= 1000 \times 1.163 \times (45 - 14)$$
$$= 36053 \text{ Wh/day}$$

2.

$$q_{sol.} = E_c \eta_{sol}$$
$$= 5 \times 0.48 = 2.4 \text{ kWh/m}^2\text{/day}$$

3.

$$A = \frac{Q_{HW}}{q_{sol.}}$$

$$= \frac{36053}{2400} = 15.02 \text{ m}^2.$$

4. No. of solar collector modules.

$$n = \frac{A}{A_C} = \frac{15.02}{2.2} = 6.83.$$

No. of solar collector modules reqd. = 7

8.10. SOLAR THERMAL POWER PLANTS

Concentrating solar collectors are employed to heat a working fluid for power cycles. The following types of solar thermal power plants have been developed.

1. Parabolic Trough Solar Power Plant
2. Central Receiver Solar Power Plant

3. Dish/Stirling Engine Power Plant
4. Solar Chimney Power Plant

Deserts and semidesert regions of the earth are most suitable sites which have annual insolation of 2000 to 2400 kWh/m^2 and sunshine duration 2400 to 3500 hours/year.

8.10.1. Parabolic Trough Solar Power Plant

It is also called solar farm power plant as a number of solar modules consisting of parabolic trough solar collectors are interconnected. Every module consists of a collector as shown in Fig. 8.9. It is rotated about one axis by a sun tracking mechanism.

Thermo-oil is mostly used as heating fluid as it has very high boiling point. Water/steam working fluid can also be used. The tubes have evacuated glass enclosure to reduce the losses. The concentration ratio is between 40 and 100. The maximum oil temperature is limited to 400°C as oil degrades above this temperature. Alternately steam at 550°C can be directly generated in the absorber tube.

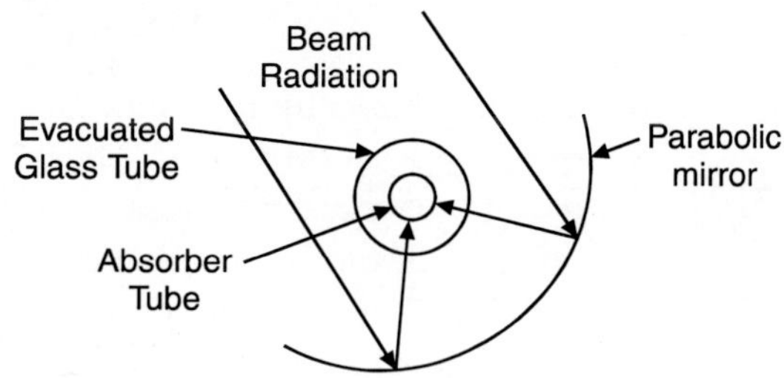

Fig. 8.9. Parabolic Solar Collector

These are commercially under operation. Fig. 8.10 shows a flow diagram of parabolic trough solar power plant. The working fluid is heated in collectors and collected in hot storage tank (2). The hot thermo-oil is used in boiler (5) to raise steam for the steam power plant. The boiler also is provided with a back-up unit (6) fired with natural gas. The cooled oil is stored in tank (3) and pumped (4) back to collector (1).

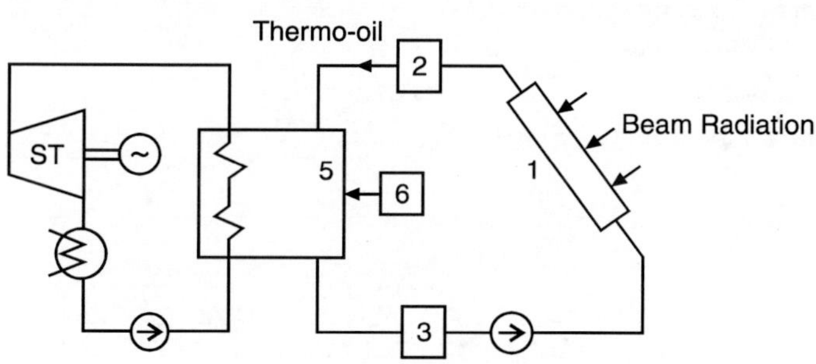

Fig. 8.10. Parabolic Trough Solar Power Plant

The specification of two such plants are given below

1. Power capacity	30 MW	80 MW
2. Thermal efficiency	37.5%	37.6%
3. Aperture area	194×10^3 m^2	464×10^3 m^2
4. Collector field efficiency	43%	53%
5. Annual energy production	93 GWh	253 GWh
6. Production cost	0.15 cents/kWh	0.10 cents/kWh

A power plant of 100 MW is planned to be installed in Rajasthan.

Performance Analysis

The heat output,

$$Q_c = n\, I_b\, A_a\, \eta_c \quad [\text{W}]$$

where A_a = Aperture area of a parabolic trough module [m^2]

I_b = Radiation intensity [W/m^2]

n = No. of collectors.

η_c = Collector efficiency

$$\eta_{mod.} = \rho\alpha_{abs} - \frac{U_c(T_{abs} - T_a)}{CI_b}$$

where ρ = Reflectivity of mirror

α_{abs} = Absorptivity of tubes.

U_c = Overall heat loss coefficient [W/m^2 –K]

T_{abs} = Temperature of absorber [K]

T_a = Ambient temperature [K]

C = Concentration ratio

I_b = Intensity of beam radiation [W/m^2]

Example 8.4. *Determine the heat output and exit temperature of thermo-oil for the following power plant.*

1. *Solar beam intensity* = 750 W/m^2
2. *Unshaded aperture area* = 240 m^2
3. *Concentration ratio* = 40
4. *Inlet temperature of oil* = 280°C
5. *Mass flow rate* = 0.6 kg/s
6. *Specific heat* = 3200 J/kg-K
7. *Ambient temperature* = 30°C
8. *Optical efficiency* = 0.74
9. *Heat loss coeff.* = 7 W/m^2-K
10. *Heat removal factor* = 0.96

Solution : (1)

$$Q_C = F_R A_a\left[I_{bc}\eta_{opt} - \frac{U_c}{C}(T_{in} - T_a)\right]$$

$$= 0.96 \times 240\left[750 \times 0.74 - \frac{7}{40}(280 - 30)\right]$$

$$= 117792 \text{ W}$$

(2) Fluid outlet temperature

$$T_{out} = T_{in} + \frac{Q_c}{m c_p}$$

$$= 280 + \frac{117792}{0.6 \times 3200}$$

$$= 341.35°\text{C}$$

8.10.2. Central Receiver Solar Power Plants

These are also called solar tower power plants. It can be divided into solar plant and conventional steam power plant. The flow diagram is given in Fig. 8.11.

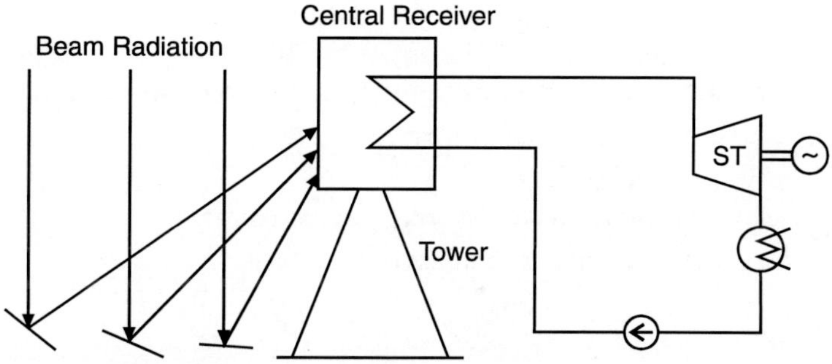

Fig. 8.11. Central Receiver Solar Power Plant

A heliostat field consists of a large number of flat mirrors of 25 to 150 m^2 area which reflects the beam radiations onto a central receiver mounted on a tower. Each mirror is tracked on two axis. The absorber surface temperature may be 400 to 1000°C. The concentration ratio (total mirror area divided by receiver area) may be 1500. Steam, air or liquid metal may be used as working fluid. Steam is raised for the conventional steam power plant.

The technical data of an experimental solar tower plant is given below :

1. Electric power output = 1.2 MW
2. Solar energy input = 5.5 MW
3. Heliostat field area = 1200 m^2
4. Tower height = 80 m
5. Working fluid = water/steam

A total heliostat area of 0.5 km^2 is required for a power plant of 20 MW_e.

Useful solar energy output,

$$E = A\, I_b\, \eta \text{ [W]}$$

where A = Total area [m^2]

I_b = Intensity of beam radiations [W/m^2]

η = Efficiency of heliostat field ≈ 55 to 80%.

Performance Analysis

The rate of useful heat output of central receiver

$$Q_c = E - \Sigma\Delta E_l - \Sigma\Delta Q_l \quad [W]$$

where E = output of heliostat field [W]

$\Sigma\Delta E_l$ = Electrical losses [W]

(Radiation losses due to reflection from the absorber surface)

$\Sigma\Delta Q_l$ = Heat losses [W]

(Due to conduction convection and radiation in the receiver walls)

The rate of useful heat output is also equal to

$$Q_c = A_a\,[I_a - F_R\, U_c\,(T_{in} - T_a)] \text{ [W]}$$
$$= \dot{m}\,(h_o - h_i) \text{ [W]}$$

The absorbed solar radiation flux,

$$I_a = \frac{(E - \Sigma\Delta E_l)}{A_a}$$

$$= I_b\, C\rho\, \tau\, \alpha \quad [\text{W/m}^2]$$

where I_b = Beam radiation intensity on heliostat surface [W/m²]

C = Concentration ratio of heliostat receiver system

ρ = Reflectivity of heliostat

γ = Intercept factor

τ = Transmissivity of transparent cover.

α = Absorptivity of absorber surface.

The efficiency, $\eta = \dfrac{Q_c}{E}$

where E = Useful solar radiation flux of heliostat [W]

8.10.3. Dish-Stirling Engine System

This is suitable for autonomous power generation

A hot gas Stirling engine receives solar irradiation from a parabolic trough concentrator (Dish). It has be very high efficiency. Glass mirrors are used for radiation concentrator. The working fluid is hydrogen or helium for Sterling engine. The maximum pressure and temperature of the engine cycle are 20 MPa and 1000K. A schematic diagram of the power plant is shown in Fig. 8.12.

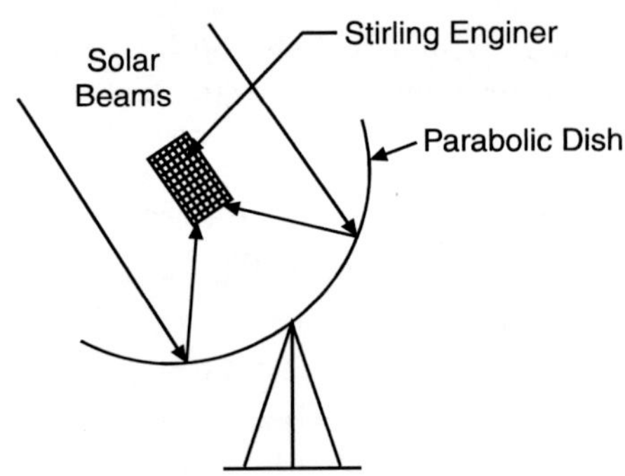

Fig. 8.12. Dish-Stirling Engine Power Plant

The specifications of such a system are given below :

1. Power output of Stirling engine = 50 kW
2. Cycle efficiency = 42%
3. Dish diameter = 17m
4. Concentration ratio = 4000
5. Optical efficiency = 89%
6. Receiver diameter = 700 mm
7. Flux density = 800 kW/m²
8. Cost = $2100/kW

8.10.4. Solar Chimney Power Plant

The air stream is heated by solar radiation absorbed by the ground and covered by a transparent cover. The hot air flows through or chimney which gives the air a certain velocity due to pressure drop caused by the chimney effect. The hot air flows through an air turbine to generate power. The schematic diagram of the plant is shown in Fig. 8.13.

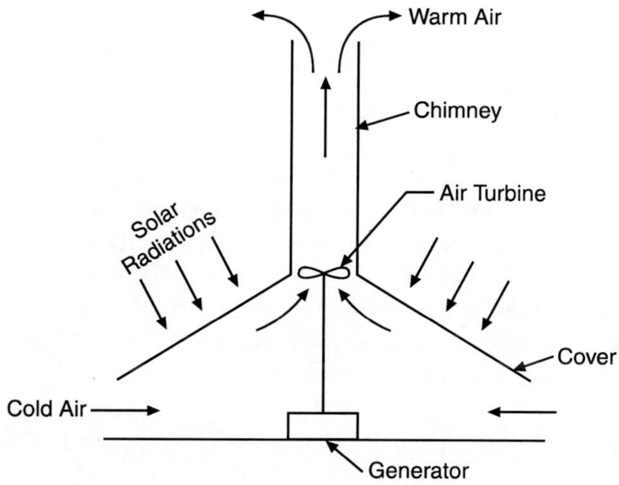

Fig. 8.13. Chimney Solar Power Plant

The pressure difference due to chimney effect,

$$\Delta p = gH(\rho_c - \rho_w) \quad [N/m^2]$$

where g = acceleration due to gravity [9.81 m/s²]

H = Chimney height [m]

ρ_c = Density of cold air [kg/m³]

ρ_w = Density of warm air [kg/m³]

The pressure energy is converted into kinetic energy of air. The air velocity in the chimney.

$$V = \sqrt{2\frac{\Delta p}{\rho}} \quad [m/s]$$

The power output of air turbine is a function of air velocity and chimney area. The plant efficiency,

$$\eta = \frac{P}{IA}$$

where P = Power output of air turbine [W]

I = Intensity of global solar radiation [W/m²]

A = Area of ground receiving solar radiator [m²]

The specification of a chimney power plant in operation are :

1. Power output = 50 kW
2. Annual energy output = 42 MWh
3. Overall efficiency = 0.05%
4. Radius of solar collector = 122m
5. Chimney height = 200m
6. Chimney diameter = 10m
7. Air turbine diameter = 10m
8. Air velocity in chimney = 7.6 m/s

8.11. SOLAR PONDS

Solar ponds are also called solar salt ponds. Natural ponds convert solar radiation into heat, but the heat is quickly lost through convection in the pond and evaporation from its surface. A solar pond is designed to reduce convective and evaporative heat losses by reversing the temperature gradient with the help of nonuniform vertical concentration of salts. A greater salt concentration at the bottom than at top causes bottom water to have greater density and remains at the bottom and is also hotter. The solar energy is absorbed in deep layers and is usefully trapped.

The solar ponds are useful in two ways :

1. The conversion of solar energy to useful work as a result of the temperature difference between bottom and top layers of the pond.
2. The use of pond as a thermal storage medium.

A solar pond is a combination of solar collector and storage medium.

A schematic of an artificial solar pond and conversion system is shown in Fig. 8.14. The pond is divided into three layers.

1. Top layer : Low density convective layer.

2. Central layer : Non-convective layer and contains the required salinity and density gradient with salinity greatest at the bottom.

3. Bottom layer : High density-salinity convective storage layer.

The bottom warm water is used as the heat source and top cool water as the heat sink. The hot bottom water is pumped through an evaporator and back to the bottom. An organic fluid is evaporated and drives a turbo generator. The turbine exhaust is condensed and the condensate is fed back to evaporator. The cooling water for condenser is taken from the top cool layer of the pond.

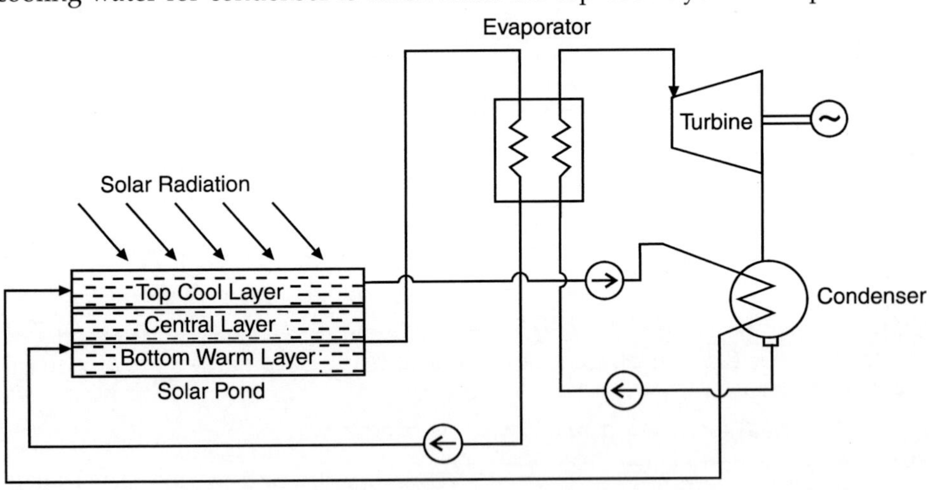

Fig. 8.14. Solar Pond Power Plant

A leak-proof pond is constructed. 330 kg of salt per square meter of pond area is added. The bottom layer can achieve a temperature of 93°C. The salinity gradient is maintained in the pond by injecting salt to the bottom layer and flushing the top layer with fresh water periodically. The efficiency is very low but these plants have large potential for tapping solar energy.

The main applications of solar ponds are :

1. Power generation
2. Space heating and cooling
3. Crop drying
4. Desalination
5. Process heat

The main limitations are :

1. Need for large land area.
2. Sunny climate
3. Availability of water.
4. Availability of salt (brackish water)

A solar pond power plant of 5 MW capacity is in operation in Dead sea, Israel.

8.12. THERMAL STORAGE SYSTEMS

In order to take care of intermittency of solar energy availability and to fill up the gap between power demand and supply of the power plant, there is need for energy storage system. There are basically two types of energy storage systems :

1. Electrical storage like batteries, superconducting coils, pumped hydro, compressed air, flywheels, etc.
2. Thermal storage including sensible heat, latent heat and chemical reaction.

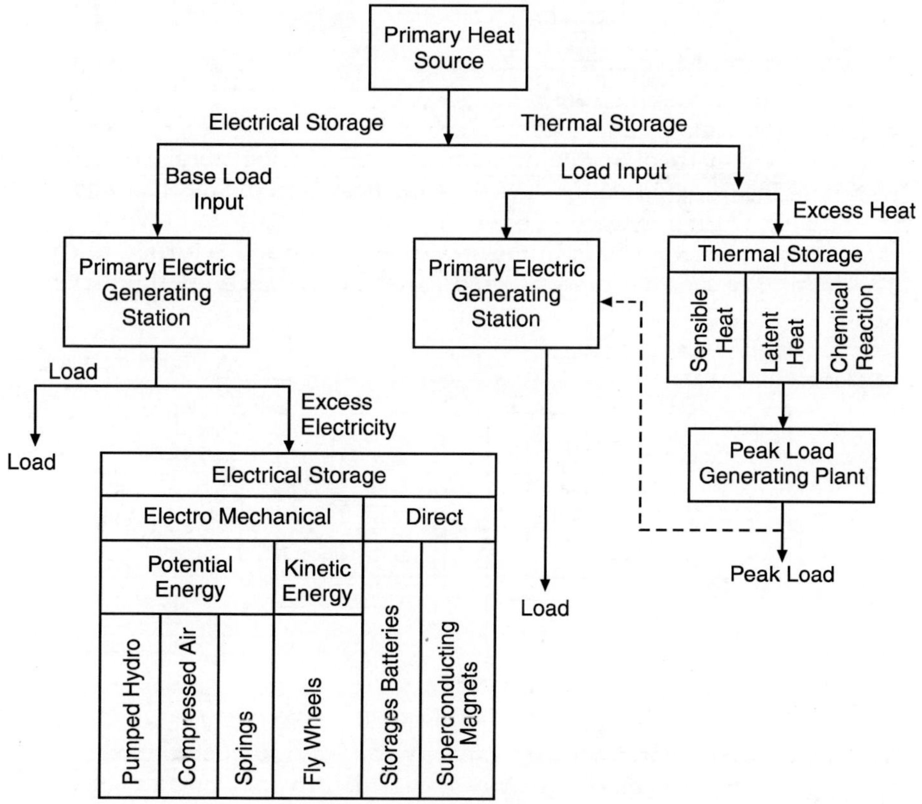

Fig. 8.15. Energy Storage Systems

Various energy storage systems are classified in Fig. 8.15. But at present thermal single-phase energy storage system is cost effective due to location of most solar power plants. There are two types of thermal storage systems suitable with solar systems.

1. Single-tank or thermocline storage system
2. Dual-tank or hot-cold storage system

8.12.1. Thermocline Storage System

The hot primary coolant is passed through a packed bed consisting of rocks + oil. The packed bed is heated and cooled coolant is returned to solar collector for reheating. During times of need, the direction of fluid flow is reversed. The primary coolant is admitted at the bottom of storage tank which gets heated by extracted heat from the packed bed and hot fluid goes to the power plant. The storage utilization factor is 0.8 for liquid-solid systems.

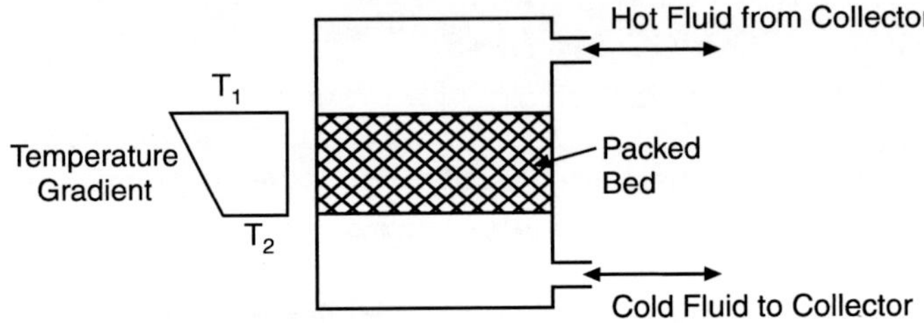

Fig. 8.16. Thermocline Storage System

8.12.2. Hot Cold System

The storage system consists of two well insulated tanks with necessary pumps, valves and piping. The storage fluid is either sodium, molten salt or oil. The tanks are made of austenitic stainless steel. During storage of extra energy cold liquid is pumped out of the cold tank, heated in the receiver and returned to hot tank. During extraction of energy, the hot liquid is drawn from the hot tank, used in raising steam in the steam generator and returned to cold tank. The amount of sensible energy stored varies by changing the fluid levels in the tanks.

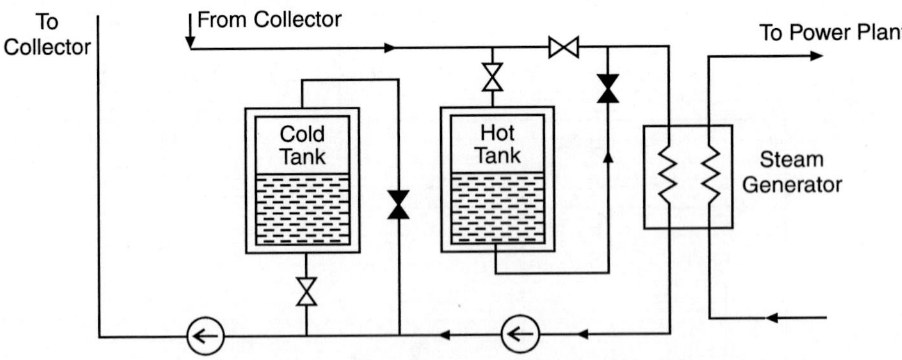

Fig. 8.17. Hot Cold Thermal Storage System

The thermal energy storage capacity (density) depends upon the storage media. Some typical values of energy storage density for a operating temperature range of 550° to 565°C are given below.

Sodium 0.08 MWh/m^3
Molten salt (nitrate) 0.22 MWh/m^3
Rock (25% void fraction) 0.15 MWh/m^3

The storage utilization factor is the fraction of stored energy which can be extracted within maximum and minimum temperature limits of storage media. This value is in excess of 0.98. The storage sizing can be made to supply enough thermal energy together with direct thermal power from receiver to operate the power plant continuously 24 h/day. Sizing depends upon many conflicting factors : variations in daily and seasonal load demands, cloud conditions, availability of backup power and cost. The cost of storage capacity will include cost of additional heliostats and large receiver to fill the storage system.

8.13. MISCELLANEOUS APPLICATIONS

Most of developing countries including India are located in those regions of the world (between 35°N and 35°S) which are most blessed with sunshine. The relative high population densities in the rural areas in these regions, and the high exposure to the sun therefore make it natural that the applications of solar energy, which lends itself to decentralized power of small size, should be given high priority when trying to solve the energy problems in the developing countries.

Although solar energy may be used in many markets such as active and passive space heating and cooling, industrial process heating, desalination and water heating and in electric generation where power requirements are small and in remote areas. The development, welfare and prosperity of rural villages depend upon the availability of cheap and abundant energy for domestic cooking, heating and lighting, community development, irrigation water, mechanised agriculture and small industrial parks based on solar energy.

If the energy and power needs of the villages can be met from solar energy, the central/zonal power grids connecting thermal, nuclear and hydro power plants can work in more efficient and reliable manner at high load factors feeding power only to industrial belts and big commercial centres. Tremendous investments and huge power losses in the long and unwieldy power lines to remote and scattered villages can be saved.

Therefore, there is an urgent need to use/develop the following solar appliances for villages

1. Solar cookers which must fit well with the age-old habits of conventional cooking and should be accepted by rural population without any reservations.
2. Small solar driven refrigeration units for preservation of food, storage of milk and other products in the rural areas where little electric power is available.
3. Solar water irrigation pumps to boost the agricultural production which is vitally important in view of growing population.
4. Solar water heaters and dryers for agricultural products.
5. Small power plants to produce electricity for domestic and street lighting and cottage industries.

8.14. SOLAR COOKERS

Eighty per cent of the basic energy needs of rural India is the provision of cooking energy, which amounts to 0.5 to 1 KWh/person/day with a peak power of 0.2 to 1.2 kW per capita. It is mostly met from traditional fuels like fire-wood and cowdung cakes. The Government of India is trying to introduce box-type solar cookers with financial subsidies but with poor success due to many technical drawbacks of the design. The temperature obtained is less than 100° C and is insufficient for chapati making and frying which are very important cooking processes. There is no provision for storage of heat and cooking has to be carried out while the sun is shining, which is an odd time for preparing breakfast and dinner. Cooking has to be carried out in the open without privacy and in the sun which is very inconvenient to the housewife.

8.14.1. Hot-Plate Solar Cooker

A small family size solar cooker has been developed to overcome the above drawbacks of solar cooking. It consists of a flat plate collector with mirror boosters. It is coupled to an oil storage insulated tank placed under the cooking platform inside the kitchen through a thermo syphon to the outside collector. The schematic diagram is shown in Fig. 8.18. It can provide sufficient flux of heat energy at high enough temperature to a hot plate through a heat pipe. The main components with specifications are as follows :

1. Solar collector ($1 m^2$)
2. Thermo syphon (32 mmNB)
3. Hot oil storage tank (100 *l* Dowtherm)
4. Thermometer (0—300°C)
5. Hot plate (2 kW)
6. Heat pipe (2 kW)
7. Control valve
8. Non-return valve.

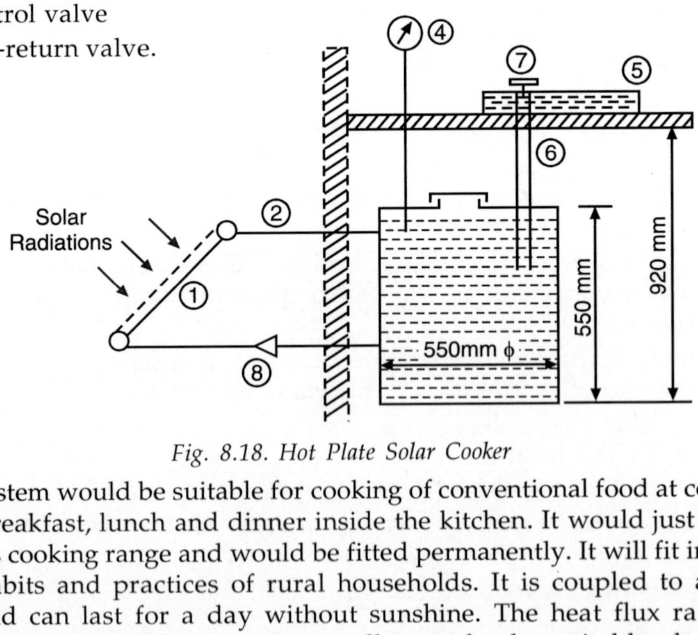

Fig. 8.18. Hot Plate Solar Cooker

The system would be suitable for cooking of conventional food at convenient time for breakfast, lunch and dinner inside the kitchen. It would just replace a bottled gas cooking range and would be fitted permanently. It will fit in with the age-old habits and practices of rural households. It is coupled to a thermal storage and can last for a day without sunshine. The heat flux rate can be conveniently controlled by turning a small control valve suitably placed at the hot plate. There are no operating and maintenance expenses and life long oil fill is provided. The prototype cost of Rs. 3000 can be substantially reduced during commercial design and manufacture.

8.14.2. Solar Ahara

An "AHARA" is a must for every household in the Indian villages in which a slow fire is maintained for almost the whole day with the help of cowdung cakes and fire wood. This is a multi-purpose cooker-cum-heater with low heat flux inputs. A versatile multi-purpose, low-cost solar ahara has been developed. It can be used as a small water heater, oil heater and ghee boiler in addition to solar cooker for rural delicacies of India such as rice, dals, sarson-ka-saag, palak and other vegetables and cereals requiring moderate cooking temperatures and low heat flux inputs to preserve taste, flavour and nutritional values of food.

The design of the ahara is based on the use of plane mirrors cut to special shape and size and arranged at suitabled angles to approach a paraboloid. The tracking is manually achieved in two axes by the use of simple pipes and joint threads. The ahara can be assembled from locally available materials and skills.

The total cost of the prototype including materials and manufacturing labour is Rs. 500 only. A temperature of 125°C is easily obtained for more than four hours on a normal sunny day.

8.15. SOLAR REFRIGERATION

A solar driven refrigerator is needed for preservation of food in rural areas where little electric power is available. Figure 8.19 depicts the flow diagram of a solar domestic refrigerator. Solid absorption zeolite refrigeration cycle can ensure long and trouble free operation life because of absence of moving parts, toxic and corrosive chemicals. The main components are :

1. 1.5 m^2 zeolite filled solar collector
2. 10 NB Non-return valve (Ball type)
3. 4.2 m^2 panel coil condenser
4. 20*l* water storage tank.
5. 0.5mm × 1.4m capillary tube
6. 0.3 m^2 panel coil evaporator and ice tank
7. 32 NB Non-return valve (Ball type)

It consists of 1.5 m^2 flat plate collector with 50 mm thick zeolite and can be fitted to 165*l* domestic refrigerator. Water vapour is driven out of collector by solar heating during the day. The vapour are condensed and stored in a tank until evenings. As the collector cools at night due to sky radiation, a vacuum in the system causes the water in the evaporator to boil constantly at a temperature of –2°C at a pressure of 4mm of Hg. The frozen surface can make 9 kg of ice in 8 hours. The water vapour returns to the collector for re-absorption into zeolite and the system is ready for the next day. The ice can be removed for use or left to cool the refrigerator at times of no sunshine. The system can be scaled up into small cold storage for preservation of food in rural areas.

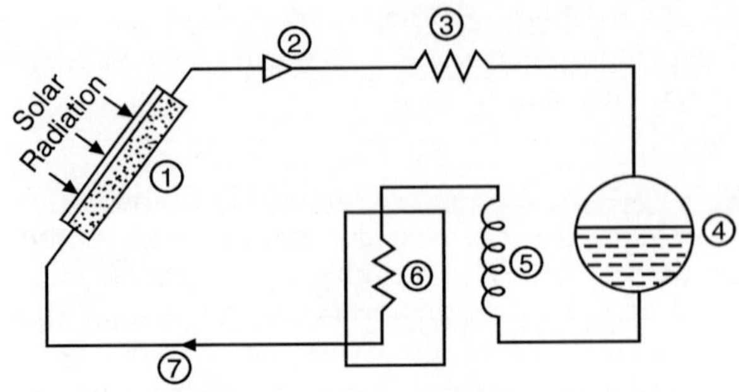

Fig. 8.19. Solid Zeolite Absorption Refrigerator

8.15.1. Solar Milk Cooler

The total milk production per day per village on an average may be taken as 2000 litres out of which the milk producers can consume 25% for self use and convert about 45% into milk products like ghee, paneer and khoa. Therefore, 500-600 litres of milk per villages is available for sale as liquid milk per day.

Rural processing of milk can add value to the product and increase rural income. The zeolite solar refrigeration system can be attached to a cooling-cum-storage tank made of single-embossed panel coils of stainless steel, the inside being plain sheet and outside made of embossed sheet. The water refrigerant circulates by thermosyphon effect in the walls and acts as a very efficient heat exchanger. The temperature difference between milk and water vapour is reduced to less then 0.5°C. As water storage is in the bottom, it makes possible to cool even small quantities of milk.

8.16. SOLAR WATER PUMPS

Organic vapour Rankine cycle system with rotary machines or screw expanders or reciprocating engine or spiral expander has been used to drive water pumps. Schematic diagram of such a pump is shown in Fig. 8.20. The main specifications are :

1. Pump rating : 1 kW
2. Refrigerant : R114
3. Collector area : $10 m^2$
4. Total cost : Rs. 12,500

The solar heat can operate the pump for 4 hours a day. When not pumping, it can be used to drive a generator or thresher or a lathe.

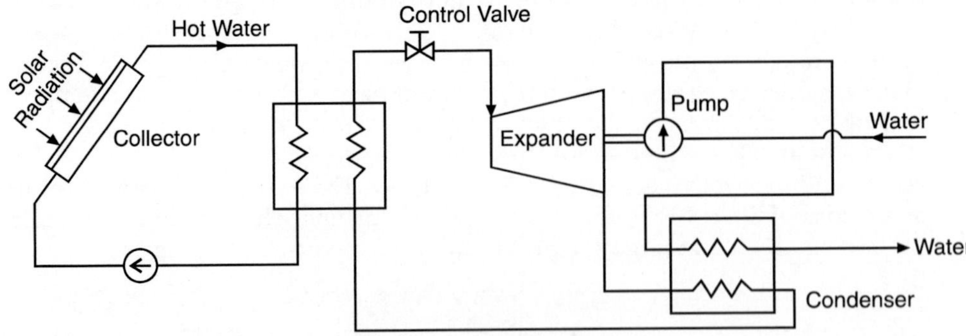

Fig. 8.20. Solar Pump

8.17. LIMITATIONS OF SOLAR ENERGY

Immense solar energy is available free of cost. Its potential is huge. But the nature of solar energy has certain limitations which affect the engineering design of the equipment and system and reduces its cost effectiveness.

1. Low Flux Density

Large surfaces are needed to collect solar energy of low flux density for large-scale utilization resulting in increased cost of the delivered energy. When the sun is directly overhead on a cloudless day, 10 m^2 of surface could theoretically provide energy at 10% efficiency of collection at the rate of 1 kW. Most of the man made solar collectors can convert only direct energy efficiently and the position of the sun, thick clouds and anthropogenic pollution let less energy pass and scatter the energy back into space and percentage of diffuse energy increases. Almost total absorption of solar energy by Ozone at wavelengths below 300nm and by carbon dioxide at wavelengths beyond 2500nm, the irradiance on the earth's surface is effectively limited to wavelengths between 300 to 2500nm.

2. Intermittency of Solar Energy

Solar energy has a regular daily cycle due to the turning of the earth around its axis, a regular annual cycle due to inclination of the earth axis with the place of the elliptic and due to the motion of the earth around the sun, and is also unavailable during periods of bad weather. The diffuse and cyclic nature of the source introduce special problems in storage and distribution of the energy, restrict the areas of application and need supplementation from other sources.

3. Global Distribution of Solar Energy

The global distribution of solar energy does not favour the industrial parts of the world, but may be of help to industrially developing countries located in the favourable radiation belt. The greatest amount of energy is available in the continent desert areas around 25°N and 25°S of the equator and falls off towards both the equator and the poles. Mostly, these are flat deserts that are practically unusable for agriculture or any type of industrial development. These regions have little or no water which can create special problems in the generation of thermal electric power.

4. Cost Effectiveness

Although solar energy is essentially free, there is definite cost effectiveness associated with its utilization whereas for conventional energy sources, the processing cost has traditionally been borne by large industries which borrow money from a bank and then charge the consumer for each unit of energy used. Solar energy installations, especially for low temperature applications in home heating and cooling, are the total responsibility of the user. As a result of this shift in the economic risk, a solar energy installation appears to the users as a huge additional investment that they personally must make in advance before driving any benefit from it. Solar energy must be viewed as a long-term investment, and its cost must be prorated just like cost of fossil fuels. Acceptance of this view is necessary to assess the cost effectiveness of solar energy.

5. Energy Intensive

Much of the construction materials used for collecting and converting solar radiation are themselves very energy intensive. For example, large quantities of aluminium, steel, copper, concrete, glass and plastic are needed, all of which require large quantities of energy for conversion from ore to finished products. The energy consumed in the processes for producing the above finished products and in constructing the systems is a large fraction of the net energy generated during the life of the plant. This fraction is in most cases bordering on the unacceptable limits, except in special cases where solar energy is uniquely suited.

6. Environmental Pollution

Solar energy systems are not pollution free as usually expected. The mining of large quantities of mineral ores, the processing of these ores into the finished material, the manufacturing of equipment and construction of the plants are the processes that generate their own pollution and health and occupational safety risks. In addition, the low efficiency of these plants, results in large heat rejection, hence the thermal pollution.

8.18. DESIGN OF SOLAR THERMAL PLANTS

There are basically two types of solar thermal plants.

1. Low temperature solar plants

These plants are used for various thermal applications such as swimming pool heating (pool water temperature of 23 to 28°C), hot water system (45 to 60°C), space heating (heating water to a temperature of 30 to 90°C) and refrigeration and space cooling (heat is supplied to the refrigerant vapour generator at 90 to 150°C). Domestic water heating is the most widely used system.

Solar air heaters are used for drying of agricultural produce, heating of green houses and buildings.

2. High temperature solar plants

These plants are used for generation of electricity by using a thermodynamic cycle: Rankine cycle, Brayton cycle.

All solar thermal plants have the following systems/components:

1. Solar radiation data
2. Solar collector and receiver
3. Heat transport system
4. Thermal storage system

The detailed design of each component/system depends upon the specific application.

8.19. SOLAR RADIATION DATA

8.19.1. Radiation

Average cumulative daily solar radiation on a horizontal surface in MJ/m^2 per day are available for different locations. The available territorial solar energy at a given time and place is influenced not only by time of the day (hour angle, ω) or year (sun's declination angle, δ), location (latitude angle, θ) and scattering but also by cloudiness. The incident radiation on the earth's surface is influenced by the solar angles and is usually presented in terms of dimensionless air mass m_a, defined as

m_a = air mass = ratio of optical thickness of the atmosphere through which beam radiation pass to the surface to its optical thickness if the sun were at the zenith, *i.e.*, directly above (dimensionless).

In addition to three basic, angles, latitude, hour angle and sun's declintion, certain additional angles are also useful in solar radiation data-analysis.

1. Altitude angle α (solar altitude)

It is a vertical angle between the projection of the sun's rays on the horizontal plane and the direction of the sun's rays passing through a point.

2. Zenith angle (θ_Z)

It is a vertical angle between the sun's rays and a line perpendicular to the horizontal plane through a point. Therefore, θ_Z and α are complementary angles.

$$\theta_Z = \frac{\pi}{2} - \alpha$$

3. Solar azimuth angle, γ_s

It is a horizontal angle between the north and the horizontal projection of the sun's rays.

The above solar angles can be represented in terms of basic angles.

$$\cos\ \theta_z = \chi o\sigma\phi\ \cos\omega\ \cos\delta + \sin\phi\ \sin\delta$$

$$\because \qquad \theta_z = \frac{\pi}{2} - \alpha$$

$$\therefore \qquad \cos\theta_z = \sigma\iota\nu\ \alpha$$

$$\cos\gamma_s = \sec\alpha(\cos\phi\ \sin\delta - \cos\delta\ \sin\phi\ \cos\omega)$$

and $\quad \sin\ \gamma_s = \sigma\epsilon\chi\ \alpha\ \cos\delta\ \sin\omega$

The air mass,

$$m_a = (\cos\theta_z)^{-1} \text{ for } \theta_z = 0° \text{ to } 70°$$

For extraterrestrial radiations, $m_a = 0$

For sun at the zenith, $m_a = 1$ at sea level

For $\qquad \theta_z = 60°,\ m_a = 2$

Clearness Index (C_i)

All the effects of solar angles, scattering, absorption, cloudiness may be combined in one parameter called the Clearness Index.

C_i = Clearness index = ratio of the average radiation on a horizontal surface for a given period to the average extraterrestrial radiation for the same period.

The averaging could be monthly, daily or hourly. Therefore, C_i would be a monthly, daily or hourly clearness index. It varies widely from 30% to 70% in some localities on earth. Even in the day time due to bad weather, C_i could be zero.

8.19.2. Measurements of Solar Radiations

The total radiation is the sum of both the beam and diffuse components. The following measurements are made:

1. Beam radiations on a horizontal surface

A Pyrheliometer is used to measure the beam radiation on a horizontal surface.

It is a small telescope mounted on a drive mechanism and follows the sun throughout the day.

2. Diffuse radiations on a horizontal surface

A Pyranometer with a shade ring is used to measure the diffuse radiation on a horizontal surface.

3. Total radiation on a horizontal surface

A Pyranometer without a shade ring is used to measure the total or global radiations.

4. Sunshine duration

A sunshine recorder is used to measure the hours of bright sunshine during the course of a day. A lens burns a trace on a card when exposed to the sun. The length of the trace directly measures the duration of bright sunshine.

5. Beam Radiation on a normal surface

The total radiation received by a surface normal to the beam radiation is more than that on a horizontal surface. Therefore, a radiation collecting surface would be more effective when held perpendicular to the direction of the sun rays. The collector surface should track the sun by changing angle of installation.

Example 8.5. *Calculate the sun's altitude angle and azimuth angle at 7.30 am solar time on August 1 for a location at 40° north latilide.*

Solution :

$$\phi = 40°$$

$$\omega = 15 \times 4.5 = 67°.30'$$

$$\delta = 23.45 \sin\left[360 \times \frac{284 + n}{365}\right]$$

$$n = 213 \text{ for 1st August}$$

$$\delta = 17.91°$$

$$\cos\theta_z = \sigma\nu\alpha = \cos\phi\cos\delta\cos\omega + \sin\phi\sin\delta$$

$$= \cos 40° \cos 17.91° \cos 67.5° + \sin 40° \sin 18.75°$$

$$\therefore \quad \alpha = 28.46°$$

$$\sin\gamma_s = œ\chi\ \alpha \cos\delta \sin\omega$$

$$= \sec 28.93° \cos 17.91° \sin 67.5°$$

$$\therefore \quad g_s = 89°\ 1'36''$$

8.20. DESIGN OF FLAT PLATE COLLECTOR

Low temperature solar plants including various types of water and air heaters use flat plate collectors for collection and absorption of solar radiations and heating of working fluid. The design of flat plate collector consists of the following:

1. Optical design of collector
2. Thermal design of collector
3. Selection of materials for critical components such as transparent cover, absorber plate with selective coating, fluid flow channels and thermal insulation.

8.20.1. Optical Design

The flat plate collector works on the basis of green house effect. A transparent cover plate admits the short wave length radiation from the sun and surroundings.

These radiations heat the absorber plate and reradiates long wave length radiation as the plate is at low temperature. The cover plate is opaque to these radiations. The cover plate should have the following properties:

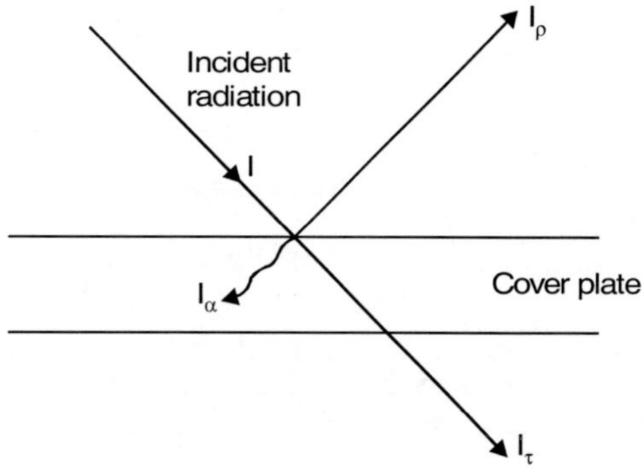

Fig. 8.21. Solar radiation.

1. Radiation received on the surface may be partly transmitted, partly absorbed and partly reflected depending upon the properties of the cover.

2. $$I = I_\alpha + I_\rho + I_\tau$$

where I = Amount of incident radiation

I_α = Amount of radiation absorbed

I_ρ = Amount of radiation reflected

I_τ = Amount of radiation transmitted

Also
$$I_\alpha = \alpha I$$
$$I_\rho = \rho I$$
$$I_\tau = \tau I$$
$$\alpha + \rho + \tau = 1$$

where $\alpha = \dfrac{I_\alpha}{I}$ = Absorptivity of cover plate

$\rho = \dfrac{I_\rho}{I}$ = Reflectivity of cover plate

$\tau = \dfrac{I_\tau}{I}$ = Transmissibility of cover plate

3. A perfectly transparent plate will have
$$\alpha = 1$$
$$\rho = \alpha = 0$$

4. Most real materials are only partially transparent. The transmittance is dependent both upon the reflection and absorption of radiation. Transmissibility can be calculated as

$\tau = 1$ □ $\rho - \alpha$

It can also be calculated as follows.

(*i*) Calculate τ_ρ considering reflection alone and taking $\alpha = 0$

(*ii*) Calculate transmissivity τ_α considering absorption only and $\rho = 0$

(*iii*) Calculate transmisivity τ taking care of both reflection and absorption as follows.

$$\tau = \tau_\rho \times \tau_\alpha$$

5. **Calculation of τ_ρ**

Transmittance of a single cover when $\alpha = 0$

$$\tau_{\rho 1} = \frac{1-\rho}{1+\rho}$$

Transmittance for a system of n covers,

$$\tau_{\rho, n} = \frac{1-\rho}{1+(2n-1)\,\rho}$$

where n = No. of covers of the same material.

6. Calculation of τ_α

Transmittance of a single cover when $\rho = 0$

$$\tau_\alpha = e^{-kL}$$

Transmittance of n covers when angle of reflection = θ_2.

$$\tau_\alpha = e^{\frac{-nkL}{\cos\theta_2}}$$

where k = extinction coefficient
= 0.01/cm for clear and white glass
= 0.32/cm for poor quality glass
L = Actual path of the radiation through the cover plate.

Example 8.5. *A flat plate collector has two glass covers each 3 mm thick, having refractive index of 1.526 and an extinction coefficient (k) of 0.0161/mm. Calculate the transmittance taking into account the absorption and reflection for normal incidence.*

Solution:

$$\tau_{\alpha(0)} = e^{-nkL}, \text{ for normal incidence}$$
$$= e^{-2 \times 0.0161 \times 3} = 0.915$$

$$\rho_{(0)} = \left|\frac{n-1}{n+1}\right|^2 = \left[\frac{1.526-1}{1.526+1}\right]^2 = 0.0434$$

$$\tau_{\rho(0)} = \frac{1-\rho}{1+(2n-1)\rho}$$

$$= \frac{1-0.0434}{1+(2\times 2-1)\,0.0434} = 0.846$$

$$\therefore \quad t_{(0)} = \tau_{\alpha(0)} \times \tau_{\rho(0)} = 0.915 \times 0.846 = 0.7741$$

7. Transmittance Absorptance Product

The radiation passing through the cover system strikes the absorber plate. Most of the radiation is absorbed but some part is reflected back.

Τηε ραδ ιατιον αβσορβεδ = $\tau \propto$

The radiation reflected back = $(1 - \alpha)\,\tau$.

This reflected radiation is mostly diffuse radiation which is again reflected back to absorber plate by the cover plate. This process continues. The sum of total radiations absorbed will be more than $\tau \propto$.

Total radiation absorbed

$$<\tau\alpha> = \tau\alpha \sum_{n=0}^{n=\infty}\left[(1-\alpha)\,\rho_d\right]^n = \frac{\tau\alpha}{1-(1-\alpha)\,\rho_d}$$

where, ρ_d = Diffuse radiation reflected by cover plate
= 0.16 for incidence angle of 60°.

Example 8.6. *Find the absorptance-transmittance product if transmittance for two glass cover systems is 0.4187, ρ_d = 0.24 and absorptance of absorber plate is 0.94.*

Solution :

$$\langle\tau\alpha\rangle = \frac{\tau\alpha}{1-(1-\alpha)\rho_d} = \frac{0.4187 \times 0.94}{1-(1-0.94)\,0.24} = 0.399$$

8.20.2. Thermal Design

The radiations striking the absorber plate raise its temperature. Under the plates, fluid carrier tubs are bonded through which working fluid flows and gets heated. The energy balance for the whole collector can be written to find out the useful heat output of the flat-plate solar collector.

$$Q_C = A[I_C F_R (\tau\alpha)_e - F_R U_C (T_{in} - T_a)]\ [W]$$

where A = surface area of collector or absorber [m^2]

I_C = intensity of solar radiation incident on the collector [W/m^2]

F_R = heat removal factor of the collector

$\langle\tau\alpha\rangle_e$ = effective product of transmissibility τ of the transparent cover and absapticity α of the absorber.

U_C = overall heat loss coefficient of collector [W/m^2]

T_{in} = Fluid inlet temperature [°C]

T_a = Ambient air temperature [°C]

Here, absorbed energy = $A\, I_C\, F_R \langle\tau\alpha\rangle_e$

Effective heat loss = $A\, U_C (T_p - T_a)$

where T_p = Temperature of absorber plate

1. Evaluation of overall loss Coefficient (U_C) :

The energy loss from the collector plate consists of radiation and convection to the cover and edges and conduction through the back insulation. Thermal network of a flat plate collector is shown in Fig. 8.22.

Insulation Resistance (R_1)

$$U_1 = \frac{1}{R_1} = \frac{k}{x}$$

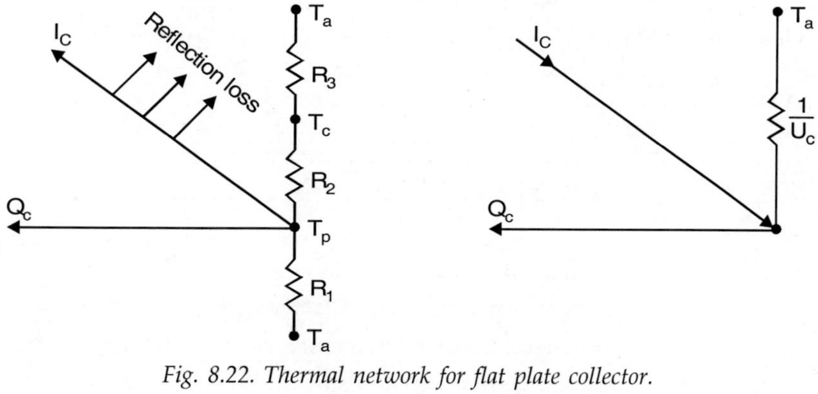

Fig. 8.22. Thermal network for flat plate collector.

k = insulation thermal conductivity

x = insulation thermal thickness

Heat loss resistance from absorber plate (R_2)

This constitutes convective resistance and radiation resistance.

$$R_2 = \frac{1}{h_p + h_r}$$

h_p = heat transfer coefficient between two inclined parallel plates, *i.e.*, absorber plate and cover plate

h_r = radiation heat transfer coefficient

$$= \frac{\sigma(T_p + T_c)(T_p^2 + T_c^2)}{\frac{1}{\varepsilon_p} + \frac{1}{\varepsilon_c} - 1}$$

where σ = Stefan Baltzmann's constant

T_p = Temperature of absorber plate (metal)

T_c = Temperature of cover plate (glass)

ε_p = Infra-red emissivity of the absorber plate

ε_c = Infra-red emissivity of the cover plate

Heat loss resistance from top cover to surroundings (R_3)

$$R_3 = \frac{1}{h_r + h_w}$$

h_r = radiation conductance from top cover to sky

$$= \varepsilon_c\, \sigma\, (T_c + T_{sky})(T_c^2 + T_{sky}^2)\, \text{Ÿ}\, \frac{(T_c - T_{sky})}{(T_c - T_a)}$$

where $T_{\text{sky}} = T_a - 6$ [°C]

The overall loss coefficient,

$$U_c = \frac{1}{R_1 + R_2 + R_3}$$

2. Evaluation of Heat Removal Factor (F_R)

Heat removal factor (F_R) is defined as the ratio of actual useful energy collected by the fluid to the useful energy collected if the entire collector aborber surface were at the temperature of the fluid entering the collector.

$$F_R = \frac{\dot{m}\, c_p}{U_c A}\left[1 - e^{\frac{-U_c A F'}{\dot{m} c_p}}\right]$$

where $\dot{m}$ = mass flow rate of fluid [kg/s]

c_p = heat capacity of fluid [kW/kg-K]

F' = collector efficiency factor

F_R can also be found out with the help of fig. 8.5.

8.21. MATERIAL SELECTION

The performance and durability of a flat plate collector depends upon the proper selection of material for critical components.

1. Cover Plate

The purpose of cover plate is

(*i*) to transmit maximum solar energy to the absorber plate.

(*ii*) heat loss from absorber plate to environment should be minimum.

(*iii*) to protect absorber plate from direct exposure to weathering.

(*iv*) to receive maximum solar energy for the maximum duration in a day.

The most important properties are :

(*i*) transmissibility

(*ii*) strength

(*iii*) durability

(*iv*) non-degradability.

The most common materials used are

1. Tampered glass with low iron content.
2. Plastics such as Lexian.

2. Absorber Plate

The most important properties of absorber plate should be

(*i*) high thermal conductivity

(*ii*) high strength

(*iii*) good corrosion resistance.

The common materials used are

1. Copper due to high conductivity and resistance to corrosion
2. Aluminium
3. Steel
4. Thermoplastics

3. Fluid Flow Channels

(*i*) Tubes soldered or welded to the bottom of absorber metal plate

(*ii*) Roll-Bond panel made of formed copper or aluminium sheets to provide fluid channels.

4. Thermal Insulation

Thermal insulation of 5 to 10 cm thickness is placed behind the absorber plate to prevent heat losses from the bottom surface. The insulation material used are

(*i*) mineral wool

(*ii*) glass wool

(*iii*) heat resistance fibre glass

5. Selective Coating

An effective way to reduce thermal losses from the absorber plate is by using selective absorber coating. This coating will make absorber plate a perfect absorber of solar radiator of short wave lengths and perfect reflector or poor emitter of thermal radiation of long wave lengths. Some common selective coatings with their properties are given below

	Coating	Type	Absoptance (α)	Emittance (ε)
1.	Black chrome	Electoplated	0.96	0.10
2.	Black Nickel	Electroplated	0.90	0.10
3.	Black copper	Copper oxide	0.87 – 0.92	0.07 – 0.35
4.	Black Anodize	Aluminium oxide	0.94	0.07
5.	Solar foil	Black chrome over copper	0.96	0.10
6.	Nextel	Paint	0.98	0.89

Other properties of selective coating should be

1. Non-degradability due to exposure to humidity.
2. Ability to withstand temperatures and thermal shocks.
3. Resistance to atmospheric corrosion and oxidation.
4. Reasonable cost.

8.22. OPTICAL DESIGN OF FOCUSSING (CONCENTRATING) COLLECTORS

High temperature solar plants are solar-electric conversion systems where the collectors gather the sun's energy and direct it onto receivers that contain the working fluid of the thermodynamic cycle. The main designs are:

1. Solar-thermal central-receiver system

A large field of reflecting mirrors called heliostats receive and redirect the sun's energy and concentrate it on a central receiver mounted on top of a tower. The heliostats are individually guided, since they cover a large field so that each plain mirror focusses the solar radiations on the central receiver at all hours of sun light. In the receiver the concentrated solar energy is absorbed by a circulating fluid. The fluid could be water, which vaporises into steam that is used to drive a turbogenerator in a Rankine cycle, or an intermediate fluid that transports the heat to the steam cycle.

2. Parabolic Trough System

Long troughs of cylindrical parabolic shape are lined with mirrors to collect and concentrate the solar radiation onto a focal linear conduit through which the primary fluid flows (Fig. 8.9). The fluid from different receivers is combined for the thermal-electric conversion.

3. Dish-Stirling Engine System

A mirrored parabolic dish tracks the sun and focusses the captured energy on a receiver which is mounted at the focal point of the parabola (Fig. 8.12).

The focussing collectors involve additional optical losses. The optical efficiency,

$$\eta_{opt} = \rho\gamma\tau\alpha_a C_{opt}$$

(*a*) Reflectance Loss

It depends upon the reflectivity of mirror ρ and depends upon the nature and smoothness of the reflector surface.

(*b*) Intercept Factor (γ)

The energy received by the absorber depends upon the part of the reflected radiation that is intercepted by the receiver. The intercept factor depends upon :

(*i*) The shape, size and orientation of reflector

(*ii*) The shape, size and positioning of receiver relative to the reflector/concentrator.

(*c*) Concentration ratio (C)

The ratio of effective area of the aperture to area of the solar energy absorber is called concentration ratio.

$$C = \frac{A_a}{A_r}$$

The collector efficiency increases with the increase of concentration ratio.

Example 8.7. *Determine the intercept factor and concentration ratio for a paraboloidal concentrator and receiver for an axially symmetrical system with aperture radius R and radius of absorber in focal plane, r, and R/r = 0.01 and normal flux distribution coefficient, h = 60.*

Solution: The intercept factor,

$$\gamma = 1 - e^{-\frac{h^2}{(r/R)^2}} = 1 - e^{\frac{-(60)^2}{(0.01)^2}} = 0.296$$

The concentration ratio,

$$C = \left(\frac{R}{r}\right)^2 = \left(\frac{1}{0.01}\right)^2 = 10{,}000$$

8.23. HELIOSTATS

1. These are reflecting mirrors to reflect the sun rays on the receiver.
2. It consists of a mirror, mirror support structure, pedestal, foundation and control and drive mechanism.
3. A glass heliostat is divided into 10 to 14 panels for ease of manufacture and transportation. The rectangular panels are 1.2 × 3.6m with 1.5 to 3 mm thickness of low iron glass sheets to minimize absorption.
4. The cost of heliostats may be 45% of total capital cost of a solar thermal power plant.
5. Plastic heliostat have lower cost, lower mass but lower in reflectance.
6. A north heliostat field is used for higher efficiency.
7. During normal operation, the heliostats are in sun-tracking mode.
8. Active reflected beam sensors are used to control the drive motors. Pre-programmed computer control can also be used.

Heliostat Losses

There are additional energy losses between the incident energy on the heliostat field and the receiver.

1. Shadowing. This is caused by one heliostat causing a shadow on the reflective surface of another at certain times of the day.
2. Cosine Loss. In order to adjust the angle of beam reflected to the receiver, the heliostat may not be perpendicular to the sun rays. The area of solar flux intercepted by the heliostat is cosine of the angle between the surface and perpendicular to the beam.
3. Blocking. Reflected light from one heliostat may be partly blocked by the back side of another.

4. Reflective loss. Some radiation may be absorbed by the glass and silvering and some may be scattered due to dirt on the mirror surface.
5. Attenuation. Absorption and scattering by water vapour, haze, fog, smoke and particulates in the atmosphere between heliostat and receiver.

8.24. RECEIVER

1. Central receivers sit atop tall towers and receive radiation energy fluxes from 300 to 700 kW/m^2.
2. The receiver should intercept, absorb and transport this energy to a heat-transfer fluid.
3. The panels of parallel tubes with headers at each end absorb the solar energy on their outside surface and conduct it to heat transfer fluid inside. The panels are supported at top to allow thermal expansion downwards.
4. The coolant tubes are 20 to 56 mm diameter and 1.2 to 6.4 mm thickness.

The efficiency of a receiver is reduced due to following losses.

1. Spillage. The reflected rays may miss the receiver due to heliostat tracking errors, wind effects, steering back loss, etc. These losses are less than 5 percent in a well designed system.
2. Reflection. The energy scattered back from the receiver heat transfer surface can be minimized by painting these surfaces with high-absorptivity paint. These losses are less than 5 percent in a well designed system.
3. Convection. There is a loss by wind convection from the hot surfaces of the receiver.
4. Radiation. There are radiation losses from the hot receiver surface to the environment.

   The convection and radiation losses can be 5 to 15 percent.
5. Conduction. About 1 percent energy is lost due to conduction of heat to structural members, insulation, etc.

8.25. HEAT TRANSPORT SYSTEM

The heat transport system is composed of primary coolant piping, pumps, etc. The primary coolant may or may not be same as power plant working fluid.

1. Water-steam

Steam may be generated at 70 to 140 bar and 540 to 600 °C as in a once-through boiler or drum type boiler. It is extension of conventional technology to solar system.

2. Liquid Metal

Liquid metals have low vapour pressures and hence low operating temperatures. Sodium may be operated as a single-phase coolant upto 540°C as used in fast-breeder nuclear reactor.

3. Molten salts

Molten salts have also low vapour pressures, high volumetric heat capacities and high freezing temperatures. Nitrate salt mixtures can be used for solar central receiver system.

4. Gases

Air and helium can be used upto 840°C. These can be used to generate steam for a Rankine cycle or may be used directly in a Brayston-cycle power plant gas turbine of open type (with air) or in a closed cycle (with helium) or in a combined cycle.

5. Heat Transfer Oil

Low corrosive properties and low vapour pressures of some oils can be used as primary coolant of raising steam in a boiler. However, the danger of decomposition of oil and fouling of piping can cause operational problems.

TUTORIAL SHEET NO. 8.1.

1. Prove that direct solar energy can meet the needs of entire pollution of the world.
2. What is a solar constant ? What is the difference between extraterrestrial and terrestrial solar radiations ? Give reasons for the difference.
3. How can amount of solar radiations falling on an tilted flat surface be estimated ?
4. Describe the construction of different types of solar collectors. Discuss various materials used.
5. How is the performance of a flat plate solar collector evaluated ?
6. Compare the performance of various types of solar collectors. How is focussing accomplished in paraboloid dish collector?
7. Describe a solar water heating plant.
8. What are various types of solar thermal power plants ? Discuss in details.
9. Describe the construction and performance of a solar tower power plant.
10. Write notes on
    (*a*) Stirling engine system
    (*b*) Solar chimney system
11. Describe the principle and applications of solar ponds.
12. Give a detailed classification of energy storage systems. Describe thermocline and Hot-cold thermal storage systems.
13. Describe the application of solar energy for the following rural applications
    (*a*) solar cooking
    (*b*) solar refrigeration
    (*c*) solar water pumping
14. What are the main limitations of solar thermal energy ? How these limitations affect its widespread utilization?
15. Define various solar angles and their effects on the collection of solar radiations on a tilted flat surface.
16. How is the optical efficiency of a solar collector be estimated?

❑❑❑

Chapter **9**

OCEAN THERMAL ENERGY

Introduction ; Availability ; Ocean Temperature Differences ; Conversion Technologies ; Open (Claude) Cycle ; Modification of Open Cycle ; The Closed OTEC Cycle ; Limitations of OTEC Power Plants ; Special Applications.

9.1. INTRODUCTION

The oceans and seas constitute some 70 percent of the earth's surface area. These absorb solar radiation which causes moderate temperature gradients from the water surface downwards, especially in tropical waters. This ocean thermal energy is immense. The temperature gradient can be utilized in a heat engine to generate power. This is called ocean temperature energy conversion (OTEC). But because the temperature difference is small, even in the tropics, OTEC systems have very low efficiencies and consequently have very high capital cost.

9.2. AVAILABILITY

The availability of ocean thermal energy can be estimated as follows :

Total extraterrestrial solar energy received by the earth

$= 1.516 \times 10^{18}$ kWh/year

or 5.457×10^{18} MJ/year

Average clearness index $= 0.5$

Fraction of area of oceans $= 0.7$

$\therefore$ Total terrestrial ocean thermal energy $= 1.516 \times 10^{18} \times 0.5 \times 0.7$

$= 0.53 \times 10^{18}$ kWh/year

or 1.9×10^{18} MJ/year

Average terrestrial ocean thermal energy is nearly equal to 50% of solar constant $= 1353 \text{ W/m}^2 \times 0.5$

$= 676 \text{ W/m}^2$

This total terrestial incident energy is not absorbed by water because some of it is reflected back to the sky. The energy absorbed by the ocean water can be estimated from the annual evaporation of water. Although the same amount is replenished back by rainfall & runoff from the land.

Annual evaporation $= 120$ cm

or $120 \text{ cm}^3/\text{cm}^2$

or $1.20 \text{ m}^3/\text{m}^2$.

Average water surface temperature	= 20°C
The latent heat of vaporization	= 2454 kJ/kg
Sea Water density	≈ 1000 kg/m^3
∴ The annual energy absorbed	= 1.20 × 1000 × 2454
	= 3 × 106 kJ/m^2 per year

This is equivalent to 95 W/m^2

Absorbed energy as percentage of incident energy

$$= \frac{95}{676} \times 100 = 14\%$$

This figure may be higher in the tropics and lower in arctic waters.

9.3. OCEAN TEMPERATURE DIFFERENCES

Solar energy absorption by the water takes place according to Lambert's Law of absorption which states that each layer of equal thickness absorbs the same fraction of light that goes through it. If I is the intensity of radiation and y is the depth of water,

$$\frac{dI(y)}{dy} = \mu I$$

or

$$I(y) = I_o e^{-\mu y}$$

where I_o = Intensity of radiation at the surface where y = 0.

μ = Extinction coefficient or absorption coefficient

For very fresh clear water, μ = 0.05 m^{-1}

For turbid fresh water, μ = 0.27 m^{-1}

For very salty water, μ = 0.50 m^{-1}

The intensity falls exponentially with the depth (y) and depending upon the value of μ, almost all of the absorption occurs very close to the surface of deep waters. Therefore, maximum temperature occurs just below the surface due to heat and mass transfer at the surface itself.

Water density decreases with an increase in temperature. However, pure water at 3.98°C has the maximum density.

There will be no thermal convective currents between the warmer, lighter water at top and coolar, heavier water at a depth. Similarly, heat transfer by thermal conduction between the water layers at the surface and a depth is too low to alter the picture and mixing is retarded. The warm water stays at the top and the cool water stays at the bottom.

In tropical waters there are essentially two infinite heat reservoirs : a heat source at the surface at about 27°C and a heat sink at about 4°C at some depth of one kilometer : Both the reservoirs are maintained annually by solar incidence. These temperatures vary with latitude and season.

9.4. CONVERSION TECHNOLOGIES

The maximum possible efficiency of a heat engine working between two temperature limits cannot be more than that of a Carnot cycle operating between the same temperature limits.

The Carnot efficiency,

$$\eta_c = \frac{T_1 - T_2}{T_1}$$

where T_1 = source temperature

T_2 = sink temperature

If t_1 = 27°C and t_2 = 7°C

$$\eta_c = \frac{27 - 7}{(27 + 273)} = 6.67\%$$

For a 300K (27°C) surface temperature an OTEC system cannot have efficiency greater than $\Delta T/3$ where ΔT is the temperature difference between surface and deep waters. Taking into consideration the temperature and pressure drops in the plant equipment such as turbine, condenser, heat exchangers and pumps, the efficiency of a real OTEC power plant is less than two percent.

Although there are no fuel costs, the capital costs are extremely high due to large size of power plant.

There are two basic designs for OTEC systems

1. Open cycle also called Claude Cycle.
2. Closed cycle also called Anderson Cycle.

9.5. THE OPEN OR CLAUDE CYCLE

The seawater plays the role of heat source, working fluid, coolant and heat sink. The flow diagram and *T-s* diagram are shown in Fig. 9.1 and 9.2 respectively.

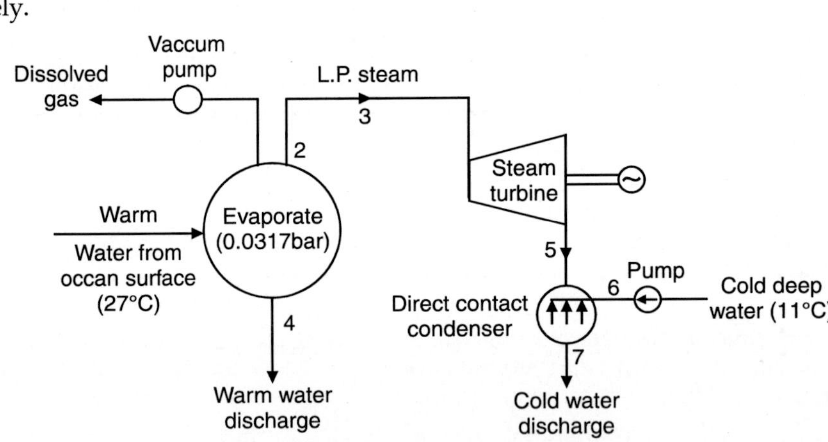

Fig. 9.1. Open Cycle OTEC Power Plant.

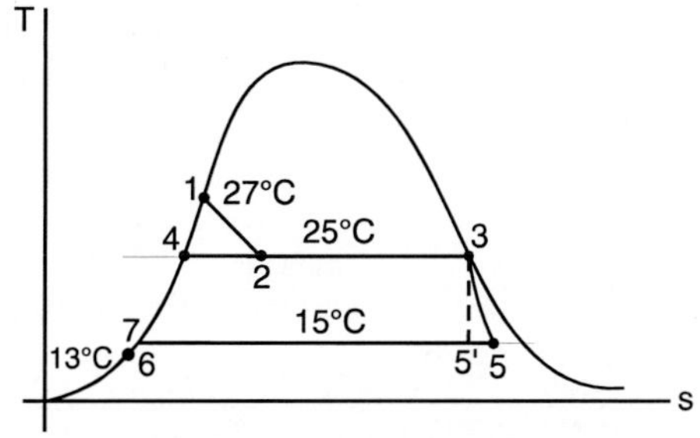

Fig. 9.2. T-s Diagram of Open Cycle OTEC Plant.

Warm water at 27°C is admitted into an evaporator where pressure is maintained slightly below the saturation pressure corresponding to warm water temperature. The steam from evaporator is superheated at pressure of 0.0317 bar. The water undergoes volume boiling, water is partly flashed into steam. Process 1-2 is a constant enthalpy throttling process. The low pressure in the evaporator is maintained by a vacuum pump which also removes the dissolved noncondensable gases from the evaporator. The steam expands in a specially designed steam turbine. The steam at turbine exit is condensed in a direct contact condenser where cold deep water at 13°C is mixed with this steam and discharged into the ocean.

A temperature difference of 10°C (25° – 15°) is available for the cycle work. The plant has to handle very large quantities of warm and cold water which requires large pumping power and large heavy cold water pipes.

Example 9.1. *A Claude cycle of 100 kW gross capacity operates on the parameters given in the table. The turbine has a polytropic efficiency of 0.80 and the turbine-generator has a combined mechanical-electrical efficiency of 0.90. Calculate the surface and deep-water flow rates and the gross cycle and plant efficiencies.*

Saturated Steam Data

Point in the cycle	Temp. °C	Pressure	Specific volume m^3/kg		Enthalpy kJ/kg		
		bar	v_f	v_g	h_f	h_g	h_{fg}
Water at condenser inlet	13	0.01497	0.0010007	88.18	54.60	2525	2470
Condenser	15	0.01704	0.0010010	77.97	62.97	2528	2465
Turbine inlet	25	0.03166	0.0010030	43.40	104.80	2547	2442
Surface water	27	0.03564	0.0010036	38.82	113.20	2550	2437

Solution : **1. Evaporator**

$$h_1 = h_f\,(27°\text{C})$$
$$= 113.2 \text{ kJ/kg}$$
$$v_1 = 0.0010036 \text{ m}^3/\text{kg}$$
$$h_2 = h_1 = 113.2$$
$$= (h_f + x_2\, h_{fg}) \quad \{\text{at } 25°\text{C}]$$
$$= 104.8 + x_2 \times 2442$$
$$\therefore \quad x_2 = 0.00344 \quad \text{or} \quad 0.344\%$$

Warm water mass-flow rate per unit turbine mass-flow rate

$$m'_w = \frac{m_1}{m_3} = \frac{m_2}{m_3}$$

$$= \frac{1}{x_2} = \frac{1}{0.00344} = 290.7$$

2. Turbine

$$h_3 = h_g \text{ (at 25° C)}$$
$$= 2550 \text{ kJ/kg}$$
$$s_3 = 8.5570 \text{ kJ/kg-K}$$
$$s_{5,s} = s_3 = 8.5570$$
$$= (s_f + x_{5,s} \cdot s_{fg}) \text{ at 15°C}$$
$$= 0.2244 + x_{5,s} \times 8.5562$$
$$\therefore \quad x_{5,s} = 97.39\%$$
$$h_{5,s} = (h_f + x_{5,s}\, h_{fg}) \text{ at 15°C}$$
$$= 62.97 + 0.9739 \times 2465$$
$$= 2463.6 \text{ kJ/kg}$$

Adiabatic reversible turbine work $= h_5 - h_{5,s}$

$$= 2550 - 2463.6$$
$$= 86.4 \text{ kJ/kg}$$

Actual turbine work, $w_T = 86.4 \times \eta_p$

$$= 86.4 \times 0.8 = 69.1 \text{ kJ/kg}$$
$$h_5 = h_3 - \text{actual work}$$
$$= 2550 - 69.1$$
$$= 2480.9 \text{ kJ/kg}$$
$$A\, x_5 = 0.9809$$
$$v_5 = 76.48 \text{ m}^3\text{/kg}$$

Turbo generator output $= 69.1 \times 0.88 = 60.8$ kJ/kg

3. Condenser

$$h_6 \approx h_g \text{ at 13°C}$$
$$= 54.60 \text{ kJ/kg}$$
$$v_6 = 0.0010007 \text{ m}^3\text{/kg}$$
$$h_7 = h_f \text{ at 15°C} = 62.97 \text{ kJ/kg}$$

Cold water mass-flow rate per unit turbine mass-flow rate

$$\dot{m}_c = \frac{h_5 - h_7}{h_7 - h_6}$$

$$= \frac{2480.9 - 62.97}{62.97 - 54.60} = 288.9$$

4. The Cycle

$$\text{Turbine mass-flow rate } \dot{M}_T = \frac{\text{Turbine work}}{w_T}$$

$$= \frac{100}{69.1} = 1.447 \text{ kg/s}$$

$$\text{Turbine volume flow rate at throttle} = \dot{M}_T v_3 = 1.447 \times 43.40 = 62.8 \text{ m}^3/\text{s}$$

$$\text{Turbine volume flow rate at exhaust } \dot{M}_T v_5 = 1.447 \times 76.48 = 110.7 \text{ m}^3/\text{s}$$

$$\text{Warm-water mass flow rate } \dot{M}_w = \dot{M}_T \dot{m}_w = 1.447 \times 290.7 = 420.6 \text{ kg/s}$$

$$\text{Warm-water volume flow rate } \dot{V}_w = \dot{M}_w v_1 = 420.6 \times 0.0010036 = 0.422 \text{ m}^3/\text{s}$$

$$\text{Cold-water mass-flow rate, } \dot{M}_c = \dot{M}_T \dot{m}_c = 1.447 \times 288.9 = 418 \text{ kg/s}$$

$$\text{Cold water volume flow rate, } \dot{V}_c = \dot{M}_c v_6 = 418.0 \times 0.0010007 = 0.418 \text{ m}^3/\text{s}$$

$$\text{Gross cycle efficiency} = \frac{w_T}{q_A} = \frac{h_3 - h_5}{h_5 - h_7} = \frac{69.1}{2487} = 0.0278 = 2.78\%$$

$$\text{Gross plant efficiency} = 0.0278 \times 0.9 = 0.0250 = 2.5\%$$

Note : The gross power, 100 kW, and the gross plant efficiency 2.5%, do not take into account pumping and other auxiliary power inputs to the plant.

9.6. MODIFICATIONS OF OPEN OTEC CYCLE

Certain changes are proposed in the open OTEC cycle to make it economically viable.

1. Controlled Flash Evaporator
2. OTEC plant for cogeneration of electricity and fresh water.

9.6.1. Controlled Flash Evaporation (CFE)

A CFE chamber is a vertical vessel of 2.5m height which is filled with a large number of vertical parallel chutes. The water flows on both sides of the chutes vaporises without violent process of bubble formation and bursting. Pure vapour free of entrained solids flows downwards. Both brine and vapour cool down as they travel down the chutes. The pure low pressure steam enters the steam turbine and expands to the condenser. The deaeration required is minimal in CFE. The high quality of steam going to the turbine is well suited to electrical and fresh-water cogeneration. The corrosion problem in steam turbine is alleviated.

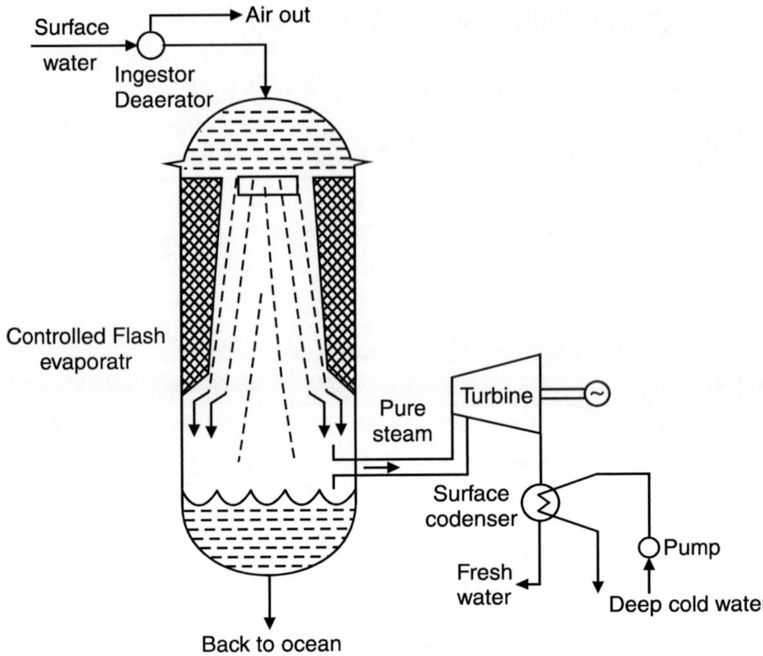

Fig. 9.3. Controlled Flash Evaporator and OTEC Plant for Cogeneration of Electricity and Fresh Water.

9.6.2. Cogeneration of Fresh Water and Electricity

A surface condenser either shell and tube or plate type is used. The deep cold water is pumped through the condenser and discharged back to the ocean. The turbine fresh water condensate is used for various uses. When combined with a controlled flash evaporator, the quality of condensate is a potable water with salt content 1 to 5 ppm. There is slight reduction in efficiency due to slightly higher pressure and temperature in the condenser as compared to direct contact condenser.

9.7. THE CLOSED OTEC CYCLE

This is also called Anderson Cycle and uses Ammonia, Propane or Freon as the working fluid for the Rankine Cycle.

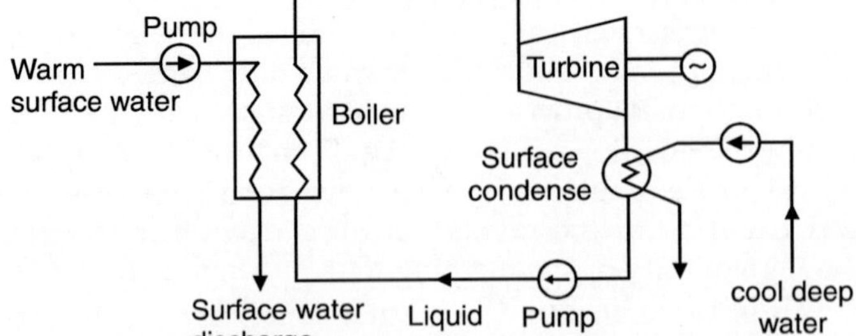

Fig. 9.4. Closed Cycle OTEC Power Plant.

High pressure of working fluid results in low specific volumes and compact plant and less costs. The problems of evaporator are also avoided. However, a large size heat exchanger has to be used as boiler to keep the temperature differences small.

An Anderson OTEC power plant uses propane with 20°C temperature difference between warm surface and cool deep water. The propane is vaporized in the boiler at 10 bar and exhausted in the condenser at 5 bar. It employs thin plate-type heat exchangers to minimize the amount of material and cost. The heat exchangers are placed at depths in the ocean where static pressure of water is roughly equal to the pressure of the working fluid.

9.8. LIMITATION OF OTEC POWER PLANTS

Problems facing commercial development of OTEC systems are enormous.

1. Low temperature differences result is very low plant efficiencies, very large plant size and huge capital cost.
2. For open systems, the problems associated with the evaporator design, operation and maintenance are many.
3. Steam turbines capable of generating 10 MW or more using low pressure steam have yet to be developed.
4. For closed systems, heat exchangers of very large size have to be designed and built.
5. Pumps handling large amounts of water have to be developed.
6. A 100 MW OTEC power plant may require 1 km long and 30m diameter pipe.
7. The whole plant has to be stationed and moored at large depth.
8. The deployment of the electrical cables required to carry the power to shore must be carefully done because it is subjected to severe stresses from its own weight and ocean currents & eddies.
9. An OTEC power plant will have to be capable of withstanding sever ocean storms over its lifetime, corrosion by seawater salts, erosion due to large volume flows, biofouling due to algae growth and encruotation by various marine life such as barnacles.

9.9 SPECIAL APPLICATIONS

(*i*) OPEC Plant Ship : The commercial-size OTEC systems are very large and suffer from various problems when placed in a stationary position. OTEC Plant ship concept can be used. The OTEC plant can be built as a specially designed ship. The grazing plant ship would move at low speed (about 0.5 knots) in search of warmest surface water. The electricity generated can be used on board to manufacture energy-intensive products like aluminium, magnesium, nickel, various alloys, semiconductors, etc. deep sea bed mining, chemical processing of products like ammonia.

(*ii*) Island OTEC Plants : The case for OTEC is more promising for islands that have low population densities, that are dependent upon imported fossil fuels for generating electricity, and that have large warm-water resource in relation to their energy needs.

***(iii)* Hydrogen Energy :** The power generated in the OTEC plants can be used for hydrogen production by means of water electrolysis. In this case, the investment costs will be lower compared to electrolysis powered by electricity generated in conventional thermal and nuclear power plants. The energy transmission to the consumer locations shall be by means of hydrogen transport.

India is planning 1 MW OTEC plant off Chennai coast.

TUTORIAL SHEET NO. 9.1.

1. What is Ocean Thermal Energy ? Discuss its availability.
2. How can ocean temperature differences be estimated ?
3. What is OTEC ? Determine the overall efficiency of an OTEC plant if surface warm water temperature is 30°C and deep cool water temperature is 8°C. It can be assumed that the relative efficiency factor of the power plant is 50%. [3.63%]
4. *Describe open cycle Claude OTEC plant with the help of flow diagram and T-s* diagram.
5. What modifications can be carried out to the open cycle OTEC plant to make it economically feasible ?
6. Explain a closed cycle OTEC plant with the help of flow diagram.
7. What are the limitations of OTEC plants ?
8. Discuss the special applications of OTEC plants.
9. Discus the theory and working principle of ocean thermal energy conversion system. Also list various plants based upon it.

❑❑❑

Chapter **10**

WAVE AND TIDAL WAVE ENERGY

Introduction ; Mechanism of Wave Motion ; Power from Wave Energy ; Wave Energy Conversion by Floats ; Wave Energy Conversion Devices ; Tidal Waves ; Power production in Single-Pool Tidal Systems ; Modulated single pool Tidal System ; Two-pool Tidal System.

10.1. INTRODUCTION

Ocean and sea waves are caused indirectly by solar energy. Uneven solar heating and subsequent cooling of the earth's crust causes winds which alongwith rotation of the earth cause waves. Wave energy when active is very concentrated, therefore wave energy conversion into useful energy can be carried out at high power densities. A large variety of devices have been developed for harvestuing of wave energy but these are complicated and fragile in the face of gigantic power of ocean storms.

Advantages

The main advantages of power from wave energy are :

1. It is a free and renewable energy source.
2. Wave-power devices donot require large land masses.
3. It is relatively pollution free.
4. After removal of power, the waves are in placid state.
5. A string of devices have to be used wherever there is large wave activity. In addition to producing electricity, the system protects coast lines from the destructive action of large waves, minimize erosion and help create artificial harbour.

Disadvantages

There are several disadvantages also.

1. Waves lack dependability.
2. There is relative scarcity of accessible sites of large wave activity.
3. The conversion devices are relatively complicated in construction.
4. The devices have to withstand the enormous power of stormy seas.
5. There are unfavourable economic factors such as large capital investment and costs of repair, replacement and maintenance.
6. Problems of biological growth of marine organisms are relatively unknown.

Arabian sea off India and Pakistan have suitable sites for wave energy harvesting.

10.2. MECHANISM OF WAVE MOTION

A progressive wave has a free surface and is acted upon by gravity and is shown in Fig. 10.1.

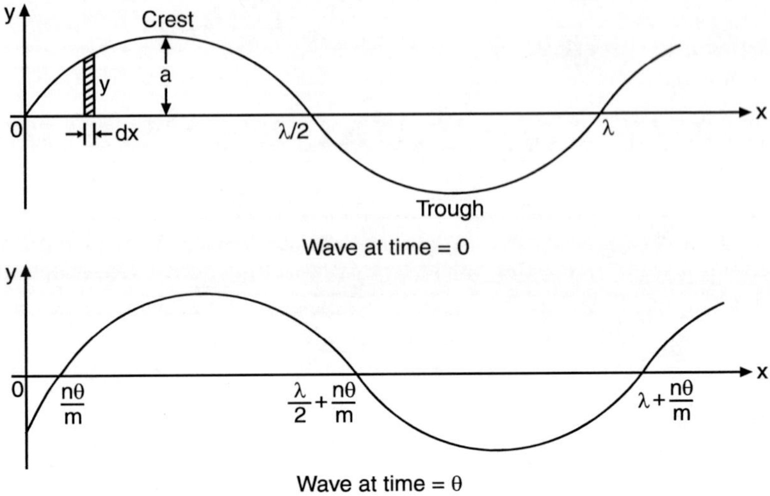

Fig. 10.1. A Progressive Wave

It has the following parameters

$$\lambda = \Omega\ \alpha\varpi\varepsilon\lambda\varepsilon\nu\gamma\tau\eta\ [\mu\ ] = C\tau$$
$$a = \text{amplitude [m]}$$
$$2a = \text{height (from crest to trough) [m]}$$
$$\tau = \text{period [s]}$$
$$f = \frac{1}{\tau} = \text{frequency } [s^{-1}]$$
$$c = \Omega\ \alpha\varpi\varepsilon\ \pi\rho o\pi\alpha\gamma\alpha\tau\iota o\nu\ \varpi\varepsilon\lambda o\chi\iota\tau\psi\ [\mu\ /\ \sigma] = \lambda/\tau$$
$$n = \text{phase rate } [s^{-1}]$$
$$= \frac{2\pi}{\tau}$$

The period τ and wave velocity c depend upon the wave-length and the depth of water. The relationship between wavelength and period can be approximated as :

$$\lambda = 1.56\ \tau^2$$

where λ is in metres and τ in seconds.

Figure 10.1 shows the cross-sections of the wave at time 0 and θ. The sinusoidal simple harmonic wave can be expressed by

$$y = a \sin\left(\frac{2\pi}{\lambda}x - \frac{2\pi}{\tau}\theta\right)$$

or $$y = a \sin(mx - n\theta)$$

where y = height of wave above mean level $[m]$

θ = time, $[s]$

$$m = \frac{2\pi}{\lambda}\ [\text{m}^{-1}]$$

$$(mx - n\theta) = 2\pi\left(\frac{x}{\lambda} - \frac{\theta}{\tau}\right) = \text{phase angle [dimensionless]}$$

The wave profile at time θ has the same shape as at time 0. It is displaced from it by a distance $x = \frac{\theta}{\tau} = \theta(n/m)$

When $\theta = \tau$, $x = \lambda$. The wave profile assumes its original position. The wave motion is continually lateral, *i.e.*, in the x direction with a speed of propagation $c = \frac{\lambda}{\tau}$.

The motion of water does not follow the wave motion. A given particle of water rotates in place in an elliptical path in the plane of wave propagation. A wave, therefore, possesses potential energy and kinetic energy.

10.3. POWER FROM WAVE ENERGY

The total energy of a wave is the sum of its potential energy and kinetic energy.

10.3.1. Potential Energy

The potential energy is due to elevation of water above the mean level ($y = 0$). Consider a differential volume $y\,d$× (see Fig. 10.1). It has a mean height of $y/2$.

The potential energy,

$$dPE = m\frac{yg}{2g_c} = (\rho y\,dx\ L)\frac{yg}{2g_c}$$

$$= \frac{\rho L}{2} y^2\,dx\,\frac{g}{g_c}$$

where m = mass of liquid in volume $y\,dx$ [kg]

g = gravitational acceleration = 9.81 m/s^2

g_c = conversion factor = $1\frac{kg.m}{Ns^2}$

ρ = Water density [kg/m^3]

L = arbitrary width of the two-dimensional wave perpendicular to the direction of wave propagation [m]

$$\therefore \qquad PE = \frac{\rho La^2}{2}\frac{g}{g_c}\int_0^{\lambda}\sin^2(mx - n\theta)\,dx$$

$$= \frac{\rho La^2}{2m}\frac{g}{g_c}\left[\frac{1}{2}mx - \frac{1}{4}\sin 2mx\right]_p^{\lambda}$$

$$PE = \frac{\rho L a^2}{2m}\frac{g}{g_c}\left[\frac{m\lambda}{2}\right] = \frac{1}{4}\lambda a^2 \lambda L \frac{g}{g_c}$$

The potential energy density per unit area is $\frac{PE}{A}$, where $A = \lambda L$ in J/m^2.

$$\frac{PE}{A} = \frac{1}{4}\rho a^2 \frac{g}{g_c}$$

10.3.2. Kinetic Energy

The kinetic energy of the wave is the energy of the liquid between two vertical planes perpendicular to the direction of wave propagation x and placed one wavelength apart. From hydro dynamic theory,

$$KE = \frac{1}{4}\rho a^2 \lambda L \frac{g}{g_c}$$

The kinetic energy density is

$$\frac{KE}{A} = \frac{1}{4}\rho a^2 \frac{g}{g_c}$$

10.3.3. Total Energy and Power

The potential energy and kinetic energy of a progressive sine wave are identical. Therefore, total energy

$$E = \frac{KE}{2} + \frac{PE}{2}$$

The total energy density,

$$\frac{E}{A} = \frac{1}{2}\rho a^2 \frac{g}{g_c}$$

The power P (energy per unit time) for a wave is the product of energy and frequency. The power density

$$\frac{P}{A} = \frac{1}{2}\rho a^2 f \frac{g}{g_c} \quad [\text{W/m}^2]$$

Example 10.1. *A 2m wave has a 6s period and occurs at the surface of water 100m deep. Find the energy and power densities of the wave. Take water density = 1025 kg/m^3.*

Solution : Wavelength, $\lambda = c\,\tau = 1.56\,\tau^2$

$$= 1.56\,(6)^2 = 56.16 \text{ m}$$

Wave velocity, $c = \frac{\lambda}{\tau} = \frac{56.16}{6} = 9.36$ m/s

Wave height, $2a = 2$m

Amplitude, $a = 1$m

Wave frequency, $f = \frac{1}{\tau} = \frac{1}{6}$ s

Energy density, $\frac{E}{A} = \frac{1}{2}\rho a^2 \frac{g}{g_c}$

$$= \frac{1}{2} \times 1025 \times (1)^2 \times \frac{9.81}{1} = 5027.6 \text{ J/m}^2$$

Power density, $\frac{P}{A} = \frac{E}{A}.f = 5027.6 \times \frac{1}{6} = 837.9 \text{ W/m}^2$

10.4. WAVE-ENERGY CONVERSION BY FLOATS

The wave motion is horizontal and motion of water is primarily vertical. The water motion is converted into mechanical power by the use of floats. A large float is driven up and down by water within stationary guides. This reciprocating motion is converted to mechanical and then to electrical power.

Float Wave-Power Machine

Figure 10.2 shows Martin's float wave-power machine. A square float moves up and down with the water. Four vertical manifolds that are part of a platform guide the float. A piston is attached to the float which compresses air in a cylinder which is stationary with the platform. The piston-cylinder arrangement acts as a reciprocating air compressor. The downward motion of the piston draws air into the cylinder via inlet check valve. The upward motion compresses the air and send it through an outlet check valve to the four underwater floatation tanks via the four manifolds. The four floatation tanks serve the dual purpose of bouyancy and air storage tanks and four manifolds serve the purpose of float guides and discharge air pipes.

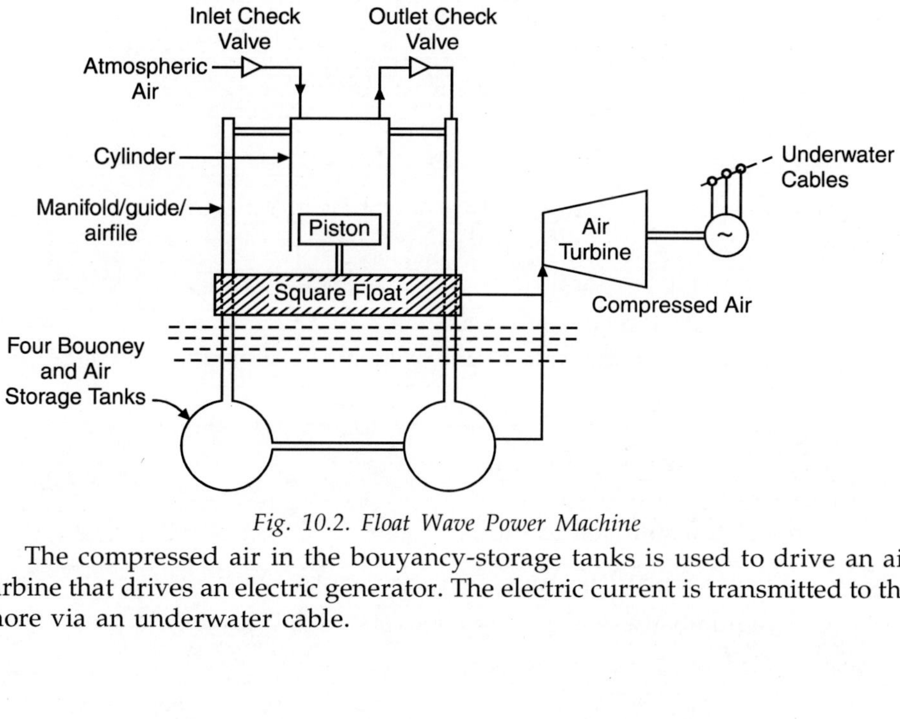

Fig. 10.2. Float Wave Power Machine

The compressed air in the bouyancy-storage tanks is used to drive an air turbine that drives an electric generator. The electric current is transmitted to the shore via an underwater cable.

Working Air Cycle

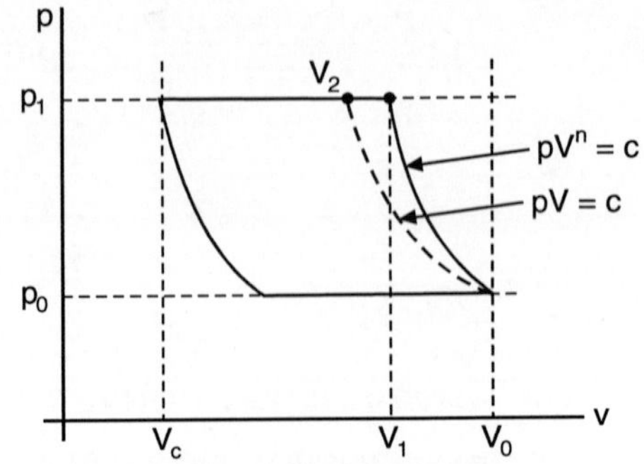

Fig. 10.3. p-v diagram

The air enteres the compressor at atmospheric pressure, volume and temperature p_o, v_o and T_o. It leaves the cylinder at p_1, v_1 and T_1. The air is cooled in the storage tanks back to T_o. Therefore, compression can be assumed as isothermal process.

$$\therefore \quad V_2 = \frac{v_o \, p_o}{p_1}$$

$$p_2 \approx p_1$$

The air expands in the turbine by adiabatic process. The turbine work,

$$\frac{W_T}{m} = C_p T_o \left[1 - \left(\frac{r_c}{r_e} \right)^{\frac{\gamma-1}{\gamma}} . r_c^{\left(\frac{1-\gamma}{\gamma} \right)} \right]$$

where m = mass of air (kg)

C_p = specific heat of air [J/kg-K]

$r_c = \dfrac{p_1}{p_0}$ = pressure ratio in compressor

$r_e = \dfrac{p_2}{p_3} = \dfrac{p_1}{p_3}$ = expansion ratio in turbine

p_3 = turbine exhaust pressure [p_a]

γ = ratio of specific heats for air

Allowing for reasonable mechanical clearance,

$$\frac{r_c}{r_e} = 1.1$$

The optimization shows a power density of 0.5 kW/m^2.

Limitations

1. It would take kilometers of linear arrays to produce 100 MW.
2. Waves not perfectly sinusoidal.
3. Aspiration of water into intake system and submersion by large waves.
4. Water entering turbine.
5. Problems of cost and corrosion. Molded plastic should be used.
6. Design to withstand storms.
6. Marine growth.
8. Power transmission to shore.

10.5. WAVE ENERGY CONVERSION DEVICES

There are a number of wave energy conversion devices developed and experimented.

10.5.1. Hydraulic Accumulator Wave Machine

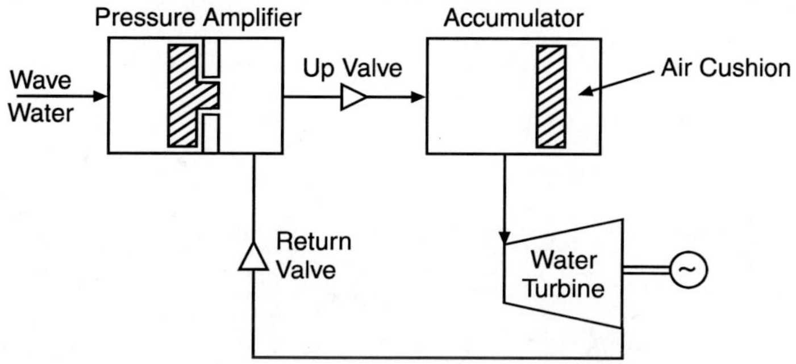

Fig. 10.4. Hydraulic Accumulator Wave Machine

The machine consists of the following main components :

1. **Pressure Amplifier**

   The waves enter the cylinder of pressure amplifier at the bottom and move the main piston. The pressure of the closed loop fluid is amplified to about 5 bar.

2. **Hydraulic Accumulator**

   The high-pressure fluid is conducted through a one-way up valve to a hydraulic accumulator. The accumulator has air cushions on the top which helps to maintain a constant pressure.

3. **Pelton Turbine**

   Part of high-pressure fluid flows through a Pelton wheel or Francis hydraulic turbine that drives an electrical generator

On the trough of the wave, the composite piston is pushed downward by the high fluid pressure. The exhaust water is sucked via return valve. The capacity of hydraulic accumulator is large enough to permit continuous turbine operation though the waves are cyclic.

10.5.2. High-level Reservoir Wave Machine

It also consists of pressure amplifier which sends high pressure water to a high level reservoir on a shore line. The potential energy of water in the reservoir is used to run a water turbine.

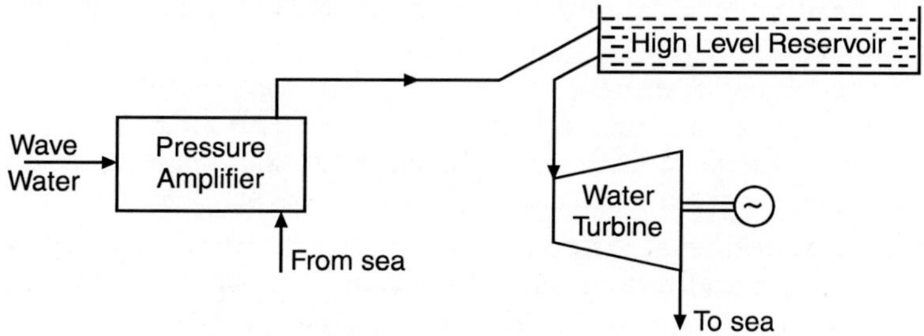

Fig. 10.5. High-level Reservoir Wave Machine

10.5.3. The Dolphin-Type Wave-Power Machine

The major components of the system are a dolphin, a float, a connecting rod and two electrical generators, Fig. 10.6.

The float has two motions. The rolling motion about its own fulcrum with the connecting rod is amplified and converted into continuous rotary motion with the help of gears. The electrical generator is driven. The other vertical motion is also amplified and converted into rotary motion to drive the gears.

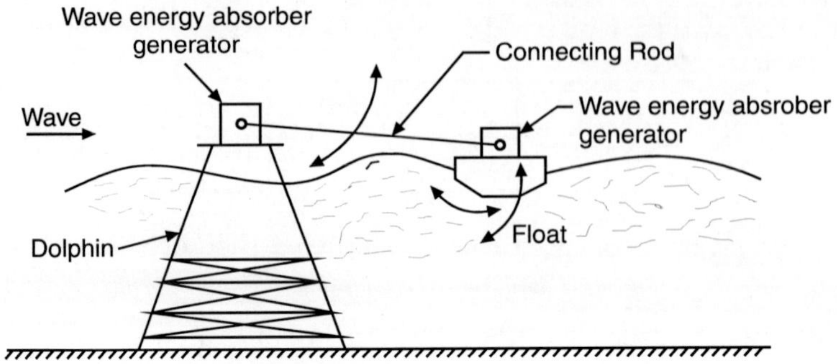

Fig. 10.6. Dolphin type Water Generator

10.5.4. Dam-Atoll Wave Machine

The waves wrap themselves around the atolls (small volcanic islands) from all sides, creating a spiral in the centre. A turbine is driven when water is discharged radially outward.

10.6. TIDAL WAVES

The periodic rising and falling of oceans associated with the position of the moon in the sky give rise to tidal waves. These can be harnessed to produce power. It has a great potential and can form a source of renewable type of energy to produce "clean" electricity without environmental pollution.

The total power of lunar tides is estimated at (2 to 5) $\times 10^6$ MW. But difficulties associated with finding suitable locations, high construction costs and non-uniform availability of power have hindred the progress of harnessing the lunar energy. Development work is going on in different countries to harness the lunar energy.

A lunar day consists of 24 hrs. 50 min and in one lunar day, water in the oceans and seas rise and fall twice. The effect is maximum when the Earth, the Moon and the Sun are in one line. The phenomenon is based on the Newton's theory of gravitation. The force of attraction of the Moon is 2.6 times than that of the Sun. This force of attraction causes acceleration of water and the Earth towards the Moon. The mass of water shifts towards the Moon and its bulge moves over the surface of the Earth as it rotates about its axis. There is tidal friction which tries to slow down the Earth's rotation resulting in tides near the coasts and waves are formed away from the coasts.

The height of tides is different at different points on the Earth and depends upon the latitude and topography of the coast. The height of tide varies from a fraction of a meter at certain points to as high as 18m near the Magellan Strait and 21m on the shores of America. The tides travel many kilometers into the coast of France. There is negligible effect on the inland seas and lakes.

Three possible locations have been identified for the development of tidal power in India : Gulf of Kutch and Gulf of Cambay in Gujarat, Sundarbans in

West Bengal. A feasibility of tidal power scheme for 600 MW installed capacity in the Gulf of Kutch is under study.

The energy of the tides varies over the lunar day and lunar month but can be predicted quite precisely. The tides are characterized by their schedule and range *R*. The tidal cycle lasts 12h 25 min. The tidal range R is defined as the difference between water elevation at high tide and water elevation at low tide. The developments in tidal electric power plants are discussed below.

10.7. POWER PRODUCTION IN SINGLE-POOL TIDAL SYSTEM

The simple single-pool tidal system has one pool or basin behind a dam. The basin is filled from the ocean at high tide and is emplied at low tide. When the ocean is at high tide, the water level in the pool is at low level, there is a filling process. When the ocean is at low tide and pool is at high level, the basin is emptied. The flow of water in both directions (during filling and emptying) is used to drive a number of reversible water turbines installed inside the dam. Electric power would be generated during two short periods in every 6h and 12.5 mm.

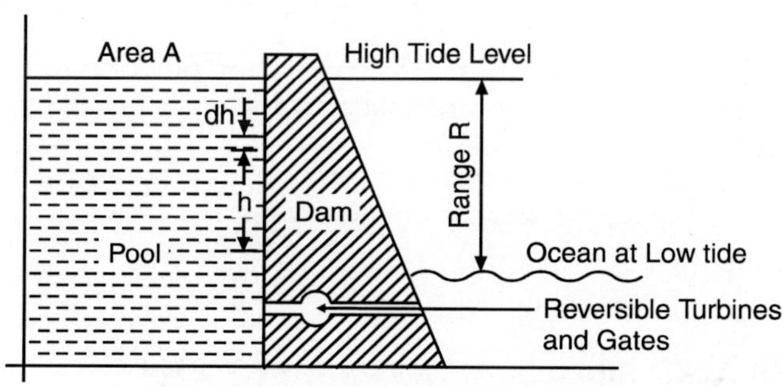

Fig. 10.7. Single Pool Tidal System

For a tidal range *R*, and an intermediate head *h* at a given time during the emptying process, the differential work done by the water is equal to potential energy at the time

$$dW = \frac{g}{g_c} dm\ h$$

But $$dm = -\rho A\ dh$$

$\therefore$ $$dW = -\frac{g}{g_c} \rho A h\ dh$$

where, W = work done by water [J]

g = 9.81 m/s^2 = gravitational acceleration

g_c = 1.0 kg/(N.s^2) = conversion factor

m = mass flowing through turbine, [kg]

h = head, [m]

ρ = water density [kg/m^3]

A = surface area of pool, considered constant [m^2]

The total theoretical work during a full emptying (or filling) period is obtained by integration.

$$W = \int_R^0 dW = -\frac{g}{g_c}\rho A \int_R^0 h\, dh$$

$$= \frac{1}{2}\frac{g}{g_c}\rho AR^2$$

The average theoretical power delivered by water is W divided by 6 h, 12.5 min or 22350 seconds.

$$P_{av} = \frac{1}{44700}\frac{g}{g_c}.\rho AR^2 \quad [\text{W}]$$

If $\rho = 1025 \text{ kg/m}^3$

$$\frac{P_{av}}{A} = 0.225\, R^2 \quad [\text{W/m}^2]$$

The actual power generated by a real tidal system will be only 25 to 30 percent of above, taking into account friction losses and inefficiencies of turbines and generators.

Note : There are two high-peak, short-duration power outputs in every tidal period. Large turbine generators are needed which remain idle much of the time. The power peaks also occur 50 minutes later every day. The peak power demand will not correspond to period of high and low tides. These difficulties have been overcome in the following tidal systems.

1. Modulated single-pool tidal system
2. Two-pool tidal system

10.8. MODULATED SINGLE POOL TIDAL SYSTEM

In simple single pool tidal system, power is generated only at high and low levels of pool only. Therefore, power generation is nonuniform. The modulated single-pool tidal system partially corrects for the above deficiency. The reversible turbines are allowed to operate during periods of pool filling and emptying processes. The turbines cease to operate only when pool level is too low for efficient operation. The no-operation period is very short. Smaller machines are needed.

The ocean and pool levels are shown in Fig. 10.8.

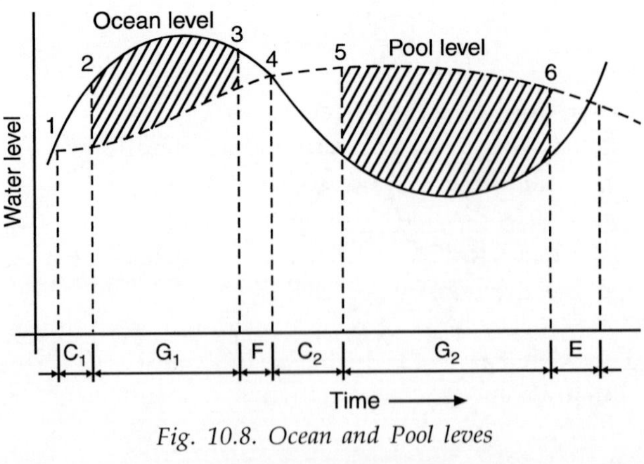

Fig. 10.8. Ocean and Pool leves

1. Period C_1

The head is too low. Both ocean and pool are at low tide level (1). All gates are closed.

2. Period G_1

When the head is sufficient (2), gates to the turbine are opened and water from the oceans is allowed through the reversible turbine to the pool. Power is generated during period G_1. Both ocean and pool levels rise.

3. Period F

The ocean level reaches its peak and begins to decrease. The pool level is still increasing upto point (3). The head is too low for power generation. The gates to the turbine are closed and by-pass gates are opened so that the pool is allowed to fill up during period F.

4. Period C_2

At point 4, all gates are once again closed and pool level remains constant. The ocean level decreases during the period C_2.

5. Period G_2

At point 5, the head is once again sufficient to allow for turbine water flow in the opposite direction. There is power generation during period G_2.

6. Period *E*

At point 6, generation ceases but the pool is allowed to empty during period *E*. The system goes back to point 1, repeating the cycle.

The power generation is not uniform but much more so than in the case of simple system.

Power plants based on the above system have been built in Russia and France.

10.9. TWO POOL TIDAL SYSTEM

The dam construction is costly but the system is not much affected by the fluctuations of the tides.

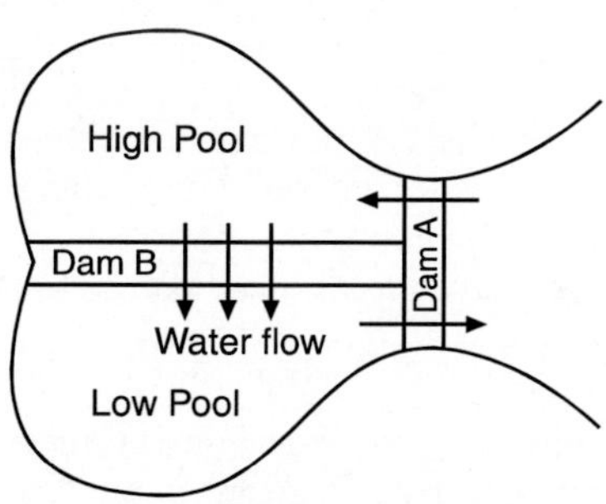

Fig. 10.9. Two Pool Tidal System

An inland basin is enclosed by a dam A which is divided into high pool and low pool by dam B. By proper gating in dam A, the high pool gets periodically filled at high tides from the ocean. Similarly, low pool gets periodically emptied at low tide. Water flows from high pool to low pool through the turbines situated in dam B. The capacities of the two pools are large enough in relation to water flow between them so that the fluctuations in the head are minimised. There is continuous and more uniform power generation.

Example 10.1. *A single basin tidal power plant has a basin area of 30 × 10⁶ m². The tide has a range of 12m. The turbine can work with head of 3 m or more. The turbine generator efficiency is 73%. Estimate the total power generated in one filling empting cycle.*

Solution. The theoretical work,

$$W = \int_R^r dw$$

$$R = \text{Range} = 12 \text{ m}$$

$$r = \text{Minimum head for turbine operation} = 3 \text{ m}$$

$$W = -\int_R^r \rho g h\, A\, dh = \rho g A \int_r^R h\, dh = \frac{1}{2} \rho g A (R^2 - r^2)$$

Power, $$P = \frac{W}{t} = \frac{\rho g A (R^2 - r^2)}{2 \times t}$$

Duration of tide is 6 hours and 12.5 minutes which is equal to 22, 350 seconds. Density of sea water, $\rho = 1025 \text{ kg/m}^3$.

$$\therefore \quad r_{theo} = \frac{9.81 \times 1025 \times 30 \times 10^6 (12^2 - 3^2) \times 3600}{2 \times 22350} \text{ kWh}$$

$$= 3280.5 \times 10^6 \text{ kWh}$$

Actual energy generated,

$$\rho_{act} = \rho_{theo} \times \eta = 3280.9 \times 10^6 \times 0.73$$
$$= 2395 \times 10^6 \text{ kWh}$$

TUTORIAL SHEET NO. 10.1.

1. What is ocean wave energy ? What are the advantages and disadvantages of harnessing of wave energy for power generation ?
2. Explain the mechanism of wave motion.
3. How can wave energy and wave power be calculated from wave motion ?
4. Explain a Float Wave Power Machine with the help of a simple diagram. What are its limitations ?
5. How are the limitations of float-wave power machines overcome in hydraulic accumulator wave machine ?
6. Write short notes on various types of wave power machines developed.
7. What are tidal waves? How can power be produced in single pool tidal system ?
8. What are the limitations of single pool tidal system ? How are these overcome in a modulated single pool tidal system and two pool tidal system ?
9. With the help of water level-time diagram. Explain the functioning of modulated single pool tidal system.

□□□

PART 111

DIRECT ENERGY CONVERSION SYSTEMS

Chapter 11

FUEL CELLS

Induction; Principle of Fuel Cell Operation; Performance Analysis; Advantages and Limitations of Fuel Cells; Performance Characteristic of H_2–O_2 Fuel Cell; Types of Fuel Cells, Applications.

11.1. INTRODUCTION

If an electric current is passed through a dilute solution of an acid or an alkali by means of two platinum electrodes, hydrogen is produced at the anode and oxygen is evolved at the cathode. If this process is reversed by removing the power supply and connecting the two electrodes through a suitable resistance, the presence of hydrogen at one electrode and oxygen at the other will produce a small current in the external circuit, water being produced as a by-product. This reverse process of electrolysis is the essence of the fuel cell technology as chemical energy stored in hydrogen and oxygen has been combined to produce electricity.

Unlike other heat engines, fuel cells donot have moving parts. Therefore, the fuel cell will be quieter and should require less maintenance and attention in operation. The manufacturing costs should also be less, as close tolerances as in pistons are not required. Because the fuel cells convert chemical energy directly to electrical energy, the conversion efficiency can be much higher; they many be cheaer to operate, and there will be less heat produced, resulting in small radiator and exhaust systems.

Although much of the interest in the fuel cells is due to their efficient use of fuel, there are considerable pollution control advantages to be gained as well. Because the fuel reacts electrochemically rather than by combustion with air, no nitrogen oxides are formed. For the same reason, emission of unburned and partly burned gaseous and particulate products are essentially nil. There is relatively little thermal pollution because less heat is lost as waste. Energy conversion via fuel cells, therefore, represents one of the best ways to achieve this goal, because it is possible, simultaneously, to obtain more work and less pollution from a rupee's worth of fuel with a fuel cell than any other energy conversion device.

With this impressive list of advantages, it is perhaps surprising that fuel cells has not been more readily accepted. Cheap, practical and long-lasting devices are still not available in the market. Fuel cells have not been widely used or developed as terrestrial energy sources because of their high costs (partly due to expensive catalysts) and short operating life time (due primarily to degradation of electrode materials). The early development of fuel cells for electrical power generation was overshadowed by the steam turbine generation and for mobile applications, it could not compete with internal combustion engines. The biggest boost to fuel cell development has been space power application where high power density and low weight are important. Fuel cells have been an integral part of the Gemini and Apollo manned space flight systems. It is only recently that some of the problems of high cost and short life time may be overcome and fuel cell batteries may be commercially viable within short period.

Hydrogen and oxygen were selected as reactants for the fuel cells used for spacecraft power supplies because of relatively high reactivity of hydrogen. Presently, hydrogen is a relatively expensive fuel. The reactants for commercial fuel cells should be as cheap and readily available as possible; for example, air as oxidant and natural gas or petroleum derivatives as fuel. As coal gasification technology matures, very satisfactory feed stream for fuel cell power plant will be available.

The output of fuel cell is low voltage dc power, cells may be connected in various series and parallel arrangements to get whatever voltage and power are desired, and highly efficient inverters are available for conversion to ac.

Fuel cells have been proven as practical power sources in certain specific applications, such as space missions and remote site operations. The development of fuel cells for widespread commercial applications is now underway. Although advantages of fuel cells over other heat engines could swing the balance in favour of fuel cells, the relatively high development costs involved means that substantial markets must be sought by finding new applications.

City life depends upon continuous and reliable power supply for comfortable living, entertainment, efficient commutation, communications, commercial and industrial activities in a pollution-and noise-free environment. The application of fuel cells for individual houses, apartments, commercial buildings, street lighting, industrial estates, communication systems, electric vehicles for rail and road transport can be studied so that total power requirements of a city can be met on fuel cell batteries only.

11.2. PRINCIPLE OF FUEL CELL OPERATION

The following three stages of energy conversion are required in a steam power plant.

1. Conversion of chemical energy of fuel into thermal energy by combustion and heat transfer in a boiler.
2. Conversion of thermal energy into mechanical energy of rotation in steam turbine.
3. Conversion of mechanical energy into electrical energy in an electric generator.

Fuel cells convert the chemical energy of fuel directly into electrical energy by an electrochemical process. Low voltage dc power is produced by using hydrogen or natural gas as fuel. The schematic diagram of a $H_2 - O_2$ fuel cell is shown in Fig. 11.1.

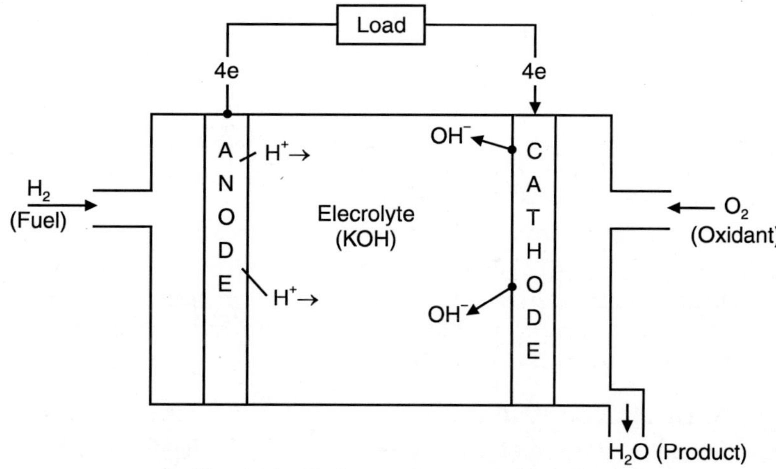

Fig. 11.1. Hydrogen-Oxygen Fuel Cell.

The working of a fuel cell may be explained with reference to the hydrogen-oxygen fuel cell using aqueous electrolyte. It consists of two porous metal electrodes with the electrolyte between their inner faces kept stirred by recirculation. The porous electrodes have a matrix of coarse pores, coated on electrolyte side with a layer of material having finner pores. The reactant gases are supplied to the electrodes as shown in Fig. 11.1. The pressure differential between the gases and the electrolyte is sufficient to displace the liquid from the coarse pores, but not from the fine pores. In order to obtain a large area of electrode-electrolyte interface (at which electron transfer takes place), the gases must either diffuse through a layer of electrolyte on the surface of the electrodes, or migrate along the surface beneath the electrolyte. Similarly the ions must diffuse through the electrolyte trapped in the fine pore layer, which is not stirred by recirculation.

Platinum and other precious metals are used as electrodes in the fuel cells for military and space applications. Platinum being expensive, porous nickel electrodes and porous carbon electrodes are generally used in fuel cells for commercial applications. A catalyst is included in the electrode to expedite the reactions. The best electro-chemical catalysts are finely divided platinum deposited on electrode material. Other cheaper catalysts like nickel for hydrogen and silver for oxygen are used where possible.

The chemical reactions taking place at the two electrodes are as follows :

(*i*) Hydrogen electrode (anode)

$$2\ H_2 = 4H^+ + 4e$$

$$4H^+ + 4OH^- = 4\ H_2O + 4\ e$$

(*ii*) Oxygen electrode (cathode)

$$O_2 = 2\ O^-$$

$$2\ O^- + 2\ H_2O + 4\ e = 4\ OH^-$$

(*iii*) Overall cell reaction

$$2\ H_2 + O_2 = 2\ H_2O$$

The H_2 molecules break up into H^+ ions at the anode. These H^+ ions combine with OH^- ions to form water and release electrons at the anode. The electrons travel to the cathode through external circuit. At the cathode the two oxygen atoms combine with the four electrons arriving by the external circuit and two molecules of water (out of the four molecules produced at the anode) to form 4 OH^- ions. These OH^- ions migrate towards the anode and are consumed there : The electrolyte remains invariant. This invariance is a critical feature of fuel cell electrolyte and requires that the composition of the electrolyte must not change appreciably as the cell operates. In other words, ions removed by the reaction at one electrode must be replaced one for one at the other electrode.

The most difficult problem in the design of a fuel cell is to obtain sufficient fuel electrode-electrolyte reaction sites in a given volume. A large effective surface is achieved by a porous electrode structure and surface tension forces are utilized to get reasonable contact stability.

11.3. PERFORMANCE ANALYSIS

The theoretical maximum electromotive force (EMF) developed in a fuel cell can be calculated with the help of Gibbs free energy as follows :

$$E = \frac{-\Delta G}{nF} \; [V]$$

where ΔG = Change in Gibbs free energy in the reaction [J/mol]

n = No. of electrons per mole of fuel which take part in the reaction (for hydrogen, $n = 2$)

F = Faraday's constant = 96487 coulombs/mole.

The change in the Gibbs free energy in a chemical reaction is given by :

$$\Delta G = \Delta H - T\Delta S \quad [J/mol]$$

where ΔH = change of enthalpy of total reaction [J/mol]

$T\,\Delta S$ = Amount of heat absorbed during a reversible process at constant temperature [J/mol]

The voltage of fuel cell depends upon temperature and pressure. For a H_2–O_2 fuel cell at a pressure of 1 atmosphere, the value of E at 25°C is 1.23V and only 1.15V at 200°C. If pressure increases, the fuel cell voltage also increases.

The maximum efficiency is achieved in a reversible fuel cell.

$$\eta_{th} = \frac{\Delta G}{\Delta H} = 1 - T\frac{\Delta S}{\Delta H}$$

where ΔG = Change in Gibbs free energy (equal to – 237.14 kJ/mol)

ΔH = Change in enthalpy (equal to – 285.83 kJ/mol) by formation of 1 mole of liquid water from H_2 and O_2 at 1 atmosphere and 25°C.

The maximum work per mole of H_2 (reactant) or per mole of H_2O (product) is given as :

$$W_{max} = \Delta G_R - \Delta G_{H_2O} \; [kJ/mol]$$

$$\Delta G_R = O \; kJ/mol \text{ for reactants.}$$

The electromotive force (EMF) is given by :

$$EMF = E_{rev} = \frac{W_{max}}{n.F} \quad [V].$$

The overall efficiency of fuel cell is its thermal efficiency and performance factor.

$$\eta_{overall} = \eta_{th} \times \text{(Loss factor)}$$

The power output of a reversible fuel cell :

$$P_{rev\cdot} = \frac{\Delta G.\dot{m}}{M_{H_2}} \quad [W]$$

where $\dot{m}$ = mass flow rate of hydrogen [kg/s]

M_{H_2} = 2.016 kg/mole = Molar mass of hydrogen.

The actual electrical power output is

$$P = P_{rev.} \times \eta_{overall} \text{ [W]}.$$

The rate of heat released which can be utilized

$$Q = P_{rv} - P \quad [W]$$

11.4. ADVANTAGES AND LIMITATIONS OF FUEL CELLS

Fuel cells have the following advantages

1. These cells have high efficiency (> 50%) in full-load and part-load operation. These are potentially ideal sources of power generation. Efficiencies of the order of 40% have already been achieved. An overall efficiency of more than 80% can be achieved in cogeneration plants if heat generated in the fuel cell can also be utilized in addition to electricity.
2. No pollution emissions. When fuel cells are used for power generation and transport sectors, NO_x will be reduced by 50% to 90% and CO_2 by 50% in comparison to present conventional technologies.
3. Water is a by-product of reaction. This is a useful product in space and remote applications.
4. There is no moving part and operation of a fuel cell is noiseless. This can be a great advantage for military and other strategic applications.

Limitations

The main limitations hindering the growth of fuel cells are :

1. High capital cost of fuel cells is the main limitation against commercialization. This limitation can be overcome by new material developments as the present material being used are very costly. Search for new applications and markets can help to reduce the costs by taking advantage of scale of production.
2. Low life span of fuel cells due to heavy corrosion of electrodes. This can be overcome again by new material technologies.

Example 11.1. *Find the reversible voltage for hydrogen-oxygen fuel cell, having the following reaction :*

$$H_2 + \frac{1}{2}O_2 = H_2O\ (l)$$

Given : $(\Delta G°)_{25} = -\ 237.3 \times 10^3$ *J/g-mol of* H_2.

Solution:

$$E = -\frac{\Delta G}{nF}$$

where n = 2 electrons per mole of H_2.

$$E = \frac{273.3 \times 10^3}{2 \times 96,500} = 1.23 \text{ V}.$$

Example 11.2. *Calculate the following for a hydrogen-oxygen fuel cell.*

(a) The voltage output of the fuel cell

(b) The cell efficiency.

(c) The electrical work output per mole of H_2 consumed and per mole of H_2O produced

(d) The heat transfer to the surrounding.

The following reactions take place.

At the anode

$$H_2 \longrightarrow 2H^+ + 2\bar{e}$$

At the cathode

$$2\bar{e} + \frac{1}{2}O_2 + H_2O = 2OH^-$$

The cell operates at 25°C. Take the following data :

$$\Delta H^\circ_{298K} = -\ 286 \times 10^3 \text{ kJ/kg mol.}$$

$$\Delta G^\circ_{298K} = -\ 237.3 \times 10^3 \text{ kJ/kg mol.}$$

Solution:

$$n = 2$$

(*a*)

$$E = -\frac{\Delta G}{nF} = \frac{237.3\times10^3}{2\times96500} = 1.23 \text{ V.}$$

(*b*)

$$\eta_{max} = \frac{\Delta W_{max}}{-(\Delta H)_{298\,K}} = \frac{-(\Delta G^\circ)_{298\,K}}{-(\Delta H^\circ)_{298\,K}}$$

$$= \frac{237.3\times10^3}{286.0\times10^3} = 83\%$$

(*c*) The electrical work output per mole

$$\Delta W_r = nFE = 2 \times 96500 \times 1.23$$

$$= 237.4 \times 10^3 \text{ kJ/kg mol. of } H_2$$

$$H_2 + \frac{1}{2}O_2 = H_2O.$$

One male of H_2O is produced for each mole of H_2.

$\therefore$ ΔW_r per mole of water = 237.4×10^3 kJ/kg mol of H_2O

(*d*) The heat transfer

$$Q = T\Delta S = \Delta H^\circ_{298K} - \Delta G^\circ_{298K}$$

$$= (-\ 286 + 237.3) \times 10^3$$

$$= -\ 48.7 \times 10^3 \text{ kJ/kg mole.}$$

The negative sign shows that the heat is rejected from cell to the surrounding.

Example 11.3. *A hydrogen-oxygen fuel cell operates at 25°C. Calculate the voltage output of the cell, the efficiency and the electrical work output per mole of H_2 consumed and per mole of H_2O produced if $\Delta H^\circ_{298K} = -\ 285838$ kJ/kg mole and $\Delta G^\circ_{298K} = -237191$ kJ/kg mole.*

Also compute the heat rejected by the cell.

Solution : $H_2 + 2OH^- = 2H_2O + 2\bar{e}$

$\therefore \qquad n = 2$

The cell voltage

$$E = -\frac{\Delta G^{\circ}_{298\,K}}{nF}$$

$$= \frac{237.191 \times 10^3}{2 \times 96.500 \times 10^3} = 1.229 \text{ V.}$$

$$\eta_{max} = \frac{\Delta G^{\circ}_{298\,K}}{\Delta H^{\circ}_{298\,K}}$$

$$= \frac{-237.191 \times 10^3}{-285.838 \times 10^3} = 83\%$$

The electrical work,

$$W_e = -\Delta G^{\circ}$$

$$= -\ 237.191 \text{ kJ/kg mole. of } H_2$$

or $\qquad -\ 237.191\ kJ/kg\ mole\ of\ H_2O$

There is one mole of H_2O produced per mole of H_2 consumed.

Heat rejected by the cell

$$\Delta Q = T\Delta S$$

$$= \Delta H^{\circ}_{298K} - \Delta G^{\circ}_{298K}$$

$$= -\ 285838 + 237191$$

$$= -\ 486.5 \text{ kJ/kg mole.}$$

Example 11.4. *Calculate the open-circuit voltage, the maximum work and thermal efficiency of a $H_2 - O_2$ fuel cell operating at 25°C and 1.013 bar. The product H_2O is in liquid state.*

Determine the power output in reversible and actual processes and overall cell efficiency when performance factor is 0.7. The hourly hydrogen consumption is 1.3 kg/hr.

Solution: (1) The Gibbs free energy change and production enthalpy of H_2O (liquid) at 1.013 bar and 25°C are :

$$\Delta G_{H_2O} = -\ 237.14 \text{ kJ/mol.}$$

$$\Delta H_{H_2O} = -\ 285.83 \text{ kJ/mol.}$$

The free energy change and production enthalpy for the reactants are zero. For reactants $\Delta H = \Delta G = 0$ kJ/mol.

(2) $$W_{max} = -\Delta G$$

$$= 237.14 \text{ kJ/mol.}$$

(3) $$E = \frac{W_{max}}{nF}$$

$$E = \frac{237.14 \times 10^3}{2 \times 96500} = 1.229 \text{ V}$$

(4) $$\eta_{th} = \frac{\Delta G}{\Delta H}$$

$$\eta_{th} = \frac{237.14 \times 10^3}{285.83 \times 10^3} = 83\%$$

(5) $$\eta_{overall} = \eta_{th} \times \eta_{per.} = 0.83 \times 0.7 = 58\%$$

(6) Maximum power,

$$P_{max} = \frac{\Delta G . m}{M_{H_2}}$$

$$= \frac{\frac{237.14 \times 1.3}{3600}}{2.016} = 42.48 \text{ W}$$

$$M_{H_2} = 2.016 \text{ kg/mol.}$$

$$m = \frac{1.3}{3600} kg/\text{sec.}$$

(7) Actual power output

$$P = P_{max} \times \eta_{overall} = 42.48 \times 0.58 = 24.64 \text{ W.}$$

(8) $$Q = P_{max} - P = 42.48 - 24.64 = 17.84 \text{ W}$$

11.5. PERFORMANCE CHARACTERISTICS OF $H_2 - O_2$ FUEL CELL

The Carnot efficiency of any heat engine increases with the source temperature for a given sink temperature.

$$\eta_c = \frac{T_1 - T_0}{T_1}$$

The heat source temperature T_1 if varied from 400K to 1400K, the performance curve in plotted is Fig. 11.2. The heat sink temperature T_0 = 300K. If the fuel cell temperature is equal to source temperature, the efficiency decreases linearly with temperature. The theoretical efficiency of $H_2 - O_2$ fuel cell as function of temperature is plotted in Fig. 11.2.

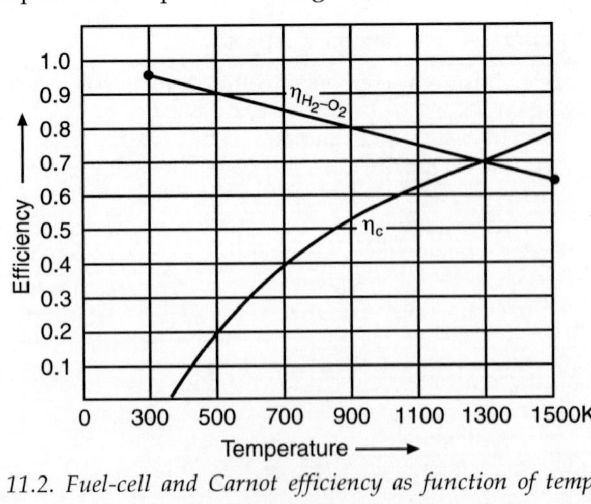

Fig. 11.2. Fuel-cell and Carnot efficiency as function of temperature.

The relationship between current and voltage and power density for $H_2 - O_2$ fuel cell are plotted in Fig. 11.3.

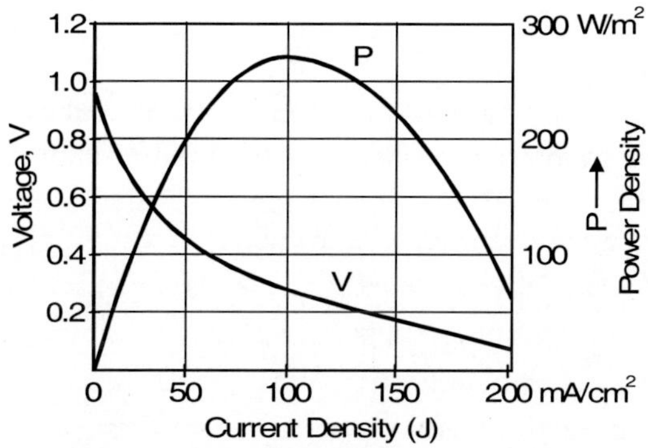

Fig. 11.3. Current-voltage-power characteristics of H_2-O_2 Fuel Cell

11.5.1. Polarization in Fuel Cells

There is a significant drop in the voltage in a fuel cell with the increase of current density. This energy loss is called polarization or over potentials.

There are three types of polarization as shown in voltage-current curve.

I. Activation Polarization

This is related to the activation energy barrier for the electron transfer process at the electrode. A drop of potential at the electrode gives a greater driving force to overcome this energy barrier and thus allows more current to flow. This polarization may be reduced by :

(*i*) using better electrode catalysts,

(*ii*) increasing surface area of electrodes,

(*iii*) raising the operating temperature.

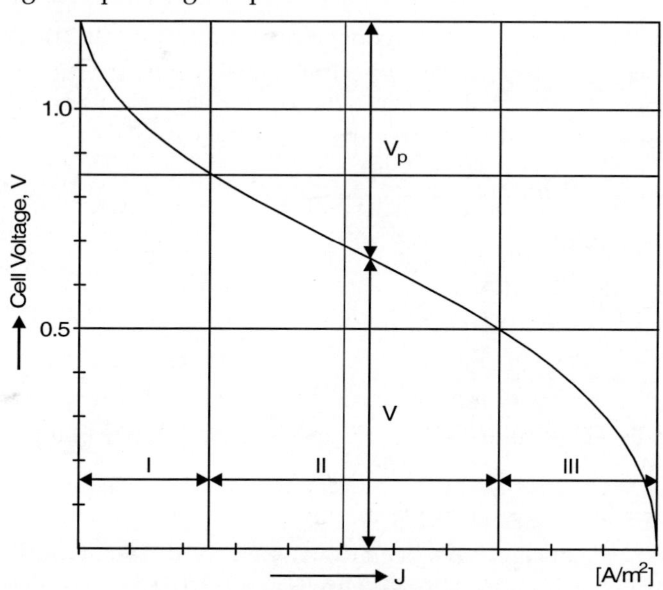

Fig. 11.3(a). Polarization in Fuel Cells

II. Resistance (ohmic) Polarization

The internal resistance of a fuel cell consists of

1. Electrode resistance,
2. Bulk electrolyte resistance,
3. Interface contact resistance between electrode and electrolyte.

The reduction in internal resistance is the main design criterion for low resistance polarization. This can be reduced by

(*i*) using more concentrated electrolyte,

(*ii*) closer spacing of electrodes,

(*iii*) increase of operating temperature.

III. Concentration Polarization

This can be divided into two parts.

1. *Electrolyte side polarization:* This is due to slow diffusion of ions into the electrolyte.

2. *Gas-side polarization:* This is due to slow diffusion of reactants into porous electrode.

This loss can be reduced by increasing temperature.

The no-load or open circuit voltage (e.m.f),

$$E = V + V_p$$

where V = Operating voltage at a given current density

V_p = Polarization loss.

All the losses in a fuel cell may be included under voltage efficiency,

$$\eta_v = \frac{V}{E}$$

For an ideal fuel cell, $\eta_v = 100\%$

Most of the energy converters convert heat energy into electricity. The efficiency is defined as the ratio of work output and heat input.

The work output of energy converter operating at the thermodynamic reversible potential of the cell is the free energy of the cell.

$$\Delta W_{max} = -\Delta G = -nFE$$

The heat input is the enthalpy change of the reaction = ΔH.

The ideal efficiency of a fuel cell,

$$\eta_{ideal} = \frac{\Delta G}{\Delta H}$$

$$= -\frac{nFE}{\Delta H}$$

Generally ΔG is quite close to ΔH and hence efficiency of a fuel cell would be close to unity under ideal conditions.

11.6. TYPES OF FUEL CELLS

Many types of fuel cells have been developed for different applications & various technical specifications are given in Table 11.1. The power conversion rates depend upon the type of fuel and fuel cell. Some typical values are given in Table 11.2.

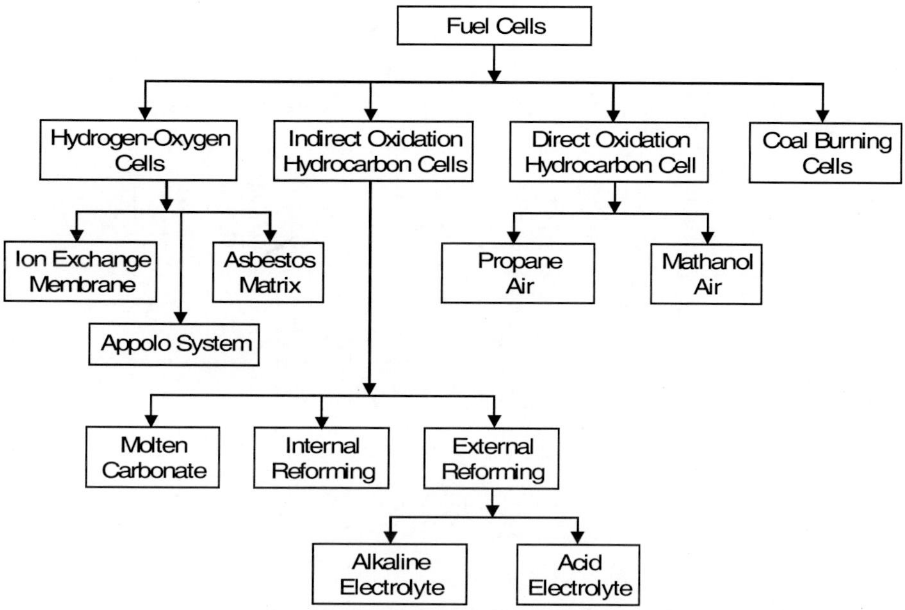

Fig. 11.3(b). Classification of fuel cells.

The pictorial representation of various types of fuel cells are shown in Fig. 11.3(*a*).

Table 11.2. Heat Rates

S.No.	Fuel	Fuel Cell	Fuel to ac power heat rate [kJ/kWh]
1.	Hydrogen	Alkaline	6500
2.	Impure hydrogen	Acid	7750
3.	Syngas	Molten Carbonate	8200
4.	Methanol	Acid	6500
5.	SNG	Molten Carbonate/Acid	7500
6.	Naptha	Molten Carbonate/Acid	7500

11.6.1. Proton Exchange Membrane Fuel Cells

The electrolyte is a thin polymer film. The electrodes are made of porous carbon with platinum catalyst.

The reaction at the anode :

$$H_2 \longrightarrow 2H^+ + 2e^-$$

The polymer membrane allows protons (H^+) to pass through. The free electrons (e^-) pass through the external circuit and reach cathode.

Table 11.1. Technical Specifications and Applications of Fuel Cells

Sl No.	Type of Fuel cell	Fuel	Electrolyte	Capacity [kW_e]	Efficiency [%]	Working Temperature [°C]	Applications	Remarks
1.	Proton Exchange Membrane (PEM) or solid polymer	H_2, CH_4 $CH_3\,OH$	Polymer Membrane	upto 250	60 (H_2) 40 (CH_4)	60 – 80°	Cars, buses, home, CG plants	H_2 is produced in a reformer. CO sensitive
2	Solid oxide (SoFC)	H_2, CH_4, coal gas	$Zr(y)O_2$	10 – 25	50 – 65	800 – 1000	Domestic, small plants, CG plants	No reforming needed.
3.	Alkaline (AFC)	Pure H_2	KOH	20	60	60 – 90	space erafts, boats	CO_2 sensitive
4.	Molten carbonate (M C F C)	H_2, CH_4 coal gas	Molten carbonate of Li, Na and K.	2200 (max)	48 – 60	650	Small plants, CG plants.	CO_2 must be circulated
5.	Direct methanal (D M F C)	CH_3 OH	Polymer membrane	N.A.	40	60 – 130	Small domestic plant, CG plants	No H_2 production needed
6.	Phoothoric Acid (P A F C)	$H_2(CH_4)$	$H_3\,PO_4$	1100 (max)	40	130 – 220	Small power plants, CG plants.	CO sensitive
7.	Regenerative (R F C)	H_2	N. A.	N. A.	N. A.	upto 120	Domastic, CG plants.	Low cost of material.

Note: CG = Cogeneration

The reaction at the cathode :

$$2e^- + 2H^+ + \frac{1}{2}O_2 \longrightarrow H_2O$$

Water is produced.

The total reaction :

$$H_2 + \frac{1}{2}O_2 \longrightarrow H_2O$$

The PEM fuel cells operate at 95°C and have high power density. The output can be varied very quickly. These are very suitable for transportation applications. The main applications are :

1. Light duty vehicles
2. Cogeneration (CHP) plants
3. Power supply for buildings
4. Rechargeable batteries for video cameras.

Fuel Cell Vehicles

1. Bus developed by Ballard Power System, Canada.
2. XCELLSIS Fuel Engine Inc. (USA), 1998
3. Fuel cell vehicle at Munich Airport, 2000 by Daimler/Chrystler's NECAR-4 fuel cell vehicle.
4. Mercedes-Benz A class fuel-cell powered cars, 2004 at $18000.

It is expected that by 2020, 25% of cars will be run by fuel cells.

11.6.2. Solid Oxide Fuel Cells

The electrolyte is ceramic membrane with Yttruim Stablized Zirconia (YSZ).

The electrodes are made of porous metal-ceramic. These operate at high temperature of 800-1000°C. No catalyst is needed.

The anode reaction :

$$H_2 + O^{2-} \longrightarrow H_2O + 2e^-$$

or

$$CO + O^{2-} \longrightarrow CO_2 + 2e^-$$

The cathode reaction:

$$\frac{1}{2}O_2 + 2e^- \longrightarrow O^{2-}$$

The total reaction :

$$H_2 + \frac{1}{2}O_2 \longrightarrow H_2O + 2e^-$$

or

$$CO + \frac{1}{2}O_2 \longrightarrow CO_2$$

The capacity of upto 250 kW_e are used in cogeneration plants with natural gas as fuel as well as industrial power plants. Efficiencies upto 60% can be attained. Some SOFC systems are being developed in Europe and Japan.

11.6.3. Alkaline Fuel Cells

These fuel cells have low weight, low operating temperature (90°C), high efficiency (60%). These are very attractive for space and military applications. The main drawbacks are extremely high cost of the order of 100,000 DM/kW.

11.6.4. Molten Carbonate Fuel Cells

The electrolyte is liquid solution of lithium and sodium/potassium carbonates soaked in matrix. The operating temperature is 650°C and attainable efficiency is 60%. Fuel utilization factor is 85% in cogeneration plants. The power plants of 2 MW capacity have been tested with hydrogen, natural gas and other gaseous fuels.

Anode reaction

$$H_2 + CO_3^{2-} = H_2O + CO_2 + 2e^-$$

Cathode reaction

$$\frac{1}{2}O_2 + CO_2 + 2e^- = CO_3^{2-}$$

Total reaction

$$H_2 + \frac{1}{2}O_2 + CO_2 = H_2O + CO_2$$

11.6.5. Direct Methanol Fuel Cells

A polymer membrane is used as electrolyte

Anode reaction :

$$CH_3OH + H_2O \longrightarrow CO_2 + 6H^+ + 6e^-$$

Cathode reaction :

$$6H^+ + 6e^- + 1.5\ O_2 \longrightarrow 3H_2O$$

Total cell reaction :

$$CH_3OH + 1.5O_2 \longrightarrow CO_2 + 2H_2O$$

The cell efficiency can be 40% at operating temperature of 50 to 90°C. The anode catalyst can draw hydrogen directly from liquid methanol without reforming.

11.6.6. Phosphoric Acid Fuel Cells

These are commercially most developed fuel cells. These operate at 160°C to 220°C. The efficiency is above 40% for electricity production but when used in cogeneration plants, the efficiency can increase to 85%.

These cells are suitable for autonomous power plants as well as for cogeneration plants in the capacity range from a few kilowatt to 10 MW. These can be employed for residential, commercial, institutional buildings and also for small factories. These have been also successfully used for hospitals, nursing homes, hotels, office buildings, schools and central power plants in USA.

The capital cost of phosphoric acid fuel cells in the power range of 200 kW is 4500 DM/kW when operated on natural gas. For widespread use of this fuel cell the cost has to be reduced to 1300 DM/kW and life span has to be increased from the present 15000 hours to 40,000 hours.

11.6.7. Regenerative Fuel Cells

Water can be split into hydrogen and oxygen by the use of solar energy. The gases H_2 and O_2 can be used as fuel and oxidant for producing electricity. Water produced is recirculated for electrolysis. These fuel cells can be used for energy storage also.

Example 11.5. *A galvanic cell has the following reaction :*

$$Ag(s) + HgCl(s) = AgCl(s) + Hg(l)$$

Take $E^\circ_{25^\circ C} = 0.0455 V$

$$\frac{dE^\circ}{dT} = 0.000338 \ V/K$$

Calculate, ΔG°, ΔS° and ΔH°.

Solution :

$$\Delta G^\circ = -nFE$$
$$= -(1 \times 96{,}500 \times 0.0455)$$
$$= -4400 \text{ joules}$$

$$\Delta S^\circ = nF\left(\frac{dE}{dT}\right)_p$$
$$= 1 \times 96500 \times 0.000338$$
$$= 32.6 \text{ J/K}$$

$$\Delta H^\circ = nF\left[T\frac{dE}{dT} - E\right]$$
$$= 1 \times 96500\ [298 \times 0.000338 - 0.0455]$$
$$= 5300 \text{ J}$$

Example 11.6. *Prove that maximum efficiency of a methane fuel cell is 92% and ideal cell voltage is 1.15 V. Calculate the quantity of methane to produce 100 kWh. How much heat will be generated ?*

Give that : $\Delta G^\circ_{25^\circ C} = -195{,}500$ *Cal/gmol*

$\Delta H^\circ_{25^\circ C} = -212{,}800$ *Cal/gmol.*

Solution : The fuel cell reaction :

$$CH_4 + 2H_2O \longrightarrow CO_2 + 8H^+ + 8e^-$$

$\therefore n = 8$

$$E^\circ_{25^\circ C} = \frac{-(\Delta g)^\circ_{25^\circ C}}{nF}$$

$$= \frac{-195{,}500 \times 4.186}{8 \times 96{,}500} = 1.060 \text{ V}$$

$$\eta_{max} = \frac{\Delta W_{max}}{-(\Delta H^\circ)} = \frac{(\Delta G^\circ)}{(\Delta H^\circ)}$$

$$= \frac{195{,}500}{212{,}800} = 0.92$$

$$\eta_V = \frac{\text{operating voltage}}{\text{open circuit voltage}}$$

$$= \frac{V}{V_o}$$

$$V_o = \frac{1.060}{0.92} = 1.15\text{V}$$

$$P = 100 \text{ kW}$$

Quantity of methane,

$$\dot{m} = \frac{100 \times 10^3 \times 3600 \times 36}{195,500 \times 4.186} = 6.547 \text{ kg/hr}$$

$$M_{CH_4} = 36$$

$$Q = T\Delta S$$

$$= \Delta H + \Delta W = \Delta H^\circ_{25^\circ C} - \Delta G^\circ_{25^\circ C}$$

$$= -212800 + 195500 = -116,700 \text{ kcal/kg mole}$$

11.7. APPLICATIONS OF FUEL CELLS

The applications of fuel cells may be discussed under the following groups :

1. Central power generation
2. Cogeneration plants
3. Automotive vehicles
4. Domestic power units
5. Special applications

11.7.1. Central Power Generation

The fuel cell converts chemical energy directly into electrical energy. Therefore, power generation is the most natural application of fuel cells. For this application, fuel cells must compete with steam turbine, which are remarkably efficient devices with efficiency approaching 40 percent at rated load. However, demand for electrical energy is far from constant. Over the course of a year, the actual output of a power plant may vary by a factor of four, and the daily variation in load can be almost a factor of three. To adjust to this changing demand, either the large base load plants must sometimes operate at part load, or smaller cycling or peaker units must be used during periods of high demand. Either way efficiency suffers and pollution increases. On the other hand, the fuel cell system not only has a greater efficiency at full load, but this efficiency is retained and even increases as load diminishes, so that inefficient peaking/ standby gensets may not be needed.

A fuel cell system, unlike a steam turbine need not be big to be efficient. This characteristic, taken together with two others, low emissions and capability of operation on a variety of fuels, allows fuel cell system to be operated almost anywhere. A small power plant for a community can be operated on the optimum fuel available locally with nearly the same efficiency achieved by a large central power plant. An electric supply company of a large metropolis can disperse a number of generators throughout its area and match capacity to local demand, substantially reducing the expense and other problems associated with transmission and distribution of electricity. Costs and other problems involved with local distribution of electrical energy are likely to be greater with the adoption of underground lines in urban areas.

The flow diagram of a hydrocarbon fuel cell for power generation is shown in Fig. 11.4.

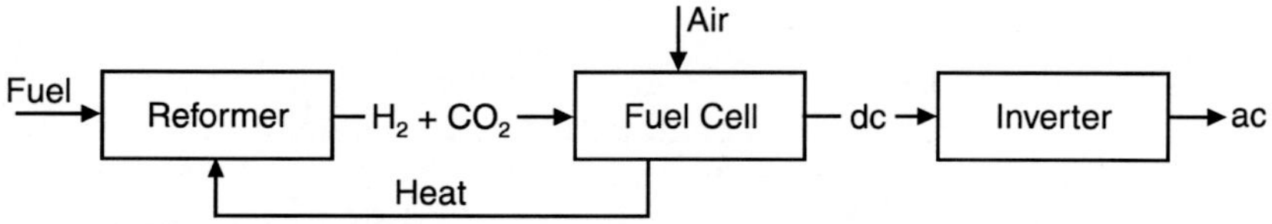

Fig. 11.4. Hydrocarbon Fuel Cell Power Unit.

Efficiency

Fuel cells have high conversion efficiency especially at low power levels as they are not subject to Carnot limitations on efficiency. It is a one-step process without moving parts and both mechanical and heat losses are absent. This indicates that these systems have the potential for very high reliability and silent and unattended operation.

The hydrocabon fuel cells operate at much the same efficiency in the 100 kW range as large multi-megawatt units see Fig. 11.5.

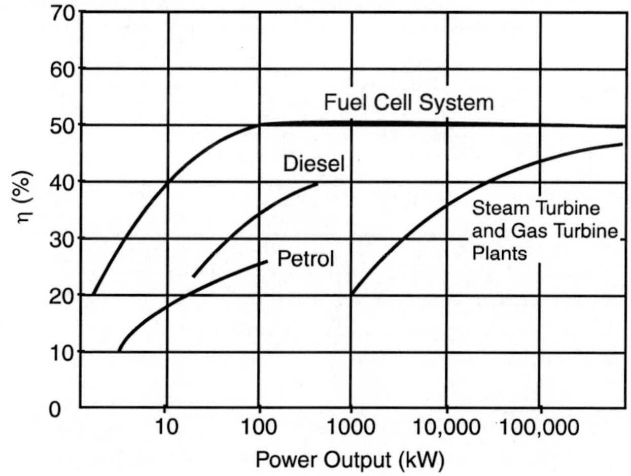

Fig. 11.5. Thermal Efficiency of Power Units

Efficiency may be further improved to about 55 percent if pure hydrogen is used in place of processed fossil fuels, and again to 60 percent when oxygen is additionally substituted for air. The efficiency of the fuel cell increases as the power level is decreased to about 40 percent load.

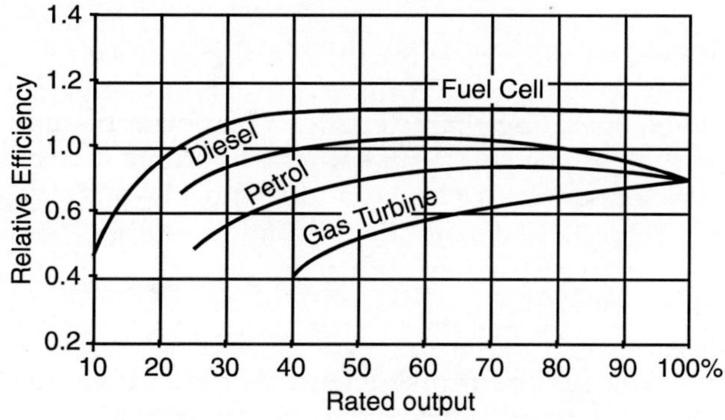

Fig. 11.6. Part Load Efficiency

Pollution

In conventional power plants, a considerable amount of NOx, SO_2, H-C and particulates are emitted due to combustion of fuel. Fuel cells emit mostly non-toxic and harmless air, CO_2, water vapour and small heat as exhaust. Thermal pollution of waterways is not a problem because fuel cells are air cooled and are not dependent on a source of water supply. There are few restrictions on site locations. Air pollutants are reduced by a factor of more than 10 over conventional systems. The environmental impact of fuel cells is compared with code requirements of conventional systems in Table 11.3.

Table 11.3. Environmental Impact of Fuel Cells (Kilogram of pollutants per million kJ heat input) (USA Federal Standards)

Emissions	Gas-fitted central stations	Oil fired central stations	Coal-fired central stations	Fuel Cell
Particulates	0.05	0.05	0.05	0.0000015
NOx	0.10	0.15	0.35	0.006—0.009
SO_2	Not required	0.40	0.60	0.000012
Smoke	20% opacity	20% opacity	20% opacity	Negligible

Scale

Fuel cell efficiency is insensitive to size. Scale-up in power output is accompanied by interconnecting fuel stacks (modules) due to the limit to the size of thin electrodes. There is a saving in cost due to mass production of identical components for scale up.

The fuel cell is a very simple device with no moving parts except a few fuel and coolant pumps. Controls are simple and automatic.

Modularity

Fuel cell units can be added to a power plant system, incrementally over a period of time and built rapidly. There is no need to tie up considerable capital in unused initial capacity which is a very serious drawback of large power plants with long construction periods.

Versatility

Fuel cells have inherent adaptability towards a broad range of applications. The direct nature of electrochemical process minimizes the effects of both scale and load level on operating characteristics. The power plant modular construction and ease with which units can be linked permits matching of capacity to demand. Increased capacity can be added quickly and simply. Because the power plant is air cooled, it is not dependent on cooling water availability and thus has few location restrictions. It has the capacity to operate on a variety of fuels as given in Table 11.2.

Small-scale fuel cell power units can be used with advantage for planning the complete power requirements of a city. For individual houses, apartments, commercial buildings and industrial sites, independent on-site power units could meet power demands from a kW to several kWs. They can be integrated into total energy packages that not only generate electricity, but provide control of temperature, humidity and cleanlines. They can also decentralize the generation of electricity and trim the expenses of transmission over networks by serving as substations. Fuels can be processed on site or supplied to individual power units through underground pipelines.

11.7.2. Co-generation Units

Co-generation presents an efficient way of utilising our limited energy resources because the same fuel source is used simultaneously to produce two forms of useful energy including electricity and heat. Cogeneration applications are process and site specific for a given industry or a commerical building. The cogeneration potential is illustrated in the Fig. 11.7.

The phosphoric acid fuel cell can be used for a cogeneration system. The block diagram for such a plant is shown in Fig. 11.8.

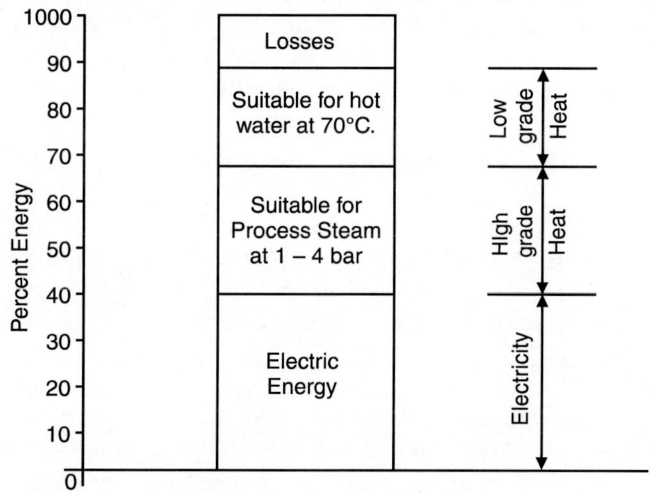

Fig. 11.7. Cogeneration Potential

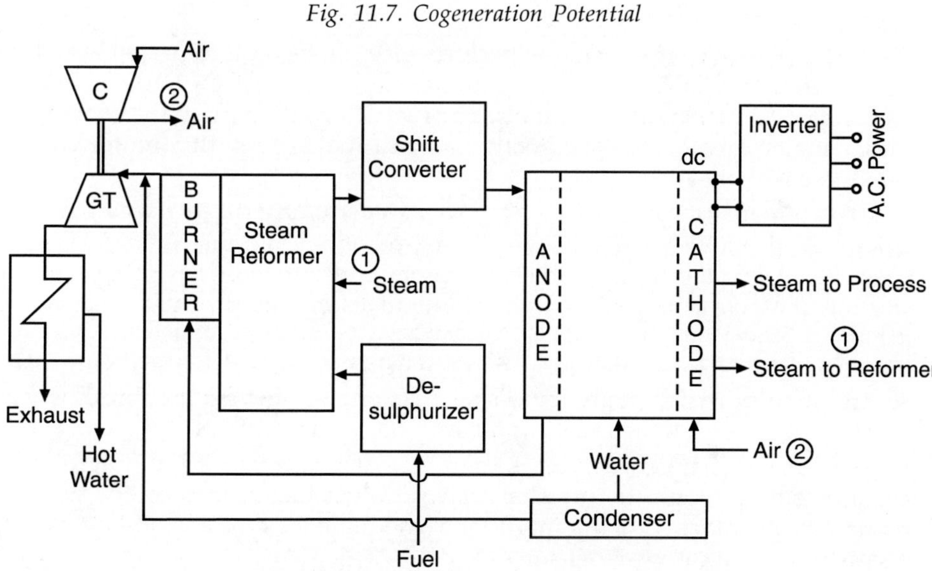

Fig. 11.8. Phosphoric Acid Fuel Cell Cogeneration Plant

It is capable of operating on a variety of fuels, including natural gas, light distillates, propane and coal-derived synthetic fuels. Synthetic fuels include hydrogen/carbon monoxide/methane mixture and methanol that can be refined, reformed and/or shifted to produce hydrogen.

The cogeneration system incorporates fuel cell stacks, turbo-compressor for supply of pressurzed air to the cell cathodes, and for recovery of pressure and heat energy from the cell exhaust streams, a desulphurizer for removal of sulphur from raw fuel, a steam reformer for conversion of hydrocarbon fuels to hydrogen and carbon monoxide, shift converters for reaction of carbon monoxide and water to produce hydrogen and carbon dioxide, and a heat recovery system for generation of process steam for industry and reformer steam from waste heat transferred from the cell.

The overall process in the fuel cells consists of the continuous-electrochemical reaction of hydrogen in the fuel stream and oxygen from air to produce electric power and by-product hot water and heat. Electric efficiency of the system, after parasitic power, is in the range of 0.38 to 0.41. If process steam generated at 6 bar is not sufficient for the industry, supplementary firing may be used.

11.7.3. Mobile units for Automotive Vehicles

Fuel cells may provide practical low-pollution vehicles with useful performance and range. Batteries, fuel cells and combination of both have been tried for electrically powered vehicles. The most practical systems will probably be hybrid power plants in which batteries provide peak power and the fuel cells act as charging units during low power periods.

In transport industry, high efficiency and low pollution make the fuel cell attractive. Fuel cells are expected to meet the requirements of high energy density handily since the amount of energy available is determined by the size of fuel tank. A fuel cell powered vehicle can have a good range without refuelling, and can be refuelled rapidly, just as can present day internal combustion engines.

Fuel cell systems of adequate performance can be built to propel local trains for the city. It would be very smooth and quiet, virtually pollution-free, and could operate on conventional fuels. The fuel cell could replace the conventional diesel engine/generator set directly or small fuel cells with a motor could be fitted to each wheel.

It is possible to meet the high power density criteria for large buses and trucks. To propel a vehicle of weight comparable to an intermediate car with speeds and accelerations usable in present traffic conditions (almost instant starting characteristic and very intermittent use), it is probably necessary to achieve a power density of about 220 W/kg, which is equivalent to 6 kg/hp. It may be possible to meet that goal by hybridising a fuel battery with one of several high power storage devices such as one of new generation flywheels.

The criteria of high power density is more difficult to achieve for small personal cars. Changed driving patterns of decreased speed and acceleration requirements can help to introduce low power vehicles. In case of a new city, roads can be planned with multi-lane and unlevel crossing to mitigate the necessity for high power densities.

11.7.4. Miscellaneous Applications

A 50 kW fuel cell stack has been developed by Plug Power LLC, Latham, New York. It produces electricity from many fuels such as gasoline, ethanol, methanol and natural gas. A fuel processor has been developed to convert raw fuel into hydrogen, which is fed into a fuel cell to produce electricity. The carbon dioxide removal system was also developed. The technology can be used for powering vehicles and in homes and apartments to provide heat and electricity.

A fuel cell of 1.5 kW capacity has been developed in Australia to meet the power requirements of a house.

TUTORIAL SHEET NO. 11.1.

1. What is a fuel cell ? Describe the principle of working of a fuel cell with reference to Hydrogen-Oxygen Cell.
2. Derive an expression for emf, free energy, potential, power output and efficiency of a fuel cell.
3. What are the advantages and disadvantages of a fuel cell ?
4. How can an energy efficient city be designed with the help of fuel cells ?
5. Explain the performance characteristics of a fuel cell.
6. What are various types of fuel cells ? Define their technical specifications and explain their applications.
7. Compare the performance of a fuel cell power plant with conventional power plant.
8. Design a cogeneration plant with the help of a phosphoric acid fuel cell.
9. Write detailed note on application of fuel cell for automotive vehicles.
10. What is an ideal fuel cell? Explain various polarizations in an actual fuel cell.

❑❑❑

Chapter 12

SOLAR CELLS

Introduction ; Photon Energy ; Theory of Solar Cells ; Solar Cell Materials ; Solar Cell Power Plants ; Energy Storage ; Applications ; Limitations

12.1. INTRODUCTION

The Solar cell is based on photovoltaic energy conversion phenomenon. It is a direct energy conversion technology. Electricity is directly produced from sunlight without the use of a working fluid such as steam or gas. There is also no need of a mechanical cycle such as Rankine or Brayton Cycle.

Photovoltaic systems are simple, convenient dependable without the need of moving parts. The basic unit of a photovoltaic system is called a solar cell. These are assembled in arrays of identical modules to produce different power capacities. The power plants may range from small residential systems installed on roof tops to large central systems.

12.2. PHOTON ENERGY

Light is radiant energy. This is transferred in discrete pieces and not continuously. The smallest unit of energy is called **quantum**. The quantum of radiant energy is called a **photon**. The photon energy is proportional to the frequency of radiation. This is called Planck's law.

$$E_p = h\nu = h\frac{c}{\gamma}$$

where E_p = photon energy

h = Planck's constant

$= 6.6256 \times 10^{-34}$ [J.s]

$= 4.13576 \times 10^{-15}$ [eVs]

ν = frequency of radiation, [herz]

c = speed of light

$= 2.997925 \times 10^{8}$ m/s

λ = wavelength of radiation, [m]

Light has dual characteristics. It consists of photons which has energy and it is also waves having a frequency and wavelength. The radiation from the sun is composed of the photons, each carrying a quantity of energy exactly equal to its frequency times Planck's constant.

Example 12.1. *Calculate the photon energy of a mono energetic radiation beam having a wave length of half a micrometer.*

Solution :

$$\lambda = 0.5 \text{ µm} = 0.5 \times 10^{-6} \text{ m}$$
$$c = 3 \times 10^{8} \text{ m/s}$$
$$h = 4.13576 \times 10^{-15} \text{ eVs}$$

The energy of a photon,

$$E_p = h\frac{c}{\gamma}$$

$$= 4.13576 \times 10^{-15} \times \frac{3 \times 10^{8}}{0.5 \times 10^{-6}}$$

$$= 2.48 \text{ eV}$$

12.2.1. Photon Flux

It is the number of photons crossing a unit area perpendicular to the beam radiation per unit line. The photon flux, ϕ_p is a useful quality in the photo-voltaic cell calculations.

The solar energy flux E'' can be calculated as :

$$E'' = \sum_i \phi_{pi}\, h\nu_i$$

where i is the frequency or energy range.

Using average values of frequency and corresponding wave length,

$$E'' = \phi_P\, h\nu_{av} = \phi_p h\left(\frac{c}{\lambda_{av}}\right)$$

The terrestrial insolation and average photon energy decrease as air mass (m) and angle of latitude (ϕ) increase.

Example 12.2. *Calculate the extraterrestrial photon flux.*

Solution : Outside the earth's atmosphere

$$m = 0$$
$$\phi = 0$$

The solar energy flux,

$$E'' = S = 1359 \text{ W/m}^2$$
$$= 0.1359 \text{ W/cm}^2$$

where S = solar constant

The average photon energy

$$h\upsilon_{av.} = 1.48 \text{ eV}$$

$$\therefore \quad f_P = \frac{0.1359}{1.48} \times \frac{1}{1.6021 \times 10^{-19}}$$

$$= 5.8 \times 10^{17} \text{ photon/s.cm}^2$$

At $m = 3$ and $\phi = 5$

$$\phi_p \approx 2.9 \times 10^{17} \text{ photons/s.cm}^2$$

When the sun's photon flux strike a *pn* semiconductor junction, there is an electric current flowing through a load connected across the junction. The electric power produced is proportional to photon flux.

12.3. THEORY OF SOLAR CELLS

A solar cell is composed of *pn* semiconductor junction as shown in Fig. 12.1. Solar cells can be manufactured from different semiconductor materials and their combinations. The voltage generated by a solar cell depends on the semiconductor material. The electric current produced depends upon the intensity of solar radiation and the cell surface area receiving the radiations. The maximum achievable power is about 100 W/m^2 of solar cell surface area. Pure semiconductor materials are needed for achieving high energy conversion efficiency.

Direct conversion of solar radiation into electricity can be studied with the help of solid-state principles. In order not to disturb the continuity of th esubject, these principles are included at the end of the chapter.

12.3.1. Types of Solar Cells

The main types of solar cells alongwith brief performance specifications are :

1. Monocrystalline silicon cells
   Band gap : 1.12 eV
   Maximum efficiency = 24%
2. Polycrystalline silicon cells
   Band gap : 1.12 eV
   Maximum efficiency : 17.8%
3. Amorphous silicon cells
   Band gap : 1.75 eV
   Maximum efficiency : 13%
4. Gallium arsenide (GaAs)
5. Copper indium diselenide (CID) Cells
6. Multi-junction Cells
7. Cadmium telluride (CdTe) Cells
   Band gap : 1.44 eV
   Maximum efficiency : 15.8%
8. Concentrator Cells
   Maximum efficiency : 32.3%

At present, silicon solar cells occupy 60% of the world market.

For single-crystal silicon, *p* is obtained by doping silicon with boron and is typically 1 μm thick; *n* is obtained by doping silicon with arsenic and is typically 800 μm thick. Thin film cells are composed of copper sulphide for *p*, typically 0.12 μm thick, and cadmium sulphide for *n*, typically 200 μm thick.

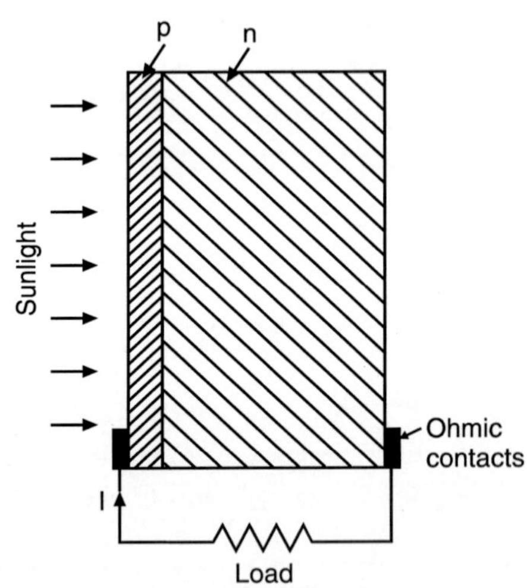

Fig. 12.1. Cross-section of a Solar Cell

12.3.2. Operation of Cell

The sun's photons strike the cell on the microthin *p* side and penetrate to the junction. There they generate electron-hole pairs. When the cell is connected to the load, the electron will diffuse from *n* to *p*. The direction of current *I* is conventionally in the opposite direction of the electrons.

12.3.3. Performance Characteristics

Typical voltage-current characteristics are shown in Fig. 12.2 at two different solar radiation levels. For each

V_o = Open-circuit voltage
I_o = Short-circuit voltage
P_m = Point of maximum power,
= $(VI)_{max}$

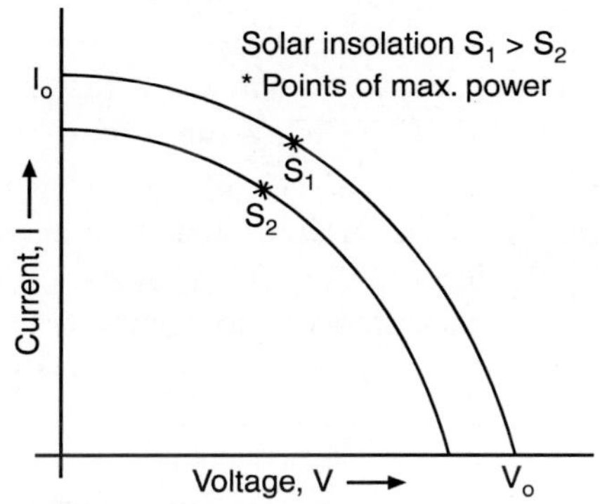

Fig. 12.2. Performance Characteristics of Solar Cells

12.3.4. Solar Energy Utilization

Table 12.1. shows the breakdown of solar energy wavelength distribution. It also shows fraction utilized by a cell and percent solar energy converted to electricity.

Table 12.1. Spectral Solar Energy Conversion

Wavelength range μm	Solar energy %age	Fraction converted by cell	Solar energy converted %age
< 0.3	0	0	0
0.3 — 0.5	17	0.36	5
0.5 — 0.7	28	0.55	15
0.7 — 0.9	20	0.73	15
0.9 — 1.1	13	0.91	12
> 1.1	22	0	0
		Total	**47**

Solar cells donot convert all solar radiation falling upon them to electricity. Weak, low-frequency (long-wavelength) photons donot possess sufficient energy to dislodge electrons. On the other hand, strong, high-frequency (short-wavelength) photons are too energetic and although they dislodge electrons, some of their energy is left over unused.

12.3.5. Solar Cell Efficiency and Losses

Efficiency is defined as the ratio of electric power output of the cell, module, or array to the power content of sunlight over its total exposed area. The maximum theoretical efficiency of solar cells is around 47 percent. Efficiencies of modules or arrays are therefore lower than those of the cells because of the areas between the individual cells. The fraction of cell to total areas is called the packing factor.

The actual efficiency of a solar cell is low because of the following losses.

1. Part of the solar energy is reflected back to the sky.
2. Some portion of solar energy is absorbed by non-photovoltaic surfaces.
3. Some solar energy is converted into heat.
4. At high temperatures there is recombination of the electron-hole pairs, Cells are usually laboratory rated at 1000 W/m^2 and 28°C. The cells normally operate at 50 to 60°C. This may reduce the efficiency by 1 to 2%.
5. The efficiency is also affected by various electrical losses.
6. There are additional losses due to mismatch between individual cells in a module and between modules in an array.
7. Mass production of modules also lower the cell efficiency.

Typical energy balance of a noncentrating silicon photovoltaic conversion array is shown in Fig. 12.3.

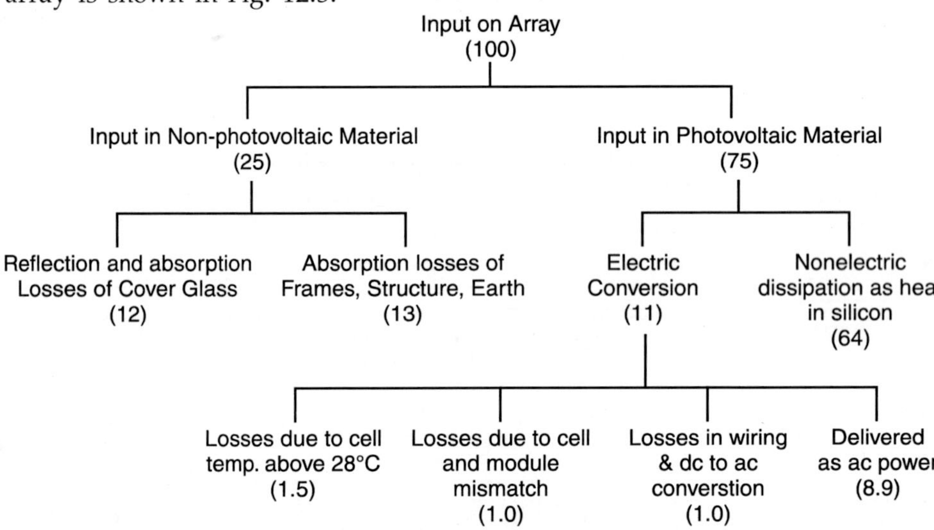

Fig. 12.3. Energy Balance of a Solar Cell

The resultant array efficiency is only 8 percent after deducting various losses.

12.3.6. Methods of Increasing Cell Efficiency

The following methods are used to improve solar cell efficiency.

1. Use of *alternative* materials.

2. *Concentration* of Incidence solar energy. The cell efficiency can be improved by the use of concentrated sunlight. The cells are located at the foci of parabolic or trough concentrators. Cell efficiency of 25 percent can be achieved with a concentration of about 500 suns. Concentrators are cost-effective.

3. *Thermophotovoltaic Systems.* Highly concentrated light is used to heat a refractory material. The hot material reradiates the solar energy to silicon cells at longer wavelengths. Such wavelengths are more effective to generate electric-

ity by silicon cells. Although there are material problems as the temperatures required are 1870 to 1925°C.

4. *Cascade Systems*. Multijunction solar systems are used. Each cell is subjected to different region of solar spectrum at which it operates most efficiently. Such a system can yield conversion efficiencies even higher than 25%.

12.4. PERFORMANCE ANALYSIS

A solar cell uses a *p-n* junction as shown in Fig. 12.1. The current and voltage relationship is given as

$$I_j = I_o\left[\left(e^{\frac{eV}{kT}}\right) - 1\right] \ [A]$$

where I_o = saturation current of diode. This is also called dark current when a large negative voltage is applied across the diode.

V = voltage across junction [V]

e = electronic charge
= 1.60219×10^{-9} Coulomb

k = Boltzmann constant
= 1.38066×10^{-23} J/K

T = Temperature [K]

The equivalent electrical circuit of a solar cell is shown in Fig. 12.4.

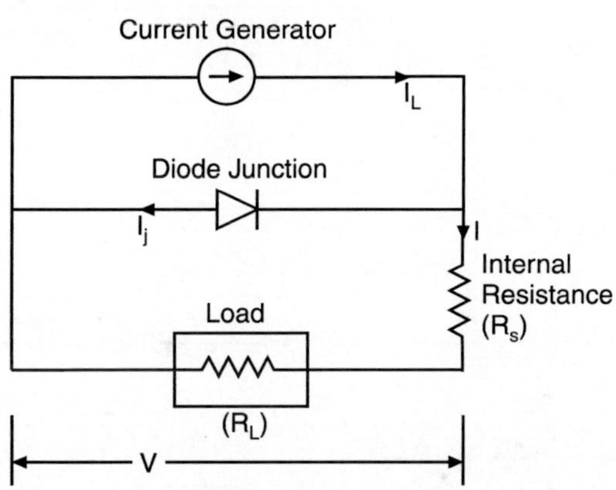

Fig. 12.4. Electrical Circuit Diagram of a Solar Cell

When light impinges on the p-n junction, electron hole pairs are created at a constant rate resulting in an electrical current flow across the junction. The net current (I) is the difference between the normal diode current (I_j) and light generated current (I_L). The internal series resistance (R_s) is due to high series resistance of the diffused layer which is in series with the junction. The photovoltaic current or light generated current I_L is branched into current I in the external circuit and I_j the diode current. The current-voltage characteristic curve is shown in Fig. 12.2. The following current-voltage relationship is valid at a constant cell temperature and a constant intensity of solar radiation.

$$I = I_L - I_j = I_L - I_o\left[\left(e^{\frac{eV}{kT}}\right) - 1\right]$$

For ideal cell $R_s = 0$. The open circuit voltage V_o for an ideal cell is

$$V_o = \left(\frac{kT}{e}\right)\ln\left[\frac{I_2}{I_0}+1\right]$$

$\therefore \quad I_L >> I_o,$

$$V_o = \frac{kT}{e}\ \ln\ \frac{I_L}{I_o}$$

The maximum power,

$$P_{max} = V_{mp} \times I_{mp}$$

where V_{mp} = voltage at maximum power point

I_{mp} = current at maximum power point

The maximum efficiency for the cell,

$$\eta_{max} = \frac{V_{mp} \cdot I_{mp}}{P_{sun}} = \frac{P_{max}}{P_{sun}}$$

$$= \left[\frac{I_L E_g}{e P_{sun}}\right]\underbrace{\left[\frac{I_{mp}V_{mp}}{I_L V_o}\right]}_{\text{Fill factor}}\underbrace{\left[\frac{eV_o}{E_g}\right]}_{\text{Voltage factor}}$$

where E_g = Forbidden energy band

FF = Fill factor

$$= \frac{I_{mp}\,V_{mp}}{I_L\,V_o} = 0.8 \text{ for a properly designed cell.}$$

$$P_{max} = I_L \times V_o \times FF$$

VF = voltage factor

$$= \frac{eV_o}{E_g} \simeq 0.5 \text{ for a silicon cell.}$$

12.5. SOLAR CELL MATERIALS

Silicon solar cells occupy more than 60% of photovoltaic market. Basic types of silicon solar cells are :

1. Monocrystalline silicon solar cells,
2. Polycrystalline silicon solar cells,
3. Thin film or amorphous silicon solar cells.

12.5.1. Monocrystalline Silicon Solar Cells

Silicon is doped with boron to produce *p* type semiconductor. A single crystal of 2m length and 10 to 15 cm diameter is drawn from the molten silicon. It is sliced into 0.3 to 0.5 mm thick wafers or discs. The upper layer of the wafers is doped with phosphorous upto 3 to 4 μm depth to produce *n* type semiconductor. This becomes a *p-n* junction. A copper layer is deposited for full area on the backside and 10% area on the front side. This metallic surfaces are used to take off the electric current.

A silicon solar cell of size 10cm × 10cm produces a voltage of 0.5V and power output of 1W at a solar radiation intensity of 1000 W/m^2. The solar cells are formed into modules by enclosing in an air tight casing with a transparent cover of synthetic glass. These modules possess high efficiency between 15 and 18% and are used in medium and large size plants.

12.5.2. Polycrystalline Silicon Solar Cell

A polycrystalline silicon block (containing many small silicon crystals) are sliced into wafers in the similar manner as in case of single crystal solar cells. The highest efficiency of solar module is 12 to 14%.

12.5.3. Thin-film Solar Cells

The crystalline solar cells are labour and energy intensive in manufacturing. The thin film cells are produced from amorphous silicon. It has the capacity to absorb more solar radiation due to irregular atom arrangement. A thin 1 mm thick solar cell is used. Gaseous $Si\,H_4$ is deposited on a glass sheet, the silicon sheet is simultaneous doped to produce *p-n* junction. The efficiency is 5 to 8%. These are very cheap to manufacture.

Thin film solar cells are also manufactured from the following materials.

1. Gallium arsenide, (GaAs)
2. Cadmium telliride (Cd Te)
3. Copper-indium-selenide (Cu In Se)

12.6. SOLAR MODULES AND ARRAYS

12.6.1. Solar Modules

Solar cells are joined in series to form a module. The number of cells will depend upon the module voltage. A 12 V solar module consists of 33 to 36 cells. The cells are mounted together under an airtight, mechanically rigid, transparent cover.

The technical specifications of Siemen's solar module M55 containing 36 solar cells connected in series are as under :

Maximum peak power	= 53W
Open circuit voltage, V_o	= 21.7V
Short-Circuit current, I_s	= 3.4A
Optimum operating current, I	= 3.05A
Optimum operating voltage	= 17.4 V
Dimensions	= 1293 × 329 × 36 mm
Mass	= 5.7 kg

The Central Electronics Ltd. Sahibabad (U.P.) have developed a solar photovoltaic panel with following specifications :

Maximum peak power	= 300W
System voltage	= 60V dc nomimal
Dimensions	= 2400 × 2500 mm

12.6.2. Solar Array or Generator

It is made by interconnecting solar modules to supply electricity. The modules are connected both in series as well as parallel. Connected in series will give a 24V solar array. In order to increase current and power output, the modules are connected in parallel.

The number of modules connected in series,

$$N_{series} = \frac{V}{V_{mod.}}$$

where V = operating voltage of solar generator

V_{mod} = nominal voltage of a module

The operating current,

$$I = \frac{P}{V} \; [A]$$

P = given power output of the array.

The number of modules connected in parallel.

$$N_{parallel} = \frac{I}{I_{mod}}$$

where I = operating current of generator

I_{mod} = nominal current of a module

Total number of modules in the solar generator

$$N = N_{series} \times N_{parallel}$$

12.7. SOLAR CELL POWER PLANTS

There are two types of power plants.

1. Autonomous power plants
   These are used for local networks
2. Grid connected power plants
   These are used for external electrical networks.

12.7.1. Autonomous Solar Power Plant

These are used for the following

1. Home power supply in rural areas.
2. Solar Water Pumps
3. Telecommunication and relay stations
4. Cattle fencing
5. National parks for street lighting
6. Tourist facilities for remote and hilly areas.

The block diagram is shown in Fig. 12.5.

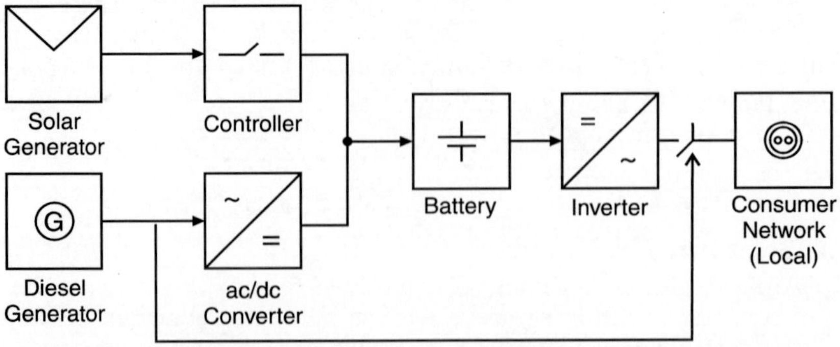

Fig. 12.5. Block Diagram of Autonomous Solar Cell Power Plant

The system consists of the following :

1. Solar generator or array
2. Storage battery
3. Charge controller
4. Inverter
5. Standby diesel engine-generator
6. Consumer network

Solar generators can be designed for required electric supply.

12.7.2. Combined Solar-Wind-Diesel Power Plant

The combined power plant of solar-wind and diesel engine is shown in Fig. 12.6. The solar generator produces electricity. A wind power plant and/or diesel generator driven by biogas from a sewage plant work as backup unit. A storage battery, an inverter and power distribution electric grid make the power supply system complete. The capacities of different power units can be as follows :

1. Solar generator = 140 kW (40 modules)
2. Wind power plant = 250 kW (Rotor diameter 25m)
3. Diesel engine = 30 kW

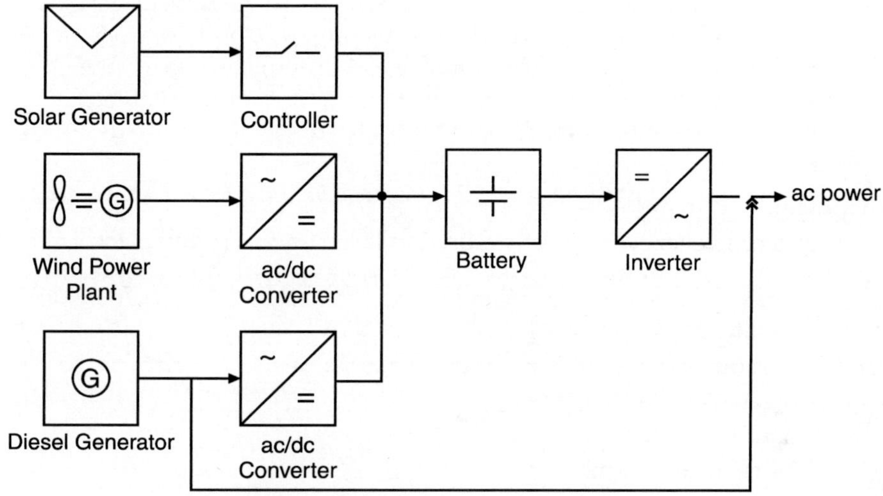

Fig. 12.6. Combined Solar-Wind-Diesel Power Plant.

12.8. ENERGY STORAGE

Solar cell power systems must share conventional power grids or must use electrical storage if their output is to last longer than sunlight. Some of the storages schemes are :

1. Battery

This is storage of electric energy by conversion to chemical energy in batteries. The most common and highly developed is the lead-acid battery. Large electric energy storage in lead-acid batteries or other batteries is not economically feasible. Other battery systems with higher energy-to-mass ratios are under development. Currently available and suitable storage batteries have capacities between 40 to 280 Ah.

A storage battery cycle operation consists of a charging process and discharging process.

The amount of energy stored during charging process,

$$P_s = C\, V_B \text{ [Wh]}$$

where C = Storage capacity [Ah]

V_B = Nominal battery voltage [V]

A storage battery with a capacity of 200 Ah and a rated voltage of 12V, can store 2400 Wh of electric energy.

The amount of energy recovered from a battery during discharging process,

$$P_d = V_d\, I_d\, t_d \text{ [Wh]}$$

where V_d = discharge voltage [V]

I_d = discharge current [A]

t_d = discharge duration [h]

In order to get highly reliable power supply a diesel engine should also be used as back-up unit in addition to a storage battery.

2. Pumped-hydro storage

This method is more suitable for large power plants. The surplus energy is used to pump water into high reservoirs during sunny periods or periods of low demands and extraction of power during evening or cloudy periods or periods of high demands by running the stored water through water turbines.

The main limitation of this system is to find sites with suitable topography near solar power plants. Such power plants are mostly located in desert like flat terrains.

3. Cryogenic Storage

The electric energy is directly stored in large underground electrical coils at liquid-helium temperatures of 4K. The electrical resistivity is almost zero.

12.9. DESIGN OF A SOLAR POWER PLANT

An autonomous solar power plant consists of :

1. PV generator or array
2. Storage battery
3. Charge control unit
4. Inverter
5. Back-up diesel generator
6. Electricity consumer network

The power balance equation for a period of plant utilization can be written as :

$$P_G + P_B + P_D - P_{IL} = P_L \quad \text{[Wh]}$$

Where P_G = Electrical energy supplied from solar generator

P_B = Electrical energy supplied from battery

P_D = Electrical energy supplied from diesel engine

P_{IL} = Internal energy losses & system consumption

P_L = Energy requirement of consumer network

The energy supplied by solar generator over a given time duration,

$$P_G = \int A_G \, S \, \eta_g \, dt \quad [\text{Wh}]$$

Where A_G = Total solar cell surface area [m^2]
S = Mean solar radiation [W/m^2]
η_G = Average solar generator efficiency
t = Time duration [h]

12.9.1. Size of Solar Array

The following input data is needed

1. Daily solar radiation at site
2. Ambient temperature
3. Network load and voltage
4. Specification of solar generator including its efficiency
5. Specification of storage battery

The daily electrical load is calculated as :

$$P_L = \sum_i n_i P_i t_i$$

where n = number of power consumers of one type
P = power requirement of consumers of one type [W]
t = daily consumption hours [h]

Total area of solar array

$$A_G = \frac{P_L}{S\eta_G} \quad [\text{m}^2]$$

where S = Daily amount of solar energy incident on 1 m^2 of solar array [W/m^2/day]

Maximum peak power output of solar generator

$$P_{max} = A_G \, S \, \eta_G \quad [\text{W}]$$
$$S = 1000 \text{ W/m}^2$$

Number of solar panels (modules) required,

$$N_m = \frac{P_{max}}{P_m}$$

where P_m = Maximum power output of a module [W]

12.9.2. Solar Panel Tilt

The tilt angle θ of a solar panel is selected so that the generator receives maximum solar radiation at the time of maximum load.

$$\theta = \phi \pm 15°$$

where ϕ = Latitude of site in degrees

12.9.3. Storage Battery Capacity

The performance parameters of a storage battery are :

1. Capacity [Ah]
2. Rated voltage [V]

3. Rated current [A]
4. Efficiency

The battery efficiency is the energy recovered to energy charged. The usual values of efficiency are 75 to 90%.

The capacity of a storage battery can be calculated as :

$$C = \frac{F P_L N_D}{V_B D_{\max} \eta_B} \quad \text{[Ah]}$$

where F = Reserve factor = 1.2

P_L = Daily power load [Wh/day]

N_D = Number of required storage days

D_{max} = Max. allowable depth of battery discharge = 0.8

V_B = Rated battery voltage [V]

η_B = Battery efficiency

For large power storage, batteries are connected in series and parallel in a storage bank.

Number of batteries to be connected in parallel is the ratio of total energy capacity and energy capacity of each battery.

The number of batteries to be connected in series is the rated voltage of the network load divided by the rated voltage of a battery.

The total number of batteries is the product of batteries in parallel and batteries in series.

Example 12.3. *An autonomous solar power plant is to be designed to supply power to a village school. The power supply is at 12V dc. The building loads are :*

1. *Fluorescent tubes* = *30 × 8W for 4.5 hrs/day*
2. *Radio with cassette player* = *2 × 25W for 3 hrs/day*
3. *Colour TV sets* = *2 × 80W for 4 hrs/day*
4. *Computers* = *4 × 24W for 7 hrs/day*

The daily solar radiation is 5.4 kWh/m²/day and the average panel efficiency is 11%. Solar panels are available with peak power rating of 160W. Determine :

1. *Daily power requirement*
2. *Solar panel area*
3. *Total peak power output*
4. *Number of solar panels needed.*

Solution : 1. Daily load

$$P_L = \sum_i (n_i P_i t_i)$$

$$= 30 \times 8 \times 4.5 + 2 \times 25 \times 3 + 2 \times 80 \times 4 + 4 \times 24 \times 7$$

$$= 2542 \text{ Wh/day}$$

2. Solar panel area required

$$A_G = \frac{P_L}{S \, \eta_G}$$

$$A_G = \frac{2542}{5.4 \times 1000 \times 0.11} = 4.28 \text{ m}^2$$

3. Total peak power output

$$P_{max} = A_G \, S \, \eta_G$$
$$= 4.28 \times 1000 \times 0.11$$
$$= 470.8\text{W}$$

4. Number of solar panels required,

$$= \frac{470.8}{160} = 2.94 \simeq 3$$

Example 12.4. *An acid-lead battery bank is required for a solar generator serving a building with a load of 2542 Wh/day for 3 days. The rated voltage of a battery is 12V.*

Determine the battery storage capacity for battery efficiency of 83%, max. depth of discharge 0.8 and reserve factor of 1.2. Also calculate the number of batteries connected in parallel if the capacity of a single battery is 180 Ah.

Solution : 1. Battery bank capacity

$$C = \frac{F P_L N_D}{V_B D_{\text{max}} \eta_B}$$

$$= \frac{1.2 \times 2542 \times 3}{12 \times 0.8 \times 0.83} = 1148.5 \text{ Ah.}$$

2. Number of batteries to be connected in parallel,

$$= \frac{C}{C_1} = \frac{1148.5}{180} = 6.38 \approx 7$$

12.10. APPLICATIONS OF SOLAR PHOTOVOLTAIC SYSTEMS

The solar photovoltaic systems are classified as per field of application, type of system, rated capacity, etc.

I. Autonomous system

1. Amorphous silicon solar cells of very small capacity are used in watches, pocket calculators, etc.
2. Small capacity solar systems from 50W to 50 kW capacity are used for remote houses and villages for lighting for domestic use, street lighting telecommunications, community development, water pumping, etc.
3. Roof-mounted solar system of 1 kW to 5 kW can be used for residential houses.
4. In developing countries, solar systems can find applications for drinking water supply, irrigation, vaccine refrigeration, milk chilling, rural power supply.
5. In developed countries, solar generators can be used for lighting, recreation centres, radios, TV sets, small refrigerators, etc.
6. Solar generators can be used for boats, lighting towers, radio buoys, traffic signals, parking lights.
7. Developing countries have high solar energy potential and solar genera-

tors are used for water pumping, domestic power supply, hospitals, schools, farm houses.

8. Other applications can be for catholic protection of oil pipelines, weather monitoring, railway signalling, battery charging.

II. Solar Water Pumps

The major application of photovoltaic systems lies in water pumping for drinking water, irrigation in rural areas, cattle stock watering. The solar generators required for this purpose should have power capacity from 10W to 10 kW. The water discharge of pump ranges between 1 to 40 m^3/h with delivery head from 2.5m to 120m. These generators usually work without storage battery.

The specifications of solar photovoltaic water pumping system standardized by Central Electronics Ltd. (U.P.) are given below :

Solar panel	:	300 W (peak) 2.4 m × 2.5m (size)
System voltage	:	60V DC nominal
Motor-pump set	:	DC motor-centrifugal pump of monoblock configuration
Peak water output	:	1.5 to 2.0 litres/sec.
Average water output at 5m head	:	30 to 40 m^3 per day
Maximum suction head	:	6m
Maximum total head	:	10m

The solar array consists of 5 rows of 4 modules each. Each module consists of 36 cells each 76mm diameter connected in series to generate about 16 peak watts of power at incident radiation of 100 mW/cm^2 at 28°C. Four such modules are connected in series and five such series connected module chains are kept in parallel. The complete array is mounted on an angle iron structure which can be provided with storage batteries and power conditioning equipment.

The total cost of the system including solar photovoltaic array, on/off switch, motor-pump set, accessories including suction and delivery valves and piping is about Rs. 40,000 which is highly subsidised and installed at a cost of Rs. 5000 only for a small or marginal farmer.

III. Central Power Generation

A solar photovoltaic power plant of 16 MW capacity has been developed to supply electricity to 2300 households Solar power plants with a total capacity of 100 MW are planned for future installation.

IV. Space Satellite Power Stations

The space satellite power station (SSPS) has been conceived as the future solar power supply system. Large surface area of photovoltaic panels will be mounted on a space satellite which will be synchronically moving with the earth orbit so that it will look stationary from every point on the earth. The energy generated will be converted into microwave energy and transmitted to the earth. It will be collected with antennas of a few square kilometer area and then converted into commercial-frequency electric power.

Satellite-mounted solar power plants can be designed for power outputs from 3 to 20 GW. The plant will require solar battery of 20 km^2 total area. The transmitting antenna of 1 km diameter and receiving antenna of 7 to 10 km diameter. The overall efficiency is expected to be 77 percent.

12.11. ADVANTAGES OF PHOTOVOLTAIC SOLAR ENERGY CONVERSION

1. Absence of moving parts.
2. Direct conversion of light to electricity at room temperature.
3. Can function unattended for long time.
4. Modular design : Voltage and power outputs can be manipulated by integration.
5. Low maintenance cost.
6. No environmental pollution
7. Very long life
8. Highly reliable
9. Solar energy is free and no fuel required.
10. Can be started easily as no starting time is involved.
11. Solar cells can be made from microwatts to megawatts. These can be used to feed the utility grid with power conditioning circuitary.
12. Easy to fabricate
13. These have high power-to-weight ratio, therefore very useful for space application.
14. Decentralized or dispersed power generation at the point of power consumption can save power transmission and distribution costs.
15. These can be used with or without sun tracking.

12.12. LIMITATIONS

1. Manufacture of silicon crystals is labour and energy intensive.
2. The principal limitation is high cost, which is being reduced through various technological innovations.
3. The insolation is unreliable and therefore storage batteries are needed. Good batteries are not available.
4. Solar power plants require very large land areas.
5. Electrical generation cost is very high.
6. The energy spent in the manufacture of solar cells is very high. The plant operation period during which photovoltaic plant recovers the spent energy varies from 4 to 7 years.
7. The initial cost of the plant is very high and still requires a long gastation period.

12.13. SOLID-STATE PRINCIPLES

Solid-state phenomenon are important for the study of photovoltaics.

12.13.1. Solid Phenomenon

A metallic crystalline solid contains atoms that have nuclei surrounded by electrons, which are tightly bound to them (the solid model). There are outer electrons which are weakly bound to the nuclei. The outer electrons are called valence or conduction electrons. These are free to migrate in the interior of the metal as these have no forces acting on them by the other free electrons or bound electrons or ionized positively charged nuclei. These free electrons.

(*a*) have equipotential field in which these move;

(*b*) each of them have a constant electrostatic potential energy, E_i.

(*c*) the electrostatic potential energy is independent of the location inside the crystal.

(*d*) donot belong to any particular atom.

(*e*) belong to the electro crystal.

(*f*) cause electric and heat conduction in the metals.

(*g*) donot have attractive forces as there are no positive ions on the surface of the metal.

(*h*) are easily moved by the electric fields inside the metal,

(*i*) encounter an energy barrier at the surface,

(*j*) require more energy to get out of the surface of the metal.

The free electrons form electron gas which is confined inside the metal. The free electrons follow **Pauli Exclusion Principle** which states that no two electrons can exist in the same state, at the same time and in the same atom.

12.13.2. Fermi Energy

The electron gas does not follow the energy distribution law of ordinary gas given **by the Maxwell-Baltzmann law. The energy distribution is given by the Fermi-** Dirac Distribution Law as given below. This law holds good for temperatures below 3000 K.

$$n(E)\,dE = \left[\frac{4\pi}{h^3}(2m_e)^{\frac{3}{2}}\right]\frac{E^{\frac{1}{2}}}{1+e^{(E-E_F)/kT}}\cdot dE$$

where, dE = energy range

$n(E)\,dE$ = number of electrons per unit volume in dE

h = Planck's constant

= 6.625×10^{-34} (J.s)

m_e = mass of the electron

= 9.13×10^{-28} g

k = Baltzmann constant

= 1.38×10^{-23} J/(molecule. K)

E_F = Fermi energy

The Fermi energy is a constant and nearly independent of temperature.

12.13.3. Energy Distribution of Electron Gas

The energy distribution of an electron gas is shown in Fig. 12.7 for different temperatures ($T_2 > T_1 > 0$). The Fermi-Dirac probability distribution is given by the function,

$$P(E) = \frac{1}{1+e^{(E-E_F)/kT}}$$

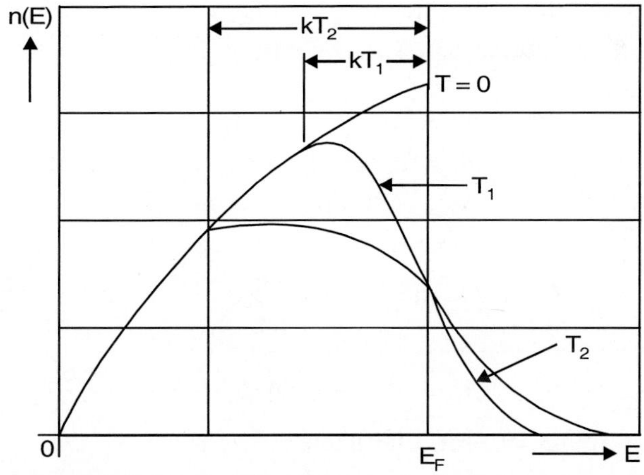

Fig. 12.7. Energy Distribution of Election Gas

The quantity $E^{\frac{1}{2}}$ gives the parabolic rise of the curve from $E = 0$. Free electron donot have zero energy at absolute zero temperature. It has finite energies upto a maximum of E_F, *i.e.*, Fermi Energy.

$$P(E) = 1 \text{ between } E = 0 \text{ to } E = E_F$$

$$P(E) = 0 \text{ for } E > E_F$$

$$P(E) = \frac{1}{2} \text{ for } E = E_F$$

$$P(E) = A\, e^{-E|kT} \text{ for } E >>> E_F$$

= Maxwell-Baltzmann distribution probability for classical gases.

A = a constant

12.13.4. Fermi Energy

The Fermi energy can be calculated from the following equation,

$$n = \int_0^{E_F} n(E)\, dE = \frac{4\pi}{h^3}(2m_e)^{\frac{3}{2}} \int_0^{E_F} E^{\frac{1}{2}} . dE$$

$$= \frac{8\pi}{3h^3}(2m_e)^{\frac{3}{2}} E_F^{\frac{3}{2}}$$

where n = total number of free electrons per unit volume

$$E_F = \frac{h^2}{2m_e}\left(\frac{3n}{8\pi}\right)^{2/3}$$

The Fermi energy can be expressed in electron volts (eV). An electron volt is a unit of energy equal to the energy acquired by or single electron charge when accelerated through a potential of 1.0 V.

$$1 \text{ eV} = 1.6021 \times 10^{-19} \text{ J}$$

$$= 4.44 \times 10^{-26} \text{ kWh}$$

One million electron volts (MeV) = 1.6021×10^{-13} J

E_F has the following value for metals.

For copper, $E_F = 7$ eV,

For sodium, $E_F = 3.1$ eV.

The average energy of a free electron at $T = 0$ is equal to $3/5\ E_F$. This value is much higher than the kinetic energy of a particle in a classical gas even at high temperatures.

$$E_F = \frac{3}{2} kT \text{ for classical gas}$$

$$E_F = 0 \text{ at } T = 0$$

$$E_F = 0.1 \text{ eV at T} = 1160 \text{ K}$$

12.13.5. Behaviour of Free Electrons

For solar cells, the behaviour of free electrons is important at temperature higher than T = 0. At moderate temperatures (see Fig. 12.7) there is slight rounding of zero temperature distribution curve. As the temperature increases, the value kT increases.

$$|E - E_F| \approx kT$$

when $E = E_F,\ P(E) = \frac{1}{2}$

The free electrons whose energies are much higher than Fermi energy (E_F) remain locked in the same energy state as those occupied at $T = 0$ irrespective of the temperature. There is a fraction of free electrons with energies within kT of the Fermi energy which occupy higher energy states than Fermi energy. These electrons can be elevated to high energies by collision with high-energy protons from the sun (solar cell) or by thermal excitations (thermionics).

12.14. THE BAND THEORY

Photovoltaic devices and the thermoelectric devices are based on band theory which distinguishes the behaviour of conductors, insulators and semiconductors. The electrical conductivities can be very different by factors as high as 10^{30}.

The valence electrons in a gas occupy a set of single-energy levels as shown in Fig. 12.8 (*a*). Where these atoms are brought together in a crystalline solid, the energy levels spread into energy bands due to interactions of atoms as shown in Fig. 12.8 (*b*). The regions between the energy bands are called forbidden bands as these cannot be occupied by electrons.

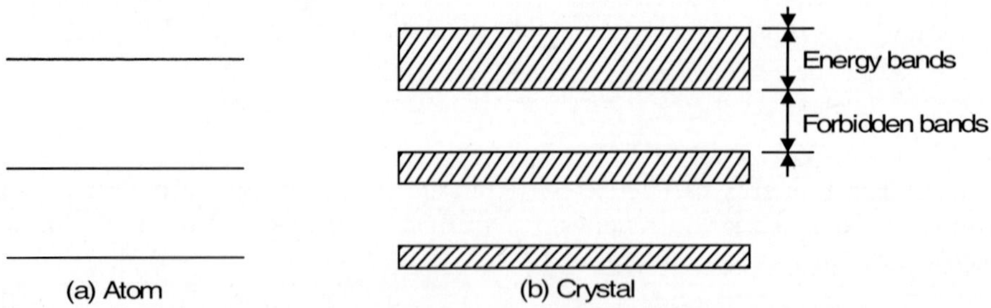

Fig. 12.8. Electron Energy Levels.

Energy Bands for Different Materials

The widths and spacings of the energy bands and forbidden bands depend upon the crystalline structure. Energy bands may be fully occupied for partially filled or completely unoccupied. The comparison for a conductor and an insulator is shown in Fig. 12.9. The occupied bands are shown as shaded.

The high conductivity of a metallic crystal is due to not completely filling of upper most band with electrons. An external electric field when applied causes the electrons in the uppermost band to gain some energy. This energy is sufficient to promote them to the continuum of available states immediately above this band. Distribution of electrons among the available states may be slightly varied by energising them. The shift is controlled by the Fermi-Dirac statics (para 12.13).

The energy bands of a conductor (sodium) and insulator (diamond) are compared in Fig. 12.9.

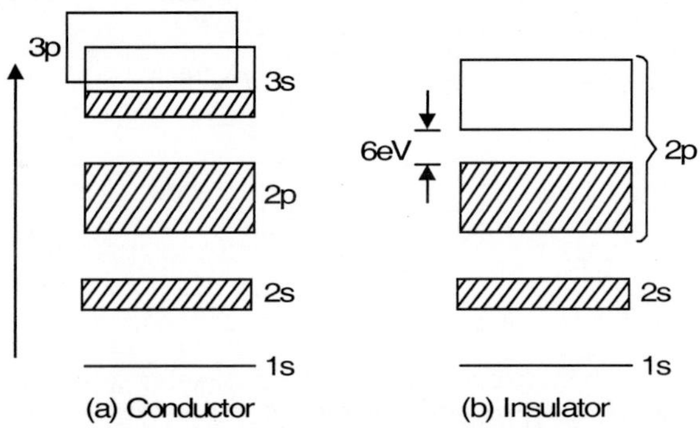

Fig. 12.9. Comparison of Electron Bands.

For sodium, the 3s band is partially filled, 3p band is unoccupied and the two bands overlap. The number of unoccupied levels readily available for the 3s band electrons is, therefore, large, which results in high electrical conductivity. Diamond, on the other hand, has two 2p energy bands, one filled, one unfilled, separated by a forbidden band that is 6 eV wide. This gap is much large than kT at room or higher temperatures. Thermal excitation, or weak electrical or other fields, will not impart sufficient energy to promote the electrons to the unoccupied 2p band and create electron flow or an electric current. The same applies to photons of visible light that donot lose energy to the electrons, which is the reason why diamond is transparent to visible light.

The uppermost filled, or partially filled energy band is called the valence band and the next empty but available energy band is called the conduction band. A substance is a conductor if these two bands are separated by a very narrow forbidden gap or if they overlap. A substance is an insulator if they are separated by a large gap.

12.15. SEMICONDUCTORS

There is a class of crystalline solids where the forbidden band between the valence and conductance band is small, *i.e.*, about 1 eV. For example silicon has the gap of 1.1 eV and germanium 0.7 eV. These solids behave like insulators at low temperatures. At very high temperatures, an appreciable number of electrons receive sufficient thermal energy and are promoted into the conduction band. The number of electrons promoted into conduction band depends upon the gap width, and temperature. The material becomes a conductor. Such a material in pure form is called an intrinsic semiconductor.

Types of semiconductors

There are two types of semiconductors: n-type semiconductors and p-type semiconductors. Some impurities are added in the intrinsic semiconductors which add some energy level in the forbidden band between the valence band and the conduction band. These semiconductors with impurities are called extrinsic semiconductors. The controlled amount of impurities are added during manufacturing.

In *n*-type semiconductors, the allowed energy level is near the bottom of the conduction band. This is called the donor level.

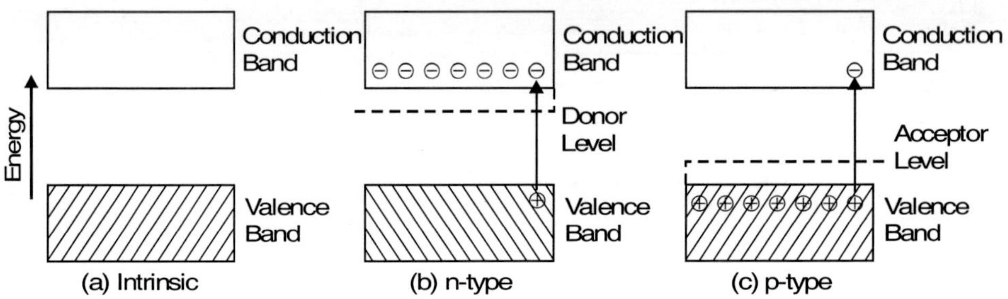

Fig. 12.10. Electron energy bands for semiconductors.

Since the gap is very narrow, there is an abundance of electrons which are excited at room temperature and promoted into conduction band. In intrinsic materials only a few electrons can be elevated from the valence to conduction bands.

In p-type semiconductors the allowed energy level is near the top of the valence band. This is called the acceptor level.

The impurity creates an electron deficiency. At room temperature electrons are easily excited from the valence band to the acceptor level. There is a deficiency created in the valence band. The absence of an electron is called a hole and may be considered a positively charged particle. The conduction in *p*-type semiconductor is due to movement of holes to the valence band.

In both types of semiconductors, the conductivity depends upon the type and amount of impurities.

12.16. p-n JUNCTION

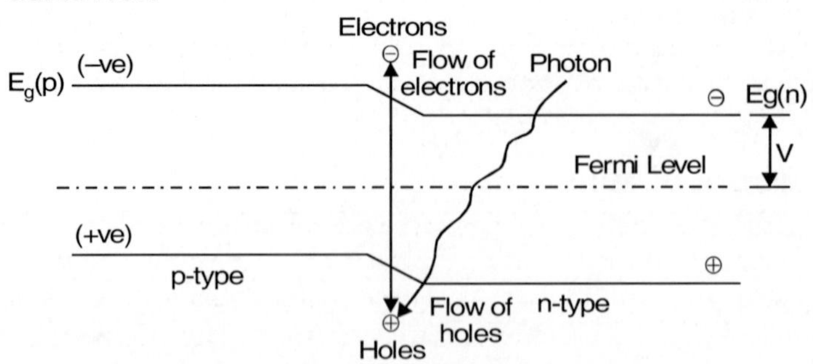

Fig. 12.11. Electric Field in p-n junction.

If a *p*-type semiconductor is brought in close contact with a n-type semiconductor a *p-n* junction is formed. Positive and negative charges are created by the absorption of photons from the sun. An electric field is created. The electrons are swept in one direction and positive holes in the other direction. Each solar cell will produce power at about 0.5V with current proportional to the area of the cell.

TUTORIAL SHEET NO. 12.1.

1. What is the principle of solar photovoltaic power generation ? What are the main elements of a solar photovoltaic conversion system ?
2. What are the advantages and limitations of photovoltaic energy conversion ?
3. What is photon energy ? Explain Planck's law.
4. How can photon flux be calculated ?
5. Describe the construction and operation of a solar cell. What are its performance characteristics ?
6. What are the losses in silicon photovoltaic conversion array ? What are achievable efficiencies ?
7. What methods can be adopted for increasing the solar cell efficiency ?
8. Derive expressions for maximum power output and efficiency of a solar cell.
9. Write a detailed note on solar cell materials.
10. How are solar modules and solar arrays designed and assembled ?
11. With the aid of suitable block diagrams, explain the following :
    (*a*) Autonomous solar power plant
    (*b*) Combined solar power plant
12. What are the various energy storage schemes used with solar power generation ?
13. Give a step-by-step procedure for the design of a solar photovoltaic power plant.
14. What are the main applications of solar photovoltaic systems? Discuss the same under following heads :
    (*a*) Autonomous systems
    (*b*) Solar water pumps
    (*c*) Central power generation
    (*d*) Space satellite power stations.
15. Discuss the basic solid-state phenomena that are important in the photovoltaic energy conversion.
16. Discuss the environmental effects of space satellite power stations.

Chapter 13

MHD POWER GENERATION

Introduction ; Principle of MHD Power Generation ; Performance Analysis ; MHD Systems ; Open-cycle Systems ; Closed-cycle Systems ; Hybridisation with Steam Power Plants ; Design Problems ; Materials of Construction ; Magnetic Field and Superconductivity ; Advantages and Limitations.

13.1. INTRODUCTION

Magnetohydrodynamic (MHD) is a direct heat-to-electricity conversion technique based on Faraday Law that when an electric conductor moves across a magnetic field, a voltage is induced in it which produces an electric current. Here the conductor is an ionized gas which is passed at high velocity through a powerful magnetic field, a current is generated and can be extracted by placing electrodes in a suitable position in the stream. It produces d.c. power directly. MHD power generation is the most promising of direct energy conversion techniques where the mechanical link can be avoided. It can overcome some of the limitations of conventional power generation by improving the efficiency from 40% to 55%, thus better utilising the fuel resources and reducing the environmental pollution.

MHD power generation has great potential for power production in excess of 1000MW. It can be used as topper for a coal-fired thermal power plant. This will increase the thermal plant efficiency by affecting direct conversion of heat to electricity. In addition to fuel economy and reduced environmental pollution, the capital cost of the power plant will also be reduced.

13.2. PRINCIPLE OF OPERATION

The arrangement of MHD power generator is shown in Fig. 13.1. The ionized gas working as electrical conductor experiences a braking force due to electromagnetic interaction. The degree of ionization required in practice is very small of the order of 0.1%. The gas is still composed of neutral particles which carry nearly all the kinetic energy of stream and are unaffected by magnetic force.

The retarding force is a complex function of collisions, cross-sections and magnetic flux density. The applied magnetic field manifests itself through the force that it exerts on the electrons in the gas. This force is then coupled to the neutral particles by the electron-ion Coulomb force and ion-neutron collisions.

In order to achieve a large power output, the gas velocity should be high (10^3m/s) and applied magnetic flux density must be as large as possible. There should be adequate gas conductivity (more than 10 mhos/m). To achieve equilibrium conditions in the pure gas by thermal ionisation, temperatures of tens of thousands of degrees are required. By seeding the gas with elements which have low ionisation potential such as alkali metals Cesium and Potassium, it is possible to achieve reasonable conductivity at temperature in the region of 2000°C. This temperature is within the limits of material technology of MHD.

The electrical power is proportional to magnetic flux density (B) and gas velocity (U) and gas conductivity (σ).

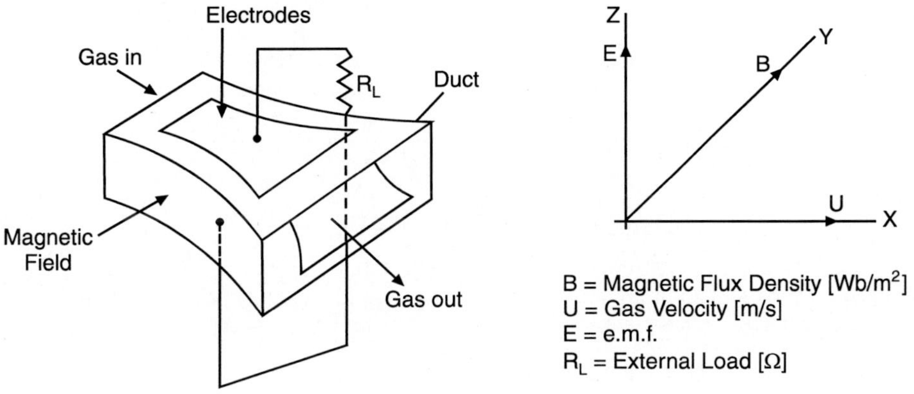

Fig. 13.1(a). MHD Power Generation.

If a particle with positive charge q is moving in a duct with plate walls P_1 and P_2 with a velocity v', the magnetic flux of density B pointing into the paper will apply a magnetic force F on the particle. The force, magnetic field and velocity are vector quantities.

$$\vec{F} = q(\vec{v}' \times \vec{B})$$

The positive charge particle moves upwards and a negatively charged particle moves downwards. If P_1 and P_2 are externally connected through a resistance, a current will flow and mechanical energy (kinetic energy) of gas is converted into electrical energy. If an electrical field of strength E is also applied, the total force on the particle,

$$\vec{F} = q\vec{E} + q(\vec{v}' \times \vec{B})$$

$$= q(\vec{E} + \vec{v}' \times \vec{B})$$

where E is the e.m.f applied opposite to direction of current.

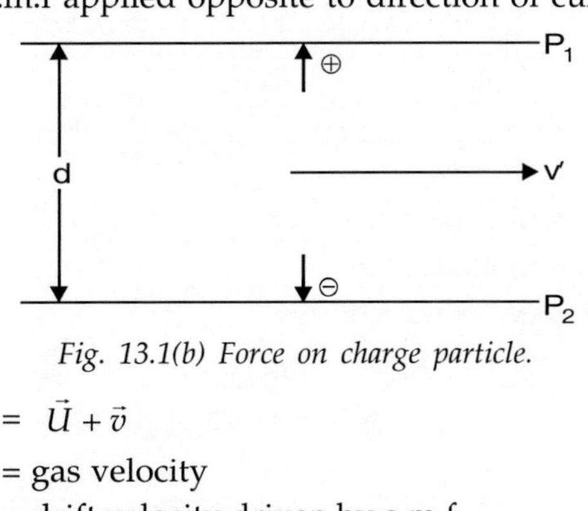

Fig. 13.1(b) Force on charge particle.

Now $\quad v' = \vec{U} + \vec{v}$

where $\quad u$ = gas velocity

v = drift velocity driven by e.m.f

$$\therefore \quad \vec{F} = q\vec{E} + (q\vec{U} \times \vec{B}) + (q\vec{v} \times \vec{B})$$

$$= q(\vec{E} + \vec{U} \times \vec{B} + \vec{v} \times \vec{B})$$

$$= q\vec{E}' + q(\vec{v} \times \vec{B})$$

where $E' = (E + \vec{U} \times \vec{B})$

The drift velocity is related to the force on the particles through the mobility, μ.

$$\therefore \quad F = \frac{qv}{\mu}$$

$$\vec{v} = \mu\vec{E} + \mu(\vec{u} \times \vec{B}) + \mu(\vec{v} \times \vec{B})$$

This is the general equation in vector form.

If gas flows in the x direction and magnetic field B is applied in the $+Z$ direction

$$U_x = U$$

$$u_y = U_z = 0$$

$$B_y = B$$

$$B_x = B_z = 0$$

Put $\mu B = \beta$

$$\therefore \quad \left.\begin{aligned} v_x &= \mu E_x - \mu v_z \\ v_y &= \mu E_y \\ v_z &= \mu E_z + \beta u + \beta v_x \end{aligned}\right\}$$

These are the scalar equations and can be solved for v_x and v_y. There will be no force acting in y direction.

$$\therefore \quad v_x = \frac{\mu(E_x - \beta E_z) - \beta^2 \mu}{1 + \beta^2}$$

$$v_y = \frac{\mu(E + \beta E_x) - \beta U}{1 + \beta^2}$$

The current density,

$$J = \frac{\sigma v}{\mu}$$

where v = drift velocity

μ = mobility

σ = gas conductivity

$$\therefore \quad J_x = \frac{\sigma(E_x - \beta E_z) - \sigma\beta B U}{1 + \beta^2}$$

$$J_z = \frac{\sigma(E_z + \beta E_x) + \sigma\beta U}{1 + \beta^2}$$

These are the basic equations for the analysis of MHD generator.

Both J_x and J_z can be extracted. In a Faraday MHD generator the current J_z is extracted and in a Hall MHD generator J_x is extracted. The flow of current in the x direction in called Hall effect.

13.3. DESIGN PROBLEMS

13.3.1. Gas Velocity

The power is proportional to square of gas velocity, the latter should be as high as possible. This can be achieved by converting thermal energy of gas emerging from the furnace into directed kinetic energy in a nozzle. This kinetic energy and thermal energy are then converted into electrical energy by the MHD process.

The conversion of thermal energy into directed kinetic energy can be brought about by conventional nozzle-expansion techniques and velocities of the order of 10^3 m/s can be obtained. For seeded inlet gas, the duct is designed to give a constant optimum Mack Number of 0.5 which ensures that maximum density of power is extracted at each section of the duct. The temperature, pressure and velocity will fall in such a way as to keep the optimum Mack Number constant. However, as the mass flow rate through the duct must remain constant, the area of the duct will vary with velocity. The area of the duct will increase and velocity will decrease with the gas flow.

13.3.2. Magnetic Flux Density

The power is proportional to magnetic flux density (B) which should be as strong as possible. This can be achieved by passing heavy currents though the coils. The energy dissipated in the magnet is proportional to the length of the duct but is independent of its cross-sectional area. It is also proportional to the resistivity of the winding material. The capital cost of the magnet is proportional to the bulk cost of the winding material.

Magnets constructed with copper windings on a soft core can carry flux density upto 3 Wb/m^2. Water cooling must be used to remove the heat generated in the windings by Joule effect. For a 100 MW MHD plant, heat dissipated is about 12 MW.

The resistivity of certain metals falls very fast with the decrease of temperature. The decrease in the power needed to generate magnetic field is much greater than additional power needed for the refrigeration equipment. This decrease in running cost is produced at the expense of capital cost of the refrigeration equipment. Magnets made of superconducting materials (NbZr, Nb_3Sb) can reduce power dissipation in the windings to zero but a refrigeration plant would be required to keep the winding at liquid-helium temperature.

13.3.3. Gas Conductivity

It is a difficult task to attain reasonable (10 to 100 mhos/m) conductivity in a gas. The gas must be made a conductor by ionising some fraction of it. Thus the gas in the MHD generator can be called plasma, *i.e.*, an ionized gas in which positive ions and electrons are present in equal number. In order to attain a reasonably high temperature plasma, it is necessary to have fuel gas with a calorific value of atleast 8500 kJ/Nm^3. At 3000°C, a small degree of ionisation equal to 0.1% is observed which gives a conductivity of 50%. By increasing the degree of ionisation to 1%, the conductivity can be raised to 80%. Hence, high degree of ionisation is not needed for practical purposes. However, it is difficult to find suitable structural materials for combustion chamber, ducts and nozzles to withstand flow of ionised gases at 3000°C.

One way of increasing the gas conductivity without the need of exceedingly high temperatures is to introduce seeding agent which will ionise more readily than the gas itself and will enhance the electrical conductivity. The best seeding material is metal cesium which has ionisation potential of 3.89 eV. Cesium is very expensive and the next seeding material being used is potassium with ionisation potential of 4.34 eV. In this case conductivity of 10 to 100 mhos/m can be achieved at temperatures of 2000°C to 2400°C, which are within the range of material technology. However, material selection has to take care of enhanced corrosiveness of plasma due to presence of seeding material.

13.3.4. MHD Duct

The main difficulty encountered in the design of MHD power generator is finding suitable material of construction for the duct to handle highly corrosive and high temperature gases and seeding materials. The development of a long-lived duct is one of the prime requirement for MHD generator. Ceramics were initially used. If uncooled walls are cooled, severe ablation and erosion takes place and extremely high thermal stresses are set up which leads to rapid disintegration of material. Temperature gradients of 10,000°C/cm may be obtained.

A more promising mode of construction is the use of water-cooled metal walls. A ceramic material, such as zinconia or alumina is sand-witched between water-cooled metal segments. This makes the wall electrically non-conducting. This results in large pressure drops due to rough surfaces on gas flow side. Water cooled peg wall or modular construction have been used successfully. But smooth walls of ceramic construction gives better flow conditions and greater efficiency. The most important insulating materials are : magnesia, alumina, calcium zirconate and strontium zirconate. For electrodes, graphite and water-cooled copper electrodes have been used.

13.4. THERMODYNAMIC PERFORMANCE ANALYSIS

A schematic diagram of MHD power cycle alongwith its T-s diagram are shown in Fig. 13.2. and Fig. 13.3. respectively.

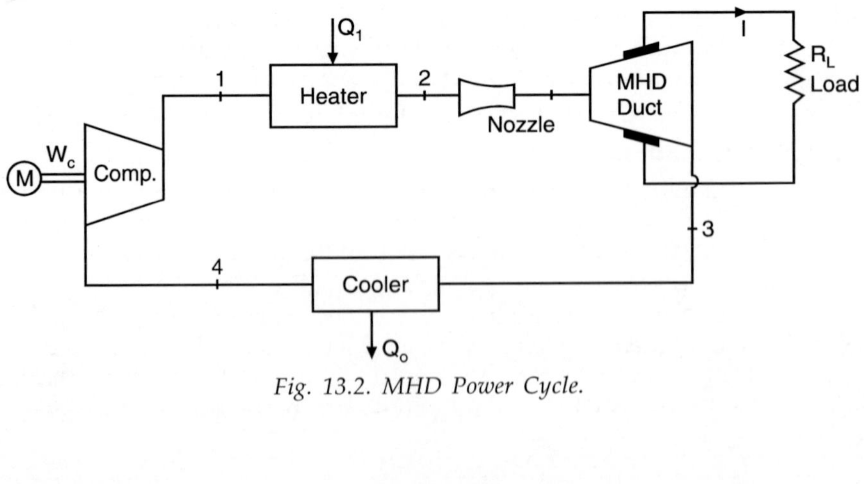

Fig. 13.2. MHD Power Cycle.

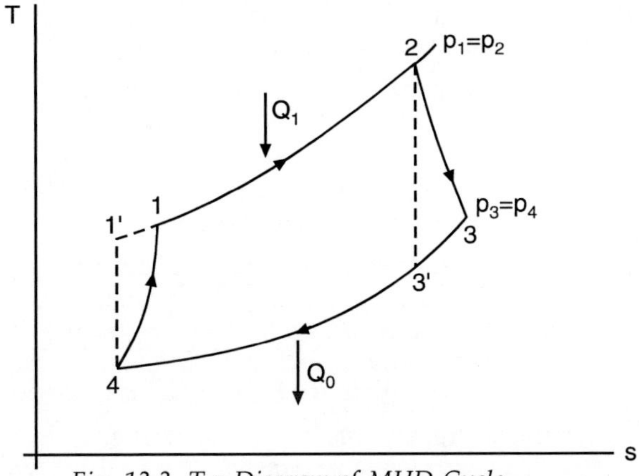

Fig. 13.3. T-s Diagram of MHD Cycle.

The total current,

$$I = \sigma(E + U \times B)$$

where
σ = Gas conductivity [mhos/m]
E = Applied electrical field [V]
U = Gas velocity [m/s]
B = Magnetic field density [W/m^2]

The work output = $W_{MHD} - W_{Comp}$

Heat input = Q_1

$$W_{MHD} = h_2^* - h_3^*$$

$$W_{Comp} = h_1^* - h_4^*$$

$$Q_1 = h_2^* - h_1^*$$

where * denotes enthalpy at stagnation values which take into account kinetic energy of flow.

$$h^* = h + \frac{U^2}{2}$$

The thermal efficiency,

$$\eta_{th} = \frac{\left(h_2^* - h_3^*\right) - \left(h_1^* - h_3^*\right)}{\left(h_2^* - h_1^*\right)}$$

The following factors can adversely affect the thermal efficiency

1. Dissipation of energy in the internal resistance of ionised gas.
2. A space charge barrier at the electrode surface.
3. Heat loss through the electrode and insulator walls.
4. Losses associated with fluid friction
5. Hall effect losses due to current induction in the direction of flow of gas.

13.5. ELECTRICAL ANALYSIS

When a current I flows across load resistance R_L with voltage V across the load, the electrical intensity across the electrode plates will be

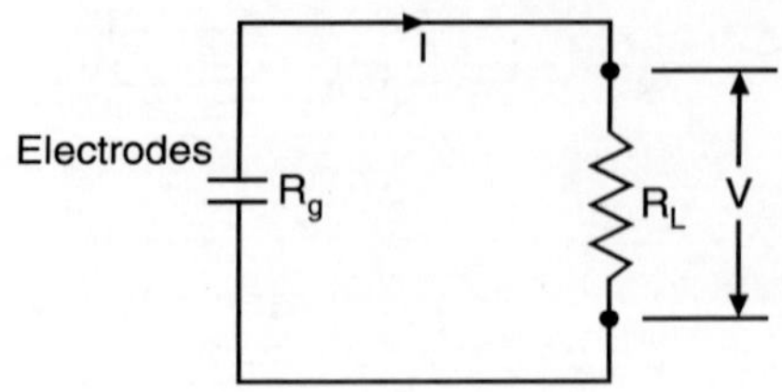

Fig. 13.4. Electrical Circuit of MHD Generator.

$$E = -\frac{V}{d}$$

where d = distance between the electrodes

The open circuit voltage,

$$E_o = B\,U\,d = IR_g$$

R_g is the internal resistance of the generator

$$R_g = \frac{d}{\sigma A}$$

where σ = gas conductivity (mhos/m)

A = electrode plate area [m^2]

Maximum power output is obtained when

$$R_g = R_L$$

$$W_{max} = E_o I = (I\,R_g)\,I = I^2\,R_g$$

$$= \left[\frac{E_o}{R_L + R_g}\right]^2 R_g$$

$$= \frac{E_o^2}{4\,R_g} = \frac{B^2U^2d^2}{4R_g} \qquad [\because E_o = B\,U\,d]$$

Now $$R_g = \frac{d}{\sigma A}$$

$$\therefore \quad W_{max} = \frac{B^2U^2d^2}{4d}\sigma A = \frac{1}{4}\sigma u^2 B^2\,dA$$

Maximum power per unit volume,

$$\frac{W_{max}}{dA} = \frac{1}{4}\sigma U^2 B^2$$

Example 13.1. *Calculate the open circuit voltage and maximum power output for the following MHD generator :*

Plate area	= 0.25 m^2
Distance between plates	= 0.50 *m*
Magnetic flux density	= 2 Wb/m^2
Average gas velocity	= 10^3 *m/s*
Gas conductivity	= 10 *mhos/m*

Solution : Open circuit voltage,

$$E_o = B\,U\,d$$
$$= 2 \times 1000 \times 0.50 = 1000\text{V}$$

Generator resistance,

$$R_g = \frac{d}{\sigma A} = \frac{0.50}{10 \times 0.25} = 0.20\Omega$$

Maximum power,

$$W_{max} = \frac{E_o^2}{4R_g} = \frac{(1000)^2}{4 \times 0.20}$$
$$= 1250 \times 10^3 \text{ W} = 1250 \text{ kW}$$

13.6. MHD GENERATOR EFFICIENCY

Possible arrangements in MHD generator are shown in Fig. 13.5.

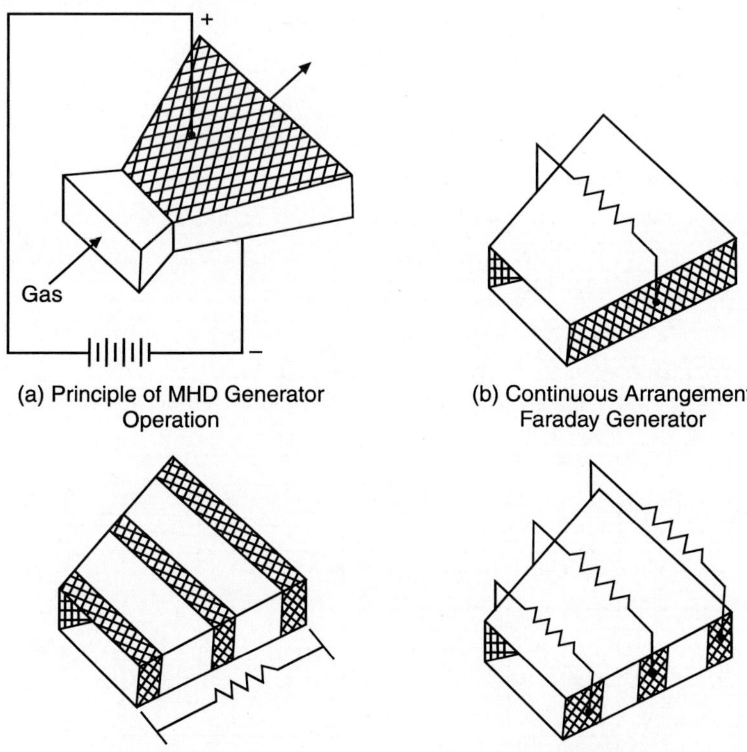

Fig. 13.5. Electrode Arrangements.

The current generated in a Faraday generator at the maximum power output is given by

$$I_{max} = UB = UB\frac{A}{\rho} \text{ [A]}$$

where U = flow velocity [m/s]

B = magnetic field density [Wb/m^2]
σ = gas conductivity [mhos/m]
A = electrode surface area [m^2]
ρ = specific resistance of ionised gas [Ωm]

The voltage at the maximum power output,

$$V_{max} = d\left(U.B - I_{\max}\frac{\rho}{A}\right) \text{ [V]}$$

where d = distance between the electrodes [m]

The maximum power output,

$$W_{max} = V_{max}\, I_{max} \text{ [W]}$$

The efficiency of energy conversion in MHD generator is defined as the ratio of useful electrical power output and the power required to transport the plasma through the MHD duct.

$$\eta_{MHD} = \frac{W_{net}}{W_{in}}$$

The efficiency of MHD generator can also be expressed as

$$\eta_{MHD} = \frac{V_{\max}}{(U.B)}$$

This can be approximated to

$$\eta_{MHD} = \frac{R_L}{R_L + R_g}$$

where R_L = external load [Ω]
R_g = internal resistance of ionized gas [Ω]

The variation of η_{max} with L/h is shown in Fig. 13.6, where L is the length of duct and h is the height of duct. η_{max} of 0.5 can be achieved with optimum design.

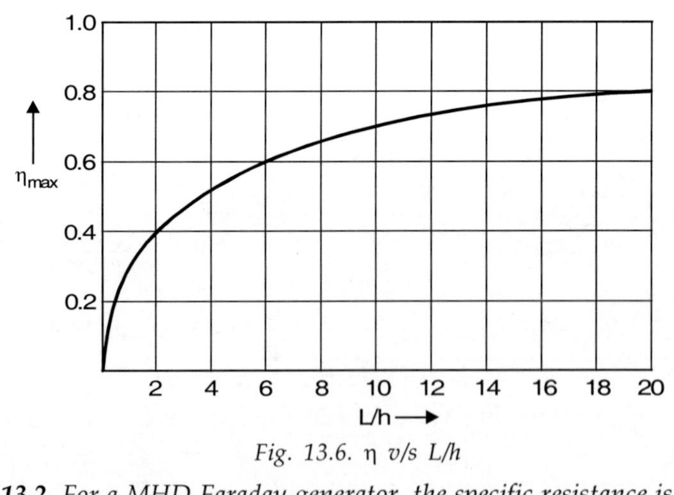

Fig. 13.6. η v/s L/h

Example 13.2. *For a MHD Faraday generator, the specific resistance is 0.08 Ωm and magnetic field density is 3.8 Wb/m^2. The electrode area is 1.2 m^2, plasma velocity is 950 m/s and the distance between the electrodes is 0.9m. Determine the net power output and conversion efficiency.*

Solution : The open circuit voltage,

$$E_o = B\,U\,d$$
$$= 3.8 \times 950 \times 0.9 = 3249\text{V}$$

The short circuit current,

$$I_{SC} = \frac{AUB}{\rho}$$
$$= \frac{1.2 \times 950 \times 3.8}{0.08} = 54150 \text{ A.}$$

The current for maximum power output,

$$I_{max} = \frac{UBA}{2\rho} = \frac{950 \times 3.8 \times 1.2}{2 \times 0.8} = 27075 \text{ A}$$

$$V_{max} = UBd - \left(I_{\max}\,\rho\frac{d}{A}\right)$$
$$= 950 \times 3.8 \times 0.9 - \frac{27075 \times 0.08}{1.2} = 1624.5 \text{ V}$$

The maximum power output,

$$W_{max} = V_{max}\,I_{max}$$
$$= 1624.5 \times 27075 = 44 \text{ MW}$$

The conversion efficiency,

$$\eta_{MHD} = \frac{V_{\max}}{UB} = \frac{1624.5}{950 \times 3.8} = 0.45 = 45\%$$

13.7. OPEN CYCLE MHD POWER GENERATION SYSTEM

An open cycle MHD power generation system is shown in Fig. 13.7. The system consists of the following units :

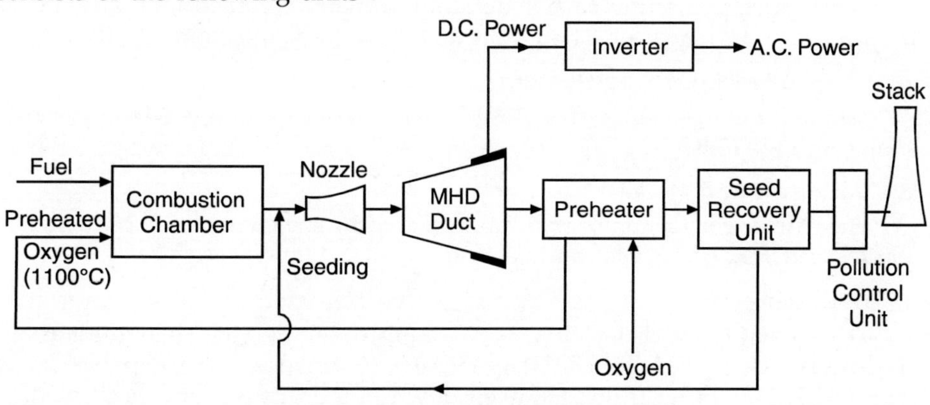

Fig. 13.7. Open cycle MHD Generator

1. Combustion Chamber

The fuel (coal, oil or natural gas) is burnt with preheated oxygen (or air) at 1100°C. The hot, pressurised working fluid at 2300°C to 2700°C is seeded with potassium carbonate (or cesium) to ionize the gas.

2. Nozzle

A convergent-divergent nozzle is used to increase the velocity to 10^3 m/s to get directed mass motion energy.

3. MHD Duct

It is a divergent channel made of heat resistant material externally water cooled. The magnetic field acts perpendicular to the direction of gas motion. The electrode pair may by connected in different ways as per Fig. 13.5 to reduce losses. The d.c power produced is converted into ac power with the help of an inverter.

4. Preheater

The gas at 1900°C enters the gas preheater where oxygen or oxygen enriched air or compressed air is heated to a temperature of 1100°C. The preheated gas helps to produce working fluid at 2300°C to 2700°C.

5. Seed Recovery Unit

The seed material is recovered for successive use for seeding of hot working fluid in the combustion chamber. The original potassium carbonate seed is converted into potassium sulphate due to presence of sulphur in the fuel. The potassium sulphate is converted back to potassium carbonate chemically in the seed recovery unit.

6. Hot Gas

The hot gases are passed through pollution control device to remove sulphur and nitrogen oxides before exhausting through a chimney.

13.8. CLOSED CYCLE MHD POWER GENERATION SYSTEM

In a closed system, helium or argon is used as working fluid which is heated in a heat exchanger. Higher temperature and better thermal efficiency are possible. However, seeding is required to attain reasonable gas conductivity at temperatures workable with available structural materials. Instead of seeding with cesium or potassium carbonate a liquid metal is mixed with an inert gas to form the working fluid. The liquid metal provides the conductivity.

13.8.1. Seeded Inert Gas System

The main components of a closed cycle seeded inert gas MHD system is shown in Fig. 13.8.

1. Combustor and Heater

The carrier gas (argon or helium) is heated by the combustion of fuel gas to 1900°C and seeded by cesium injection.

2. MHD Generator

The seeded hot working fluid is passed through the MHD generator at high speeds. The dc power from MHD generator is converted to a.c. power by the inverter. The working fluid is slowed down in the diffuser and precooled.

3. Compressor

The precooled gas is compressed for heating.

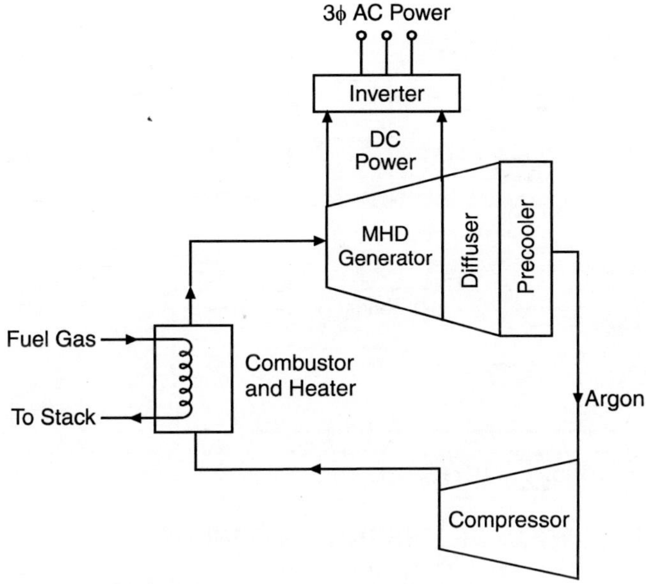

Fig. 13.8 Closed Cycle MHD System.

13.8.2. Liquid Metal System

The carrier gas (argon or helium) is pressurised and heated in a heat exchanger within the combustion chamber. The hot gas is incorporated into the liquid metal (hot sodium) to form the working fluid.

1. MHD Generator

The working fluid consisting of gas bubbles uniformly dispersed in an equal volume of liquid sodium is passed through the MHD generator with high directed velocity.

2. Breeder Reactor

The exhaust from MHD duct is passed through a condenser where potassium liquid is formed and pumped to breeder reactor. The liquid potassium is heated in vapour form and accelerated through a nozzle.

3. Separator

The vapours are separated, condensed and pumped to breeder reactor.

The schematic diagram is shown in Fig. 13.9.

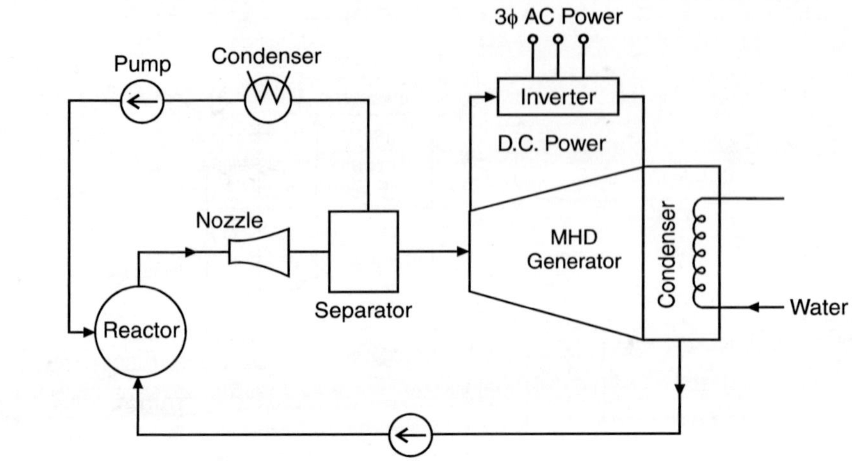

Fig. 13.9. Liquid Metal Closed Cycle MHD System.

13.9. HYBRIDISATION OF MHD POWER GENERATOR

The overall energy utilization can be improved by employing combined-cycle power plant consisting of MHD generator as a topping plant and a gas or steam turbine as a bottoming plant. The overall efficiency of about 60% can be achieved in the combined cycle. The schematic diagram is shown in Fig. 13.10.

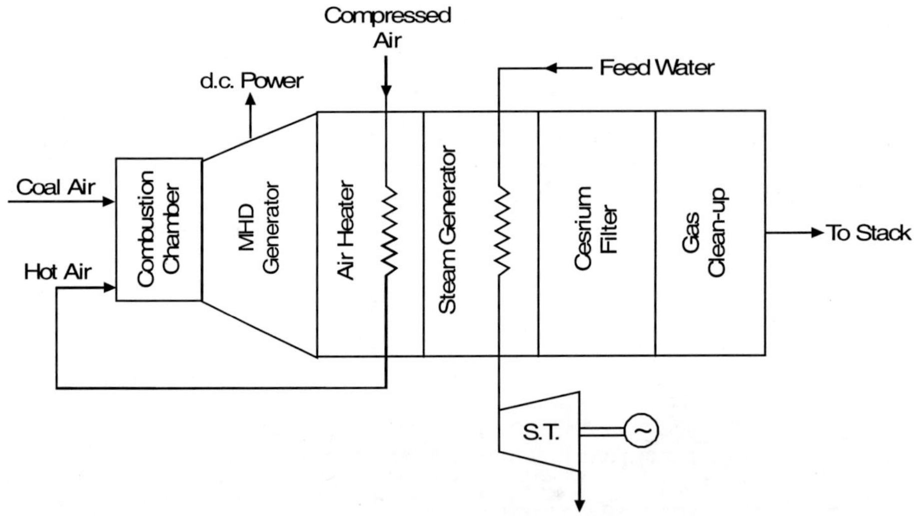

Fig. 13.10. MHD Generator-Steam Turbine Combined Cycle Plant.

If the gas entering the MHD duct at about 3000°C could be expanded to the ambient temperature of 30°C, the Carnot efficienty would have reached 90%. Unfortunately, the MHD power output is restricted because by the time the gas temperature falls to 2000°C, the electrical conductivity becomes very low with the electrons combining with ions to form neutral atoms, and the generator then ceases to operate satisfactorily. Therefore, the MHD generator is used as a topping plant and the MHD exhaust at about 2000°C is utilized in raising steam to drive turbine and generate electricity in a conventional steam power plant. If the fraction Z of the fuel energy is directly converted to electricity in the MHD generator, the remainder (1 – Z) is converted with an efficiency η' in the bottoming steam plant so that overall efficiency is

$$\eta = Z + \eta' (1 - Z)$$

If Z = 0.3 and η′ = 0.4, then η = 0.58 which is a good power plant efficiency. MHD-topped steam plants can operate either in an open cycle or in a closed cycle. A gas turbine plant can also be used as a bottoming unit. Since the combined cycle plant operates over a larger temperature difference, the efficiency will obviously be higher.

Example 13.3. *Determine the overall efficiency of a combined cycle power plant which consists of a MHD-generator with an efficiency of 40% and a steam power plant with an efficiency of 38%.*

Solution :

$$\eta = \eta_{MHD} + \eta_{ST} - \eta_{MHD} \times \eta_{ST}$$
$$= 0.4 + 0.38 - 0.4 \times 0.38 = 0.628$$
$$= 62.8\%$$

13.10. ADVANTAGES OF MHD POWER GENERATOR

1. The conversion efficiency of MHD generator in combined cycle plants can be 60–65%
2. Large amount of power can be generated.
3. It is very reliable and robust without moving parts.
4. Pollution-free power can be generated in a closed cycle system.
5. The plant can be operated on full capacity in a very short time.
6. The plant is very compact.
7. Capital costs can be lower than conventional power plants.
8. The operating costs will be lower.
9. It is a direct conversion device eliminating very large plants which reduces loss of energy and enhances reliability of operation.
10. The higher energy utilization helps savings in fuel and pollution.

13.11. INDIAN EXPERIENCE

Many countries world over are in the development of MHD technology. The specifications of a Japanese pilot plant (Tokyo) are given below :

Thermal Input :	24 MW
Electrical power :	1 MW
Mass flow of gas :	2.8 kg/s
Flow velocity :	900 m/s
Gas temperature :	2900 K
Duct Length :	1.2 m
Inlet :	8 × 10 cm
Outlet :	8 × 25 cm
Electrodes :	30 pairs
Inlet pressure :	4 bar
Magnetic flux density :	3.5 Wb/m2
Material for electrodes :	Graphite, water cooled copper

Russia has been the pioneer of this new technology. The world's first U-O2 MHD unit was designed by the Russian scientists and power engineers in 1964.

Under intergovernmental agreement, Indian scientists from BARC and power engineers from BHEL worked on a pilot plant of MHD of 5 MW capacity with close interaction with the scientists of the High Temperature Science Institute of Moscow. The pilot plant was set up at BHEL, Trichy. BARC had developed special magnet, MHD duct and power tap-off system. BHEL was responsible for the development and installation of fuel gas generator, oxygen plant, water treatment plant, high temperature regeneration air heaters and valves, the main combustion chamber, nozzles as well as seed injection and seed recovery system.

The main specifications of the pilot plant are :

Thermal input	I stage	: 5 MW
	II stage	: 15 MW
Gas temperature		: 2600°C
Flux density		: 3 Tesla
Seeding agent		: Potassium carbonate
Electrodes		: Water cooled copper

13.12. LIMITATIONS

The main limitations hindering the commercial application of MHD technology are :

1. The efficiencies attained so far have been relatively low.
2. The power output of MHD generator is proportional to the square of the magnetic field density. The electro-magnets require very large power to create strong magnetic fields. The MHD technology is waiting for development of superconductor materials which will require very little power even at ambient temperatures.
3. The combustor, MHD duct, electrodes and air preheaters are exposed to very corrosive combustion gases at very high temperatures. The life of these equipments has been very short.
4. The ash (or slag) residue from the burning coal is carried over with the combustion gases and tends to cause erosion of exposed surfaces. However, deposition of the slag on such surfaces may provide some protection.
5. There is a serious problem of separation of seed material from the flyash and reconversion of potassium sulphate to potassium carbonate.
6. Special fuel gas and preheating of air are needed to provide adequate working fluid temperatures.
7. There are serious problems associated with the fabrication of MHD duct, high temperature and high pressure heat exchangers and reactors.

TUTORIAL SHEET NO. 13.1.

1. Discuss the principle and working of MHD generation.
2. Derive equations for the power output and efficiency of an MHD power generator.
3. How are MHD systems classified ? Describe them in brief.
4. What are the advantages of MHD technology ?
5. What are the limitations of MHD conversion technology,.
6. Write short notes on :
   (*i*) Gas velocity
   (*ii*) Gas conductivity
   (*iii*) Magnetic flux density
   (*iv*) MHD duct.
7. Give names of materials for electrodes, duct and seeding materials.
8. What is hybridization of MHD power generation ?
9. Derive equation for thermal efficiency of MHD power cycle. What factors can affect the thermal efficiency ?

❑❑❑

Chapter **14**

THERMOELECTRIC POWER GENERATION

Introduction; Principle of Operation; Performance Analysis; Cascade Multistage Operation; Selection of Materials: Applications, Limitations.

14.1. INTRODUCTION

A loop of two dissimilar metals develops an e.m.f when the two junctions of the loop are kept at different temperatures. This is called Seebeck effect. This effect is used in a thermocouple to measure temperature.

Thermoelectric generator is a device which directly converts heat energy into electrical energy using the Seebeck thermoelectric effect. The device is very simple but thermal efficiency is very low of the order of 3%. Efficiency of thermoelectric generator depends upon the temperature of hot and cold junctions and the material properties. The semiconductor materials have more favourable properties which can withstand high temperatures and can give reasonable efficiency. The probability of developing peak load power stations of the order of 100 MW working at 20 percent thermal efficiency is high. Where cheap fuels are available thermoelectric generators can be developed for base load and standby power generation also. Another important application is the use of radioactive decay heat to generate power in space and other remote locations. The use of solar energy to supply heat for generating electricity can be an attractive application of thermoelectric devices if high efficiency materials can be developed.

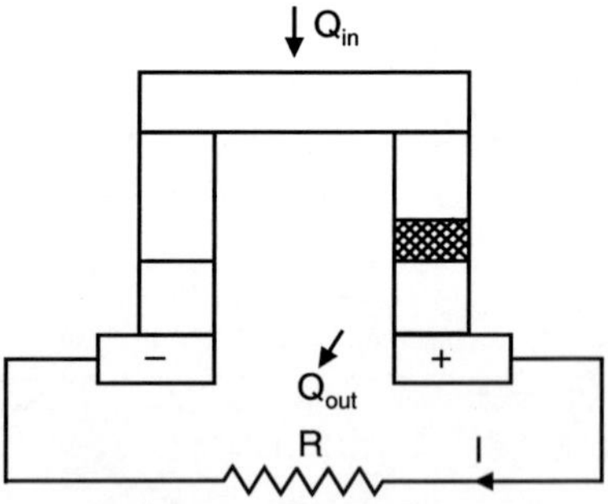

Fig. 14.1. Thermoelectric Generator.

The operation of a thermoelectric generator is shown in fig 14.1. The net useful power output is given by

$$W = I^2 R \text{ [W]}$$

where I = current [A]

R = External load resistance [Ω]

The current in the circuit is given by

$$I = \frac{\alpha \Delta T}{(R_i + R)} \text{ [A]}$$

where α = Seebeck coefficient [V/K)

ΔT = Temperature difference between hot and cold junctions [K]

R_i = Internal resistance of thermoelectric generator [Ω]

The magnitude of potential difference depends on the pair of conductor materials and on the temperature difference between the junction.

For a loop made of copper and constant an wires, the value of Seebeck coefficient α is 0.04 mV/K. For a temperature difference of 600K between the junction, a voltage of 24 mV will be developed. In order to achieve higher potential difference many generators have to be connected in parallel. For increasing the useful power output, parallel and series connections are used.

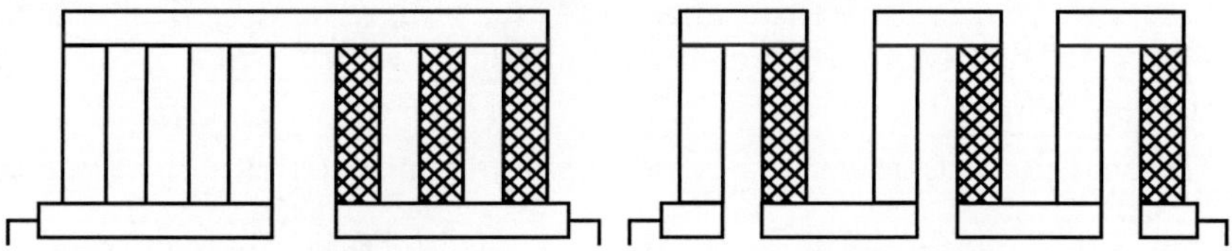

Fig. 14.2. Cascading of Thermoelectric Generators.

14.2. PRINCIPLE OF OPERATION AND PERFORMANCE ANALYSIS

The thermo-elements of a thermoelectric generator are made up of semiconductors *p* and *n* type. Heat is supplied to the hot junction and from the cold junction heat is removed. Both the junctions are made of copper, see fig. 14.3.

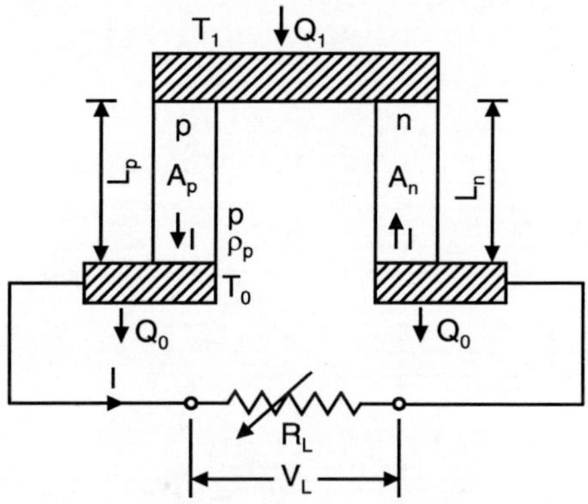

14.3. Circuit Diagram of Thermoelectric Power Generator.

Let

T_1 = source temperature [K]

T_0 = sink temperature [K]

L_p, L_n = Length of semiconductor elements [m]

A_p, A_n = cross-sectional area of thermoelectric elements [m^2]

k_p , k_n = thermal conductivity of elements [W/mK]

ρ_p , ρ_n = electric resistivity of elements [Ω-m]

k_p , k_n = thermal conductivity of elements [W/K] $= \frac{kA}{L}$

R_p, R_n = electrical resistance of elements [Ω] $= \frac{\rho L}{A}$

α_p , α_n = Seebeck coefficient [V/K]
π_p , π_n = Peltier coefficient [V]

Seebeck coefficient, $\alpha_{p,\,n} = \underset{\Delta T \to 0}{Lt} \dfrac{\Delta V}{\Delta T}$

Peltier heat, $Q_{p,n} = \pi_{p,n} I$

When a current (I) flows through the junction of two elements, Peltier heat is produced. This is called Peltier effect. The Peltier coefficient.

$$\pi_{p,n} = \alpha_{p,\,n} \cdot T$$

From I st Law of Thermodynamics as applied to upper plate (as control volume), the temperature difference ($T_1 - T_0$) will generate a Seebeck voltage, $\alpha_{pn}(T_1 - T_0)$. There will be an electrical current I which will flow through the external load R_L.

The heat $\dot{Q}_1$, will flow into hot junction and conducted into the two legs, $\dot{Q}_k$.

The Peltier heat $\dot{Q}_p$ will be produced at the junction due to current flowing through the circuit.

Joule heat $\dfrac{\dot{Q}_j}{2}$ will flow into the junction. It is assumed that half Joulean heat appears at each junction. Heat balance at the junction will give,

$$\dot{Q}_1 + \frac{1}{2}\dot{Q}_j = \dot{Q}_p + \dot{Q}_k \quad ...(i)$$

where

$$\dot{Q}_p = \pi_{p,n} I = \alpha_{p,n} I.T_1$$

$$\dot{Q}_j = I^2R = I^2(R_p + R_n)$$

$$= I^2\left[\frac{\rho_p L_p}{A_p} + \frac{\rho_n L_n}{A_n}\right]$$

$$\dot{Q}_k = k\Delta T = (k_p + k_n)(T_1 - T_0)$$

$$= \left(\frac{k_p A_p}{L_p} + \frac{k_n A_n}{L_n}\right)(T_1 - T_0)$$

Substituting the values into equation (i)

$$\dot{Q}_1 = \alpha_{p,n} I T_1 - \frac{1}{2}\left[\frac{\rho_p L_p}{A_p} + \frac{\rho_n L_n}{A_n}\right] I^2 + \left[\frac{k_p A_p}{L_p} + \frac{k_n A_n}{L_n}\right](T_1 - T_2)$$

The useful power generated,

$$W_L = I^2 R_L = \frac{V_L^2}{R_L}$$

The voltage across the load,

$$V_L = \alpha_{p,n}(T_1 - T_0) - I(R_p - R_n)$$

By Kirchhoff's Law

$$I = \frac{\alpha_{p,n}(T_1 - T_0)}{R_L + R_p + R_n}$$

Let $$m = \frac{R_L}{R_p + R_n} = \text{Resistance ratio.}$$

$\therefore$ $$m + 1 = \frac{R_p + R_n + R_L}{R_p + R_n}$$

Now $$I = \frac{\alpha_{p,n}(T_1 - T_0)}{(R_p + R_n)(1 + m)}$$

$$W_L = \frac{\alpha_{p,n}^2 (T_1 - T_0)^2}{(R_p + R_n)^2 \; (1 + m)^2} (R_p + R_n)m$$

$$= \frac{m}{(1+m)^2} \frac{\alpha_{p,n}^2 (T_1 - T_0)^2}{R_p + R_n}$$

The required heat input,

$$\dot{Q}_1 = \alpha_{p,n}^2 \frac{T_1(T_1 - T_0)}{(R_p + R_n)(1 + m)} - \frac{1}{2} \frac{\alpha_{p,n}^2 (T_1 - T_0)^2}{(1+m)^2 (R_p + R_n)} + (R_p + k_n)(T_1 - T_0)$$

The efficiency of thermoelectric generator,

$$\eta = \frac{W_L}{\dot{Q}_1} = \frac{\dfrac{m}{(1+m)^2} \cdot \dfrac{\alpha_{p,n}^2}{R_p + R_n} (T_1 - T_0)^2}{\alpha_{p,n}^2 \dfrac{T_1(T_1 - T_0)}{(R_p + R_p)(1 + m)} - \dfrac{1}{2} \dfrac{(T_1 - T_0)^2 \alpha_{p,n}^2}{(1+m)^2 (R_p + R_n)} + (k_p + k_n)(T_1 - T_0)}$$

$$= \frac{T_1 - T_0}{T_1} \cdot \frac{m}{(1+m) \dfrac{T_1 - T_0}{T_1} - \dfrac{1}{2} \dfrac{T_1 - T_0}{T_1} + \dfrac{(k_p + k_n)(R_n + R_p)(1+m)^2}{\alpha_{p,n}^2 T_1}}$$

Let the figure of merit,

$$Z = \frac{\alpha_{pn}^2}{(k_n + k_p)(R_n + R_p)}$$

The figure of merit as defined above consists of the material properties of the two semiconductors. The η will increase with increase of Z.

Let $R = R_n + R_p$

and $K = k_n + k_p$

When the product (R, K) is minimum Z is maximum and therefore η is maximum.

$$(R, K) = \left(\frac{\rho_p L_p}{A_p} + \frac{\rho_n L_n}{A_n}\right)\left(\frac{k_p A_p}{L_p} + \frac{k_n A_n}{L_n}\right)$$

Let $$\gamma_p = \frac{A_p}{L_p}$$

and $$\gamma_n = \frac{A_n}{L_n}$$

$\therefore$ $$(R, K) = \left(\frac{\rho_p}{\gamma_p} + \frac{\rho_n}{\gamma_n}\right)(k_p\gamma_p + k_n\gamma_n)$$

$$= k_p\rho_p + k_n\rho_n \frac{\gamma_p}{\gamma_n} + k_n\rho_n \frac{\gamma_n}{\gamma_p} + k_p\rho_n$$

When materials are fixed, the values of γ_n, γ_p, $\frac{A_p}{L_p}, \frac{A_n}{L_n}$ are fixed. In order to minimize the product (R, K),

$$\frac{d(R,k)}{d\left(\frac{\gamma_n}{\gamma_p}\right)} = 0 = k_n\, \rho_p - k_p\, \rho_n \left(\frac{\gamma_n}{\gamma_p}\right)^{-2}$$

$\therefore$ $$\frac{\gamma_n}{\gamma_p} = \sqrt{\frac{k_p\, \rho_n}{k_n\, \rho_p}}$$

$\therefore$ $$(R.K.)_{\min} = k_p\, \rho_p + k_p\, \rho_n \left[\frac{k_n\, \rho_p}{k_p\, \rho_n}\right]^{\frac{1}{2}} + k_n\, \rho_p \left[\frac{k_p\, \rho_n}{k_n\, \rho_p}\right]^{\frac{1}{2}} + k_n\rho_n$$

$$= \left|\sqrt{\rho_p\, k_p} + \sqrt{\rho_n\, k_n}\right|^2$$

$$Z_{\max} = \left[\frac{\alpha_{p,n}}{\sqrt{\rho_p\, k_p} + \sqrt{\rho_n k_n}}\right]^2$$

$$\eta = \frac{T_1 - T_0}{T_1} \cdot \frac{m}{(1+m) - \frac{1}{2}\frac{T_1 - T_0}{T_1} + \frac{(1+m)^2}{ZT_1}}$$

For a given pair of materials with Z maximum and T_1, T_0 are fixed, η will depend upon optimum value of m.

$$\frac{d\eta}{dm} = 0$$

$$m_{\text{opt}} = \left[1 + Z\frac{T_1 + T_0}{2}\right]^{\frac{1}{2}} = m_0$$

$$m_0 + 1 = \frac{Z}{m_0 - 1} \cdot \frac{T_1 + T_0}{2}$$

$$\therefore \qquad \eta = \frac{T_1 + T_0}{T_1} \cdot \frac{m_0 Z T_1}{(1 + m_0)^2} \left[\frac{1}{1 - \dfrac{1}{2}\dfrac{ZT_1}{(1 + m_0)^2}\dfrac{T_1 + T_0}{T_1} + \dfrac{ZT_1(1 + m_0)}{(1 + m_0)^2}}\right]$$

$$= \frac{(T_1 - T_0)\, m_0 Z}{(m_0 + 1)^2 + (m_0 + 1) Z T_1 - \dfrac{1}{2} Z(T_1 - T_0)}$$

$$= \frac{(T_1 - T_0)\, m_0 Z}{\dfrac{m_0 + 1}{m_0 - 1} \times \dfrac{Z}{2}(T_1 + T_0) + (m_0 + 1) Z T_1 - \dfrac{1}{2} Z(T_1 - T_0)}$$

$$= \frac{(T_1 - T_0)\, m_0\, (m_0 - 1)}{(m_0 + 1)(T_1 + T_0) + 2(m_0^2 - 1) T_1 - (T_1 - T_0)(m_0 - 1)}$$

$$= \frac{(T_1 - T_0)\, 2m_0\, (m_0 - 1)}{2m_0\, T_1 \left[m_0 + \dfrac{T_0}{1}\right]}$$

$$\eta = \frac{T_1 - T_0}{T_1} \cdot \frac{(m_0 - 1)}{m_0 + \dfrac{T_0}{T_1}}$$

$$\dot{W}_L = \frac{m}{(1 + m)^2} \cdot \frac{\alpha_{p,n}^2 (T_1 - T_0)^2}{R_p + R_n}$$

For a given pair of elements,

$$\frac{d\dot{W}_L}{d_m} = 0 \text{ which gives } m = 1.$$

$$\therefore \qquad \dot{W}_L = \frac{1}{4} \frac{\alpha_{p,n}^2 (T_1 - T_0)^2}{R_p + R_n}$$

$\therefore$ Maximum efficiency corresponding to max power,

$$\eta_{\text{max. power}} = \frac{T_1 - T_0}{T_1} \cdot \frac{1}{2 - \dfrac{1}{2}\dfrac{T_1 - T_0}{T_1} + \dfrac{4}{ZT_1}}$$

Example 14.1. *For a thermoelectric power generator following parameters are given :*

Temperature of the hot reservoir = 600K

Temperature of the sink = 300K

Figure of merit for the material, z = 2 ×10^{-3} K^{-1}. Determine the efficiency of the thermoelectric generator. What will be the Carnot efficiency ?

Solution :

$$m_0 = \left[1 + \frac{Z}{2}(T_1 + T_0)\right]^{\frac{1}{2}}$$

$$= \left[1 + \frac{2 \times 10^{-3}}{2}(600 + 300)\right]^{\frac{1}{2}} = \sqrt{1.9} = 1.38$$

Efficiency of the thermoelectric generator,

$$\eta = \left(\frac{T_1 - T_0}{T_1}\right)\left[\frac{m_0 - 1}{m_0 + \frac{T_0}{T_1}}\right]$$

$$= \frac{600 - 300}{600}\left[\frac{1.38 - 1}{1.38 + \frac{300}{600}}\right]$$

$$= 0.101 = 10.1\%$$

Carnot efficiency,

$$\eta_c = \frac{T_1 - T_0}{T_1} = \frac{300}{600} \times 100 = 50\%$$

Example 14.2. *The source and sink temperatures of a thermoelectric generator are 923K and 323K respectively. The geometrical parameters are A_n = 2.3 cm^2, L_n = 1.5 cm; A_p = 1.303 cm^2; L_p = 0.653cm. Using the optimum value of the product (R.K), calculate the maximum generator efficiency and efficiency for maximum power. Also calculate power output for both the cases. Use the following material properties.*

Seekbeck coefficients, $\alpha_n = -190 \times 10^{-6}$ V/K

$\alpha_p = 190 \times 10^{-6}$ V/K

specific resistivity $\rho_n = 1.45 \times 10^{-3}$ Ω-cm

$\rho_p = 1.80 \times 10^{-3}$ Ω -cm

Figure of merit $Z_n = 2 \times 10^{-3}$ K^{-1}

$Z_p = 1.7 \times 10^{-3}$ K^{-1}

Solution :

Conductivity,
$$k_n = \frac{\alpha_n^2}{\rho_n Z_n}$$

$$= \frac{(190 \times 10^{-6})^2}{1.45 \times 10^{-3} \times 2 \times 10^{-3}} = 0.0124 \text{ W/cm.K}$$

$$k_p = \frac{\alpha_p^2}{\rho_p Z_p}$$

$$= \frac{(190 \times 10^{-6})^2}{1.8 \times 10^{-3} \times 1.7 \times 10^{-3}} = 0.0118 \text{ W/cm K.}$$

$$Z_{\max} = \frac{(\alpha_n - \alpha_p)^2}{\left[\sqrt{\rho_n k_n} + \sqrt{\rho_p k_p}\right]^2}$$

$$= \frac{(-190 \times 10^{-6} - 190 \times 10^{-6})^2}{\sqrt{1.45 \times 10^{-3}\, 1.24 \times 10^{-2}} + \sqrt{1.88 \times 10^{-3} \times 1.18 \times 10^{-2}}}$$

$$= 1.85 \times 10^{-3} \text{ K}^{-1}$$

Thermal conductance,

$$K = \frac{k_n A_n}{L_n} + \frac{k_p A_p}{L_p}$$

$$= 1.24 \times 10^{-2} \times \frac{2.3}{1.5} + 1.18 \times 10^{-2} \times \frac{1.303}{0.653}$$

$$= 0.0452 \text{ W/K}$$

Resistance of the generator

$$R = R_n + R_p = \frac{\rho_n L_n}{A_n} + \frac{\rho_p A_p}{A_p}$$

$$R = \left[1.45 \times \frac{1.5}{2.3} + 1.8 \times \frac{0.653}{1.383}\right] \times 10^{-3}$$

$$R = 1.840 \cdot 10^{-3}\ \Omega$$

$$m_o = \sqrt{1 + \frac{Z}{2}(T_1 + T_0)}$$

$$m_o = \sqrt{1 + \frac{1.85 \times 10^{-3}}{2}(923 + 323)} = 1.43$$

$$R_L = m_o R = 1.43 \times 1.840 \times 10^{-3}$$

$$R_L = 2.63 \cdot 10^{-3}\ \Omega$$

Maximum efficiency,

$$\eta_{\max} = \frac{T_1 - T_0}{T_1} \cdot \frac{m_0 - 1}{m_0 + \left(\frac{T_0}{T_1}\right)}$$

$$= \frac{600}{923}\left[\frac{1.43 - 1}{1.43 + \left(\frac{323}{923}\right)}\right] = 0.1570 = 15.70\%$$

$$\eta_{max.power} = \frac{\frac{T_1 - T_0}{T_1} . m}{\frac{(1+m)^2}{T_1} \frac{(KR)}{\alpha_{pn}^2} + (1+m) - \frac{1}{2}\left(\frac{T_1 - T_0}{T_1}\right)}$$

For maximum power output , $R_L = R$

$\therefore \quad m = 1$

$$\eta_{max.power} = \frac{\frac{600}{923} \times 1}{\frac{(1+1)^2}{923} \times \frac{1}{Z_{max}} + (1+1) - \frac{1}{2}\frac{600}{923}}$$

$$\eta_{max.power} = \frac{0.65}{\frac{4}{923} \times \frac{1}{1.85 \times 10^{-3}} + 2 - 0.325}$$

$$\eta_{max.power} = 16.18\%$$

Power output, $\dot{W}_L = I^2 R_L$

$$\dot{W}_L = \frac{\alpha_{pn}^2 (T_1 - T_0) R_L}{(R + R_L)^2}$$

$$\dot{W}_L = \frac{\alpha_{p,n}^2 (T_1 - T_0)^2}{(1 + m_0)^2 \; R_L}$$

$$\therefore \quad \dot{W}_L = \frac{(-190 - 190)^2 \times 10^{-12} \times 600^2}{(1 + 1.43)^2 \times 2.63 \times 10^{-3}} = 3.318 \text{ W.}$$

$\dot{W}_{max}$ when $(R_L = R)$

$$= \frac{\alpha_{p,n}^2 (T_1 - T_0)^2 R_L}{(R + R_L)^2}$$

$$= \frac{\alpha_{p,n}^2 (T_1 - T_0)^2 R_L}{4R_L^2}$$

$$= \frac{(-190 - 190)^2 \times 10^{-12} \times 600^2}{4 \times 1.84 \times 10^{-3}} = 7.12 \text{ W.}$$

Example 14.3. *A 100 kW, 115V thermoelectric generator is designed to operate between 1500 K and 1000K. The material properties of thermal elements are :*

$$\alpha_{p,n} = 0.0012 \text{ V/K at 1250 K.}$$
$$k_n = 0.02 \text{ W/cm K}$$
$$k_p = 0.03 \text{ W/cm K}$$

$$\rho_n = 0.01\ \Omega\ cm$$
$$\rho_p = 0.12\ \Omega\ cm$$

The geometrical parameters have been optimized as

$$A_n = 43.5\ cm^2$$
$$A_p = 48.6\ cm^2$$
$$L_n = L_p = 0.49\ cm$$

The current density is limited to 20 A/cm^2. Calculate :

(i) Maximum efficiency.

(ii) No. of thermocouples in series.

(iii) Open circuit voltage.

(iv) Heat input and heat rejected at full load.

Solution :

$$\eta_{max} = \left(\frac{T_1 - T_0}{T_1}\right)\frac{m-1}{\left(m - \frac{T_0}{T_1}\right)}$$

$$m_o = \sqrt{1 + \frac{Z}{2}(T_1 + T_0)}$$

$$Z = \left[\frac{(\alpha_n - \alpha_p)}{\sqrt{\rho_n k_n} + \sqrt{\rho_p k_p}}\right]^2$$

$$= \left[\frac{0.0012}{\sqrt{0.02 \times 0.01} + \sqrt{0.03 \times 0.012}}\right]^2 = 13 \times 10^{-4}\ K^{-1}$$

$$m_0 = \sqrt{1 + \frac{13 \times 10^{-4}}{2}(1500 + 100)} = 1.635$$

(i)

$$\eta_{max} = \left(\frac{T_1 - T_1}{T_1}\right)\frac{m_0 - 1}{\left(m_0 + \frac{T_0}{T_1}\right)}$$

$$= \left(\frac{1500 - 1000}{1500}\right)\frac{1.635 - 1}{\left(1.635 + \frac{1000}{1500}\right)} = 9.1\ \%$$

(ii)

$$V = \alpha_{p,n}\ (T_1 - T_0)$$
$$= 0.0012\ (1500 - 1000) = 0.6\ V$$

$$R = R_p + R_n = \frac{\rho_p L_p}{A_p} + \frac{\rho_n L_n}{A_n}$$

$$= \frac{0.01 \times 0.49}{43.5} + \frac{0.012 \times 0.49}{48.6} = 2.34 \times 10^{-4}\ \Omega$$

$$
\begin{aligned}
V_L &= V_o - IR \\
&= V_o\,(R_p + R_n)\,I \\
&= 0.6 - (2.34 \times 10^{-4})\,(20 \times 48.6) \\
&= 0.3744 \text{ V}
\end{aligned}
$$

No. of couples in series $= \dfrac{115}{V_L} = \dfrac{115}{0.374} = 309.$

(*iii*)

$$
\begin{aligned}
V_o &= V \times 309 \\
&= 0.6 \times 309 = 185.4 \text{ V}
\end{aligned}
$$

(*iv*) Heat input at the full load

$$Q_1 = \frac{\dot{W}_L}{\eta} = \frac{100}{0.091} = 1100 \text{ kW}$$

Heat rejected at full load.

$$
\begin{aligned}
Q_0 &= Q_1 - \dot{W}_L \\
&= 1100 - 100 = 1000 \text{ kW}
\end{aligned}
$$

14.3. SELECTION OF MATERIALS

The efficiency of a thermoelectric generator depends upon suitable properties of the elements. The material should have high value of figure of merit Z and also capable of operation at very high temperature.

1. The thermal conductivity of thermoelectric element should be as low as possible. The value of k can also be reduced by introducing suitable impurities.
2. The mobility of current carriers (electrons or holes) should be as high as possible or electrical conductivity should be raised by introducing suitable impurities.
3. One element should be purely p-type and the other n-type. The semiconductor should have low ionization energy and narrow forbidden band.
4. The thermoelements should have variable impurity content so that electron concentration should depend upon the operating temperature.
5. The material should be corrosion resistant, should have high strength, elasticity so that it does not crack due to thermal stresses.
6. The bridge material should have high thermal and electrical conductivity and stable against thermal stresses.
7. The use of variable properties of thermoelectric elements becomes very important when thermoelectric pile or cascaded operation is required. The working temperatures can be different.

14.3.1. Metals

(*i*) For most metals, the value of Seeback coefficient is less than 10×10^{-6}V/K.

The ratio of electrical conductivity $\left(\dfrac{1}{\rho}\right)$ to thermal conductivity (k) for all metals can be predicted from quantum mechanics.

$$\rho\, k = \frac{\pi^2}{3}\left(\frac{k}{e}\right)^2 T = 2.44 \times 10^{-8}\ \text{T}$$

(*ii*) The figure of merit $\left[Z = \dfrac{\alpha^2}{\rho k}\right]$ at an average temperature of 500 K works out close to 8.2×10^{-6}.

(*iii*) At $T_1 = 700$K and $T_0 = 300$K, efficiency is 8.5×10^{-4} or 0.085 % which is very very low for practical applications.

(*iv*) For the best pair of metals *Bi–Sb*,

$$\alpha = 100 \times 10^{-6} \text{ V/K}$$
$$k = 9 \times 10^{-6} \text{ W/K}$$
$$Z = 1.1 \times 10^{-3}$$

For
$$T_1 = 450 \text{ K}$$
$$T_0 = 300 \text{ K}$$
$$\eta = 3.1 \text{ \%}$$

14.3.2. Semi Conductors

Semiconductor have higher values of Seebeck coefficient. The figure of merit Z exceeds 1×10^{-3} and efficiency approaches 10%. These have higher melting point, thus permitting operation at higher temperature. The efficiency of 15% or more are attainable. Therefore, semiconductors are more suitable materials for thermoelectric generators than metals.

14.3.3. Figure of Merit

Bismuth telluride, lead telluride, germanium and other semiconductors have properties suitable for thermoelectric generation. Alloying and "doping" make it possible to produce p-type and n-type materials. The figure of merit of some thermoelectric materials are given in Table 14.1.

Table 14.1. Figure of Merit for Thermoelectric Materials.

Materials	$Z(K^{-1})$
1. Bismuth telluride (doped with Sb or Se)	4×10^{-3}
2. Lead telluride (Pb Te)	1.5×10^{-3}
3. Germanium telluride (with bismuth)	1.5×10^{-3}
4. Zinc antimonide (doped with silver)	1.5×10^{-3}
5. Cesium sulphide	1.0×10^{-3}

High temperature Semiconductors

The melting point of Bi_2Te_3 is 584° C and that of *Pb* Te is 922° C. Silicates are considered the most promising of all the high temperature materials used for thermoelectrodes. With $Mn\,Si_2$, with $T_1 = 1300$ K and $T_0 = 300$ K, the thermal efficiency is as high as 9.7 % Boride and carbide base high temperature materials can give an efficiency of 15 to 20%.

14.4. CASCADING OPERATION

The materials may have best figure of merit at high temperature than at low temperatures. But certain materials have favourable figure of merit at lower temperatures. The problem can be solved by cascading where heat rejected from one part of system becomes the heat input to another part. The size of device decreases from one stage to another as the average temperature is lowered. The heat input to each stage is less than to the stage above. The overall efficiency of the system is given by :

$$\eta = 1 - \frac{Q_0}{Q_1}$$

$$= 1 - (1 - \eta_1)(1 - \eta_2) \ldots..$$

$$= 1 - (1 - \eta_s)^n$$

The greater the number of stages, the greater the efficiency. If each stage of thermoelectric cascade generator is optimized for geometry and resistance ratio; each stage efficiency can be improved.

The output voltage can be increased by putting a number of thermocouple is series (Fig. 14.2). Such an arrangement is called a thermopile.

14.5. APPLICATIONS

The thermoelectric generators are already recognized as very convenient direct energy conversion systems.

1. These are very simple in construction
2. Very compact
3. Absence of moving parts
4. Suitable for remote and space applications.
5. A promising system for waste heat recovery.

14.5.1. Nuclear Reactor with Thermoelectric Fuel Elements

A thermoelectric generator as incorporated in the fuel elements of a nuclear reactor is shown in fig. 14.4. This will help in obtaining large power outputs.

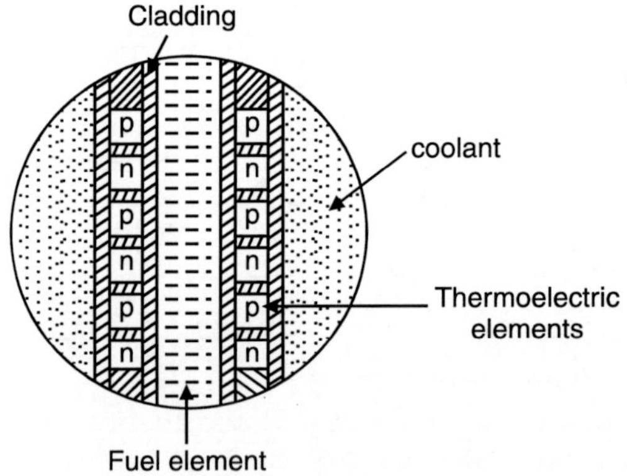

Fig. 14.4. Thermoelectric Fuel Elements.

14.5.2. Combined Thermoelectric and Steam Power Plant

Thermoelectric generator can be used as topping plant to a steam power plant. The overall efficiency of the combined plant will increase due to higher source temperature. The scheme is shown in Fig. 14.5.

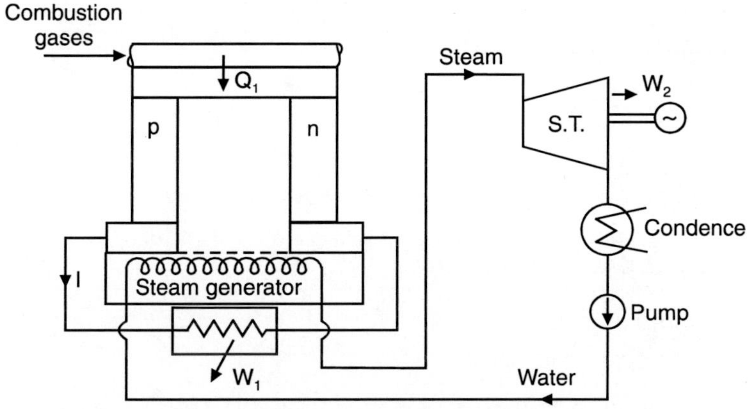

Fig. 14.5. Combined Thermoelectric and Steam Power Plant.

14.5.3. Thermoelectric Waste Heat Stack

The waste heat from gas turbines, diesel engines and stack gases can be used to produce electricity by a thermoelectric generator. The metal stack consists of a series of rings of two alternate metals connected at the inner and outer annular edges alternately. These rings are thermally and electrically insulted. A schematic diagram is shown in fig. 14.6.

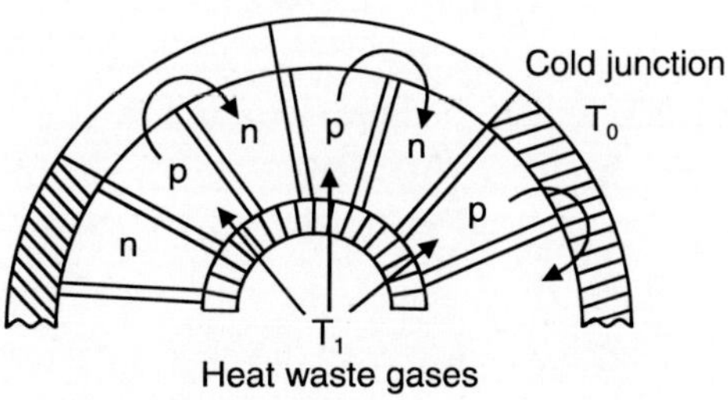

Fig. 14.6. Thermoelectric Heat Waste Stack.

14.5.4. Decay Heat of Radioactive Isotopes

The decay heat of radioisotopes has been used for the operation of small (0.1 kW) thermoelectric generators. Based on heat of decay of Strontium 85, remote generators for signalling have been used.

14.5.5. Solar Energy

A combination of thermoelectric generator and solar collector can be employed to generate electricity from solar energy.

14.6. LIMITATIONS

The main limitation of thermoelectric generators is low thermal efficiency. Lot of work is needed to develop thermocouple elements which can give reasonable efficiencies.

14.7. THERMOELECTRIC EFFECTS

Thermoelectric effects involve interchange between thermal energy and electrical energy. There are four thermoelectric effects.

(*a*) Seebeck effect

(*b*) Peltier effect
(*c*) Thomson effect
(*d*) Joule effect.

These phenomena form the basis for the design and analysis of thermoelectric generators.

Seebeck Effect

Presence of a temperature gradient in a conductor induces an electric potential gradient even when current flow is zero.

$$\left[\frac{dV/dL}{dT/dL}\right]_{I=0} = \alpha$$

where dV/dL = Induced potential gradient [V/m]
dT/dL = Applied temperature gradient [K/m]
I = Electric current

α is called the Seebeck coefficient for a single material. It has the units of volt per degree temperature difference. The value of α depends upon the materials used.

The maximum induced potential gradient is called Seebeck e.m.f.

$$\int_{L_1}^{L_2} \frac{dV}{dL}\, dL = \int_{L_1}^{L_2} \alpha \frac{dT}{dL}\, dL = \int_{T_1}^{T_2} \alpha\, dT$$

For two materials p and n the Seebeck e.m.f is

$$V_{pn} = \int_{T_1}^{T_2} (\alpha_p - \alpha_n)\, dT$$

where T_1 = Temperature of hot junction at which heat is supplied
T_2 = Temperature of cold junction at which heat is rejected

$$V_{pn} = \int_{T_1}^{T_2} \alpha_{pn}\, dT$$

where
$$\alpha_{pn} = \alpha_p - \alpha_n = \left[\frac{dV_{pn}}{dT}\right]_{I=0}$$

$$= \underset{\Delta T \to 0}{Lt} \left[\frac{\Delta V_{pn}}{\Delta T}\right]_{I=0}$$

α_{pn} is called the differential Seebeck coefficient or thermoelectric power. The variation of Seebeck coefficient with temperature is shown in Fig. 14.7. The Seebeck coefficient is positive for p type semiconductor and negative for n type semiconductor. Therefore,

$$\alpha_{p,\,n} = \alpha_p - (-\alpha_n) = \alpha_p + \alpha_n$$

The Seebeck e.m.f will be very high if thermoelectric generator is made from p type and n type semiconductors. The heat loss due to Fourier effect will also be reduced as the thermal conductivity of semiconductor is less than that of metal.

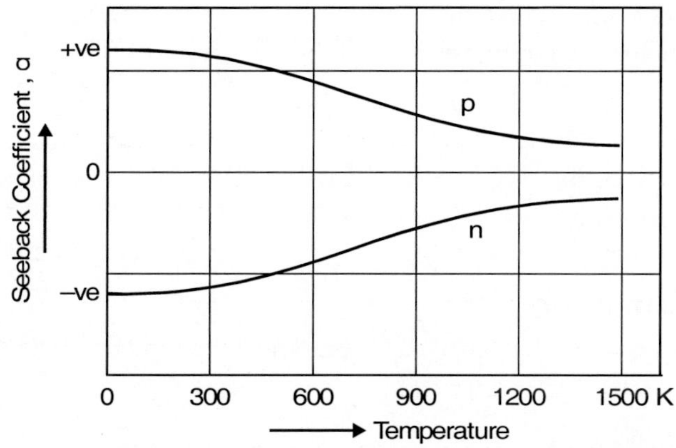

Fig. 14.7. Variation of Seeback Coefficient with Temperature

Peltier Effect

When an electric current (I) flows through the junction of two materials, Peltier heat is produced. This is called Peltier effect. The peltier effect is reversible effect because if heat is produced when current flows in one direction, the same amount of heat is absorbed at the junction if the current flow is reversed. Peltier heat per unit time,

$$Q_{p,\,n} = \pi_{p,\,n} \cdot I$$

where $\pi_{p,\,n}$ is called Peltier coefficient.

Peltier coefficient is the heat produced at the junction per unit current flow, per unit time.

$$\pi_{p,\,n} = \pi_p - \pi_n = \frac{Q_{p,n}}{I} = \alpha_{p,\,n} \cdot T$$

Thomson Effect

When an electric current flows through a material having a temperature gradient, there is production (or absorption) of heat. This phenomenon is called Thomson effect. It is also a reversible process because reversing the direction of current flow reverses the direction of heat transfer without change in magnitude.

$$\sigma = \frac{dQ_T / dL}{(dT / dL)\, I}$$

where σ = Thomson coefficient

$\frac{d\,Q_T}{dL}$ = Heat interchange per unit time per unit length of material

$\frac{d\,T}{dL}$ = Temperature gradient across the material

Thomson coefficient (σ) is defined as the Thomson heat produced per unit time per unit electric current and per unit temperature gradient.

The Thomson heat per unit time is given by,

$$\frac{d\,Q_T}{dL} = \alpha I \frac{dT}{dL}$$

The Seebeck effect, the Peltier effect and Thomson effect are more pronounced in semiconductors as compared to metals.

Joule Effect

When a current flows through a material with resistance R, an amount of heat is produced per unit time. This is called Joule heat.

$$Q_j = I^2R$$

This is irreversible process.

14.8. KELVIN RELATIONS

The relations between the Seebeck, Peltier and Thomson coefficients were derived by Kelvin.

$$\pi_{p,\,n} = \alpha_{p,\,n} \cdot T$$

$$\sigma_{p,\,n} = T \cdot \frac{d\alpha_{p,n}}{dT}$$

where α_p, α_n = Seebeck coefficients [V/K]

π_p, π_n = Peltier coefficient [V]

σ_p, σ_n = Thomson coefficient.

Example 14.4. *The Seebeck coefficient of a material*

$$\alpha = [5 \times 10^4 + 400T - T^2] \times 10^{-7}\ [V/K]$$

Calculate the Peltier and Thomson coefficients.

Solution: From Kelvin relations,

$$\begin{aligned}\pi &= \alpha T\\ &= [5 \times 10^4 + 400T - T^2] \times 10^{-7} \times T\ [V]\\ &= [5 \times 10^4 T + 400T^2 - T^3] \times 10^{-7}\ [\text{V}]\end{aligned}$$

$$\begin{aligned}\sigma &= T \cdot \frac{d\alpha}{dT}\\ &= T[0 + 400 - 2T] \times 10^{-7}\ [\text{V/K}]\\ &= [400T - 2T^2] \times 10^{-7}\ [\text{V/K}]\end{aligned}$$

Example 14.5. *Find the Thomson heat transferred to the surroundings from a wire whose end points are maintained at 373 K and 273 K respectively. A current of 10 mA is flowing in the wire and its absolute thermoelectric power increases linearly with the temperature at a rate*

$$\frac{d\alpha_p}{dT} = 5.4 \times 10^{-9}\ [\text{V/K}^2]$$

Solution: From Kelvin relations,

$$\sigma_p = T\frac{d\alpha_p}{dT} = (5.4 \times 10^{-9})\text{T}\ [\text{V/K}]$$

The Thomson heat,

$$Q_T = I\int_{T_1}^{T_2} \alpha_p\, dT$$

$$Q_T = (10 \times 10^{-3}) \times (5.4 \times 10^{-9}) \int_{273}^{373} dT = 1.74 \ \mu\text{W}.$$

TUTORIAL SHEET NO. 14.1.

1. What is Seebeck thermoelectric effect ? Describe briefly a thermoelectric power generator.
2. Define figure of merit. Derive an expression for the efficiency of a thermoelectric generator.
3. (*a*) Describe the operation of a thermoelectric generator.

   (*b*) Prove that the optimum value of resistance ratio, m that gives the maximum thermal efficiency is

$$m_o = \sqrt{1 + Z_{\max}\left(\frac{T_1 - T_0}{2}\right)}$$

where $$m = \frac{R_L}{R_p + R_n}$$

Z = Figure of Merit.

4. Write short notes on :

   (*i*) Thermoelectric effects.

   (*ii*) Seebeck and Peltier coefficients.

   (*iii*) Multistage thermoelectric generators.

   (*iv*) Selection of thermoelectric materials.

   (*v*) Figure of merit of a thermoelectric generator.
5. What is the optimum resistance ratio for (*a*) maximum power, (*b*) maximum efficiency ?
6. Briefly discuss the merits of thermoelectric power generators. What are the main limitations.
7. (*a*) Why are semiconductors preferred as thermocouple elements as compared to metals ?

   (*b*) Explain the cascade multistage operation of thermoelectric generators.
8. With the help of suitable diagrams, explain how thermoelectric generators can be used for utility power generation as well as waste heat utilization.

❑❑❑

Chapter 15

THERMIONIC POWER GENERATION

Introduction; Principle of operation; Performance analysis; Applications; Limitations.

15.1. INTRODUCTION

A thermionic generator transforms heat directly into electrical energy by utilizing thermionic emission. All metals and some oxides have free electrons which are released on heating. These electrons can travel through a space and collected on a cooled metal. These electrons can return to hot metal through an external load thereby producing electrical power.

A thermionic generator has two electrodes enclosed in a tube. The cathode is called an emitter and is heated enough to release electrons from its surface. The electrons cross a small gap and accumulate on a cooled metal anode called the collector. The space between the electrodes is maintained at high vacuum or filled with a highly conducting plasma like ionised cesium vapour to minimal energy losses. The external load R is connected through anode to cathode. The electrons return to cathode through the external load and electrical power is produced.

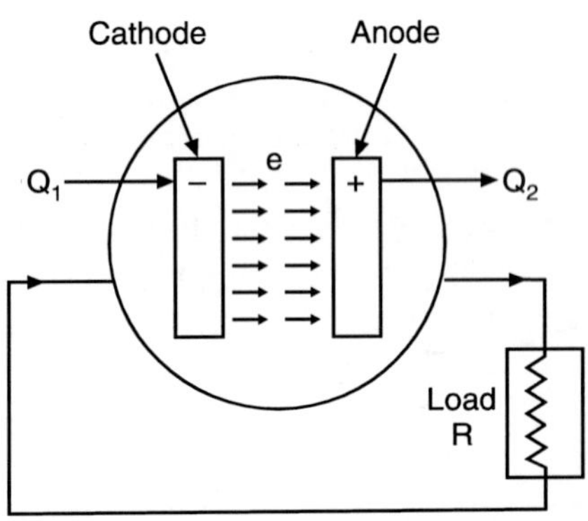

Fig. 15.1. Thermionic Generator.

A thermionic generator is like a cyclic heat engine and its maximum efficiency is limited by Carnot's law. It is a low-voltage, high current device where current densities of 20 – 50 A/cm^2 have been achieved at voltage from 1 to 2V. Thermal efficiencies of 10 – 20% have been realized. Higher values are possible in future.

Thermionic generators can be used for large power generation. The fuel elements of a nuclear reactor may be very suitable high temperature heat source for thermionic generator. This can be surrounded by the cooled anode and the in between space can be filled with ionized cerium vapour.

The energy of high temperature combustion gases can be partly converted to electricity if the riser tubes of the boiler are provided with cathode and anode of a thermionic generator with the interspace filled with ionized cerium vapour.

In principle any heat source, fossil or nuclear fuel, a radioactive material or solar energy can be used in a thermionic generator. Many applications have been suggested for remote locations on the earth and in space as the device is robust and reliable in unattended operation.

15.2. PRINCIPLE OF OPERATION

The emission of an electron from a metal surface is opposed by a potential barrier equal to the difference between the energies of an electron outside and inside the metal. Therefore, a certain amount of energy has to be spent to release the electron from the surface. This energy is called surface work function (ϕ).

The maximum electron current per unit area emitted from the surface is given by the following Richardson Dushman equation :

$$J = A_1 T^2 e^{\left(-\frac{\phi}{KT}\right)}$$

where J = current density, [A/m^2]

T = Temperature, [K]

ϕ = work function, [eV]

$K = 1.38 \times 10^{-23}$ J/molecule – K

= Boltzmann constant

$A_1 = 120$ A/cm^2 – K^2)

= Emission constant.

The work function varies between 1 and 5 eV and data is given in Table 15.1.

Table 15.1. Work Function For Different Materials

Element	Work function (eV)	Element	Work function (eV)
Ag	4.65	Mo	4.25
Al	3.60	Na	2.00
Ba	2.52	Ni	5.00
Ca	3.00	Pt	6.15
Cs	1.75	Rb	1.82
Cu	4.30	Th	3.50
K	1.90	T	4.55
Li	2.21	Zn	3.44

1 eV = 1.602×10^{-19} J.

Fermi Energy Level

The kinetic energy of the free electrons at absolute zero would occupy discrete energy levels from zero upto some maximum value defined by the Fermi energy level, ε_f. Each energy level contains a limited number of free electrons. Above absolute zero temperature, some electrons may have energies higher than the Fermi level. The energy that must be supplied to overcome the weak attractive force on the outermost orbital electrons is the work function, ϕ, so that the electron leaving the emitter has an energy level $\phi + \varepsilon_f$.

When emitter is heated, some high-energy free electrons at the Fermi level receive energy equal to emitter work function ϕ_c and escape the emitter surface. They move through the gap and strike the collector. The K.E. (ε_{fa}) plus the energy equal to collector work function ϕ_a is given up and this energy is rejected as heat from the low temperature collector.

The electron energy is reduced to the Fermi energy level of the anode ε_{fa}. This energy state is higher than that of the electron at the Fermi energy level of cathode ε_{fc}. Therefore, the electron is able to pass through the external load from anode to cathode. The cathode materials are selected with low Fermi levels as comprised to anode materials which must have higher Fermi level.

15.3. PERFORMANCE

The positively charged cathode tends to pull the emitted electrons back. The electrons already in the gas exert a retarding force on the electrons trying to cross the gap. This produces a space charge barrier Figure 15.2 shows the characteristic curve of a thermionic generator with an interspace retarding potential equivalent to δ volts above the anode work function ϕ_a.

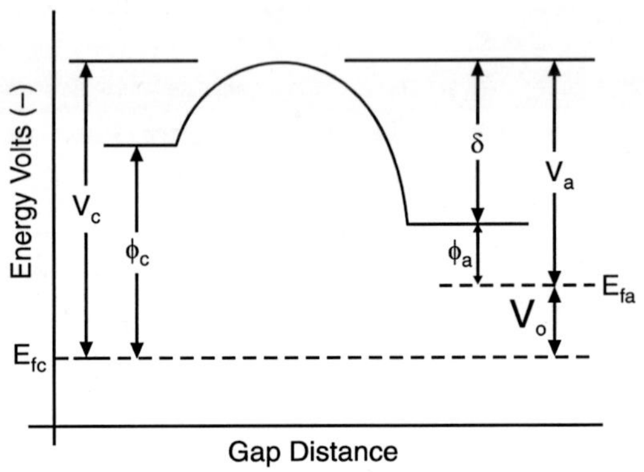

Fig. 15.2. Characteristic curve.

Potential barrier,

$$V_C > \phi_c$$
$$V_a > \phi_a.$$

The current densities are :

$$J_c = A_1 T_c^2 e^{\left(-\frac{V_c}{KT_c}\right)} \quad [A/\text{cm}^2]$$

$$J_a = A_1 T_a^2 e^{\left(-\frac{V_a}{KT_a}\right)} \quad [A/\text{cm}^2]$$

The output voltage across the electrical resistance R,

$$V_o = V_c - V_a = \phi_c - \phi_a = \frac{1}{e}(\varepsilon_{fa} - \varepsilon_{fc})$$

Each electron has to overcome the interspace potential ($V_c - \phi_c$) and work function ϕ_c when it leaves the cathode. The net energy carried,

$$Q_{1c} = J_c (V_c - \phi_c + \phi_c) = J_c V_c \quad [W/\text{cm}^2]$$

Each electron also carries away its K.E. which is equal to $2KT_c$, *i.e.*,

$$Q_{2c} = J_c \frac{2KT_c}{e} \quad [\text{W/cm}^2]$$

The back emission from the anode must similarly carry energy to the cathode. The net rate of energy supply to the cathode,

$$Q_1 = J_c\left(V_c + \frac{2KT_c}{e}\right) - J_a\left(V_a + \frac{2KT_a}{e}\right)$$

where $e = 1.602 \times 10^{-19}$ coulomb.

The power output of the generator

$$W = V_o (J_c - J_a)$$

The thermal efficiency of the thermionic generator,

$$\eta = \frac{V_o(J_c - J_a)}{J_c\left(V_c + \frac{2KT_c}{e}\right) - J_a\left(V_a + \frac{2KT_a}{e}\right)}$$

Now $\qquad V_o = V_c - V_a$

Substituting the following values.

$$\frac{V_c}{KT_c} = \beta_c\ ;\ \frac{V_a}{KT_a} = \beta_a \text{ and } \frac{T_a}{T_c} = \theta.$$

$$\therefore \qquad \eta = \frac{(\beta_c KT_c - \beta_a KT_a)(J_c - J_a)}{J_c\left(\beta_c KT_c + \frac{2KT_c}{e}\right) - J_a\left(\beta_a KT_a + \frac{2KT_a}{e}\right)}$$

$$= \frac{(\beta_c - \theta\beta_a)[1 - \theta^2 e^{(\beta_c - \beta_a)}]}{(\beta_c + 2) - \theta^2(\beta_2 + 2\theta)\, e^{(\beta_c - \beta_a)}}$$

It is found that for all values of θ, the efficiency curve peaks are very near to the value of $\beta_a = \beta_c$.

∴ If $\beta_c = \beta_a$,

$$\eta_{max} = [1 - \theta]\frac{\beta}{\beta + 2}\left[\frac{1 - \theta^2}{1 - \frac{\theta^2(\beta + 2\theta)}{\beta + 2}}\right]$$

$$\frac{1 - \theta^2}{1 - \frac{\theta^2(\beta + 2\theta)}{\beta + 2}} \approx 1$$

$$\therefore \qquad h_{max} = (1 - \theta)\frac{\beta}{\beta + 2}$$

Carnot efficiency, $\quad \eta_c = 1 - \dfrac{T_a}{T_c} = (1 - \theta)$

If β = 18,

$$\eta_{max} = 0.9\ (1 - \theta)$$

η_{max} occurs when $\beta_a = \beta_c$

or $$\frac{V_c}{T_c} = \frac{V_a}{T_a}$$

Example 15.1. *A thermionic generator has the following characteristics*

Cathode work function $\phi_c = 2.5\ V.$

Anode work function $\phi_a = 2.0\ V.$

Temperature of cathode $T_c = 2000\ K.$

Temperature of surrounding T_s = *1000 K.*

Πλασμα ποτεντιαλ δροπ $\delta = 0.1$ *V.*

Assuming the value of emissivity as 0.2 for the electrode materials used, calculate the efficiency of the generator and also compare with Carnot efficiency.

Solution: Net output voltage,

$$V_0 = \phi_c - \phi_a - \delta$$
$$= 2.5 - 2.0 - 0.1$$
$$= 0.4 \text{ V}$$

Net current in the generator

$$J = J_c - J_a$$

$$J_c = A\,T_c^2\, e^{\frac{-\phi_c}{KT_c}}$$

$$= (1.20 \times 10^6)\,(2000)^2\, e^{\frac{(1.6\times 10^{-19} \times 2.5)}{(1.38 \times 10^{-23})2000}}$$

$$= 2.4 \times 10^6 \text{ A/m}^2$$

Charge of an electron,

$$e = 1.6 \times 10^{-19} \text{ columbs.}$$

Boltzman constant,

$$K = 1.38 \times 10^{-23} \text{ J/k}$$
$$1\,eV = 1.6 \times 10^{-9} \text{ J}$$

$$J_a = AT_a^2\, e^{\frac{-\phi_a}{KT_a}}$$

$$= (1.20 \times 10^6)\,(1000)^2\, e^{\left[\frac{-(1.6\times 10^{-19})\times 2.0}{(1.38 \times 10^{-23})\,1000}\right]}$$

$$= 1.2 \times 10^2 \text{ A/m}^2,$$

$$J = J_c - J_a \approx J_c$$
$$= 2.4 \times 10^6 \text{ A/m}^2$$

Heat supplied to the cathode,

$$\frac{Q_c}{A_c} = J\left(\phi_c + \frac{2KT_c}{e}\right) + \varepsilon\,\sigma\,(T_c^4 - T_s^4)$$

$$= 2.4 \times 10^6 \left[\frac{2.5 + 2\,(1.38 \times 10^{-23})\,2000}{1.0 \times 10^{-19}}\right]$$

$$+ 0.2 \times 5.67 \times 10^{-12} \times 10^{-7}\,(2000^4 - 1000^4)$$

$$= 7.026 \times 10^6 \text{ W/m}^2$$

$$\eta = \frac{JV}{Q_c / A_c} = \frac{(2.4\times 10^6)\,(0.4)}{7.026 \times 10^6}$$

$$= 0.137 = 13.7\%$$

Carnot efficiency,

$$\eta_c = \frac{T_1 - T_2}{T_1} = \frac{2000 - 1000}{2000} = 0.50 = 50\%$$

15.4. APPLICATIONS

15.4.1. Thermionic Generator in a Nuclear Reactor

The fuel element containing the fissile material carries the cathode which is surrounded by the anode. The interspace is filled with ionized cesium gas. The anode is cooled by the coolant from outside. Some of the energy released by nuclear fission is directly converted into electricity by thermionic conversion. The remaining heat is used in conventional bottoming steam plant. The overall efficiency of the plant increases.

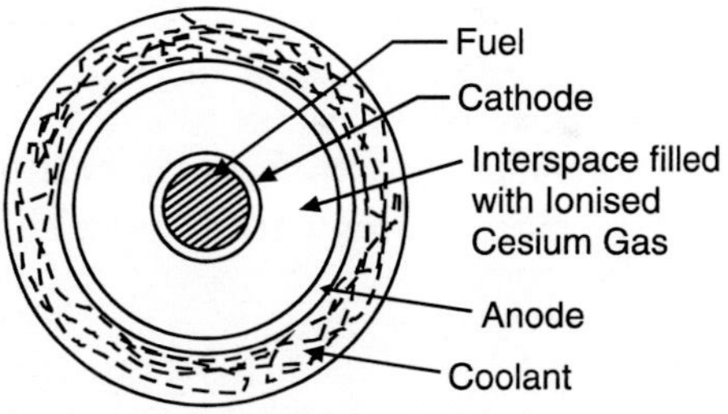

Fig. 15.3. Thermionic Generator in a Nuclear Reactor.

15.4.2. Thermionic Generator in the riser tube of a Boiler

The riser tubes of a boiler receive heat by radiation from combustion gases. The riser tube is provided by a cathode and anode of a thermionic generator. The interspace is filled with ionized cesium vapour. The use of the hot combustion gases to produce extra power before the steam cycle improves the overall plant efficiency.

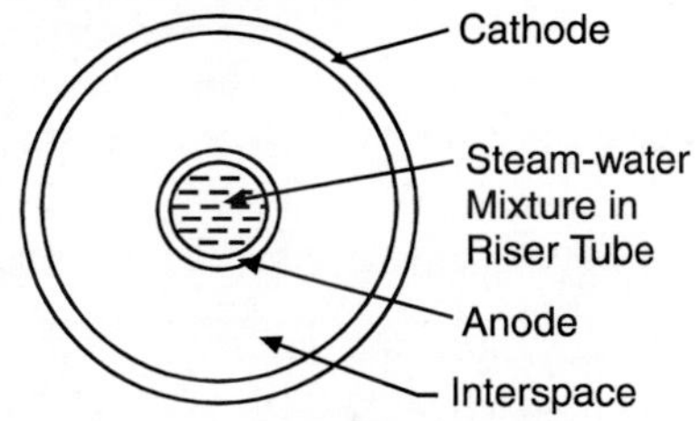

Fig. 15.4. Thermionic Generator in the Riser of a Boiler,

15.4.3. MHD-Thermionic-Steam Power Plant

The waste heat from MHD generator at about 1900°C is used to heat the cathode of thermionic generator. The heat from anode is used in the boiler of a stram power plant. The overall efficiency of the combined plant will increase.

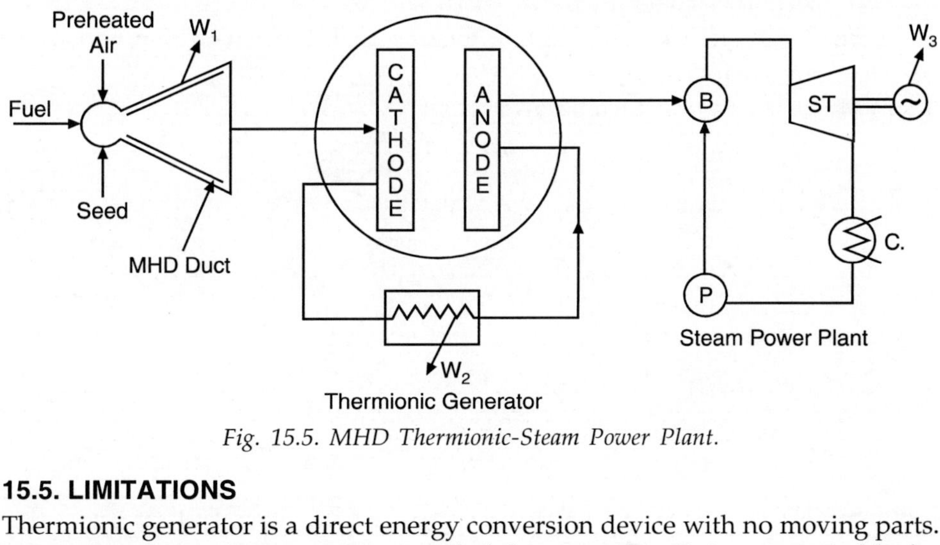

Fig. 15.5. MHD Thermionic-Steam Power Plant.

15.5. LIMITATIONS

Thermionic generator is a direct energy conversion device with no moving parts. It is very robust, no supervision needed for operation. It is very suitable for remote and space applications. It suffers from the following limitations :

1. The operating temperature of cathode is very high necessitating the use of costly materials such as tungsten or rhenium.
2. Special shields of ceramic are needed to protect cathode from corrosive combustion gases.
3. The collector may also have to be made of molybdenum coated with cesium.
4. Ionised cesium vapour has to be filled in the interspace to reduce the space charge barrier to promote electron emission from the cathode.

TUTORIAL SHEET NO. 15.1.

1. Describe the principle of operation and constructional details of a basic thermionic generator. What do you understand by "thermionic emission effect"?
2. Derive the expression for power and efficiency of thermionic generator.
3. Discuss the various applications of thermionic generator for central power production.
4. What are the main limitations of thermionic generator?
5. What do you understand by :
   (*a*) Surface work function,
   (*b*) Fermi energy,
   (*c*) Interspace retarding potential.
6. What is Richardson-Dushman equation? What is its relevance in thermionic power generation?
7. (*a*) What are the suitable materials for the emitters and collectors of a thermionic generation ?
   (*b*) What is the reason of filling the gap with ionized cesium vapour ?
8. Explain the following :
   "MHD-Thermionic-Steam power plant."

❏❏❏

PART 1V

FUTURE ENERGY SOURCES

Chapter 16

NUCLEAR FUSION ENERGY

Introduction; Principle of fusion process; Tokamak Fusion Reactor; Intertial Confinement reactor; Future Nuclear Fusion Power Plant; Advantages of fusion energy

16.1. INTRODUCTION

Of all the currently known available sources, only nuclear and solar energy have the potential of supplying large amounts of power within the available time frame. These two sources can complement each other to meet the demand economically and safely. However, considerable technical and economic problems must be solved before large-scale utilization of these sources can become possible.

Energy is produced in the sun and stars by continuous fusion reactions. Fusion of light nuclei to form a heavy nucleus releases large amount of energy. If the controlled fusion of light elements is carried out on our planet, enough energy can be generated to meet all the energy requirements of future generations.

The primary fuel for fusion is deuterium, sometimes called heavy hydrogen (H^2). There is one atom of deuterium for every 6500 atoms of ordinary hydrogen in sea water. The total energy content of deuterium as fossil fuel is about 100,000 kWh of energy per gram of deuterium. This is about 10 million times more than released per gram in the combustion of fossil fuels. The entire amount of world energy consumption of 6×10^{13} kWh per year could be supplied by the fusion of only 600 metric tons of deuterium. The amount of deuterium in the world oceans would be enough to sustain the present total world energy requirement for 100 billion years. Nuclear fusion would indeed present an ultimate solution of mankind's energy needs for all the years to come.

A nuclear fusion reaction can occur when two atomic nuclei approach very close to each other at velocities atleast large enough to overcome their mutual electrostatic repulsion called "Coulomb barrier". Fusion reaction occurs at very high temperature. Therefore, there is a problem of confinement of fusion. There are other essential difficulties lying to be overcome before this future energy source can be put to man's use.

Fusion research is indeed being taken seriously by the major industrial nations. There are serious efforts by individual countries as well as combined effort by consortium of countries. It is almost certain that large practical fusion power plants will be built in the twenty-first century. Once this technology is developed, an almost unlimited supply of energy will be available for the world's needs ushering in a better living standard for the human kind all over the world.

16.2. PRINCIPLE OF FUSION PROCESS

Energy is produced in the sun and stars by the following continuous fusion reactions :

$$4\ {}_1H^1 \longrightarrow {}_2H_e^4 + 2e_{+1}^{\circ}$$

Four nuclei of hydrogen fuse in a series of reactions to yield one nucleus of helium and protons. There is a decrease in mass of about 0.0276 amu. The energy corresponding to the change in mass is calculated from Einstein's law.

$$\Delta E = \Delta m\ C^2\ \text{(J)}$$

where Δm = charge in mass [kg]

c = Velocity of light

$= 3 \times 10^8$ m/s

$\therefore \quad DE = 9 \times 10^{16}\ \Delta m$

one amu $= 1.66 \times 10^{-27}$ kg

ΔE (in J) $= 1.49 \times 10^{-10}\ \Delta m$ (in amu)

ΔE (in MeV) $= 931\ \Delta m$ (in amu)

The energy released per nuclear reaction will be 25.7 MeV. The heat produced in these reactions maintains temperatures of the order of several million degrees in the cores of the sun and stars which serves to trigger and sustain succeeding reactions.

Artificial Fusion Reaction

The 4-hydrogen reaction shown above requires, on an average, billions of years for completion. On earth artificial fusion reaction may be accomplished when two light atoms fuse into a larger one as there is much greater probability of two particles colliding than of four. In order to cause artificial fusion reaction, the following problems must be solved.

1. The positively charged nuclei must be accelerated to high kinetic energies in order to overcome the electrical repulsive forces.
2. The temperature must be raised to hundreds of millions of degrees resulting in a plasma.
3. The plasma must be prevented from contacting the walls of the container.
4. The plasma must be confined for a certain period of time (of the order of a second) at a minimum density.
5. The fusion heat must be converted to electricity.
6. All operational problems of a fusion power plant must be solved.

There are several possible reactions between the nuclei of light elements that can be the basis for controlled thermonuclear fusion. These reactions along with energy released per reaction are given in Table 16.1. The following symbols have been used

$$D = {}_1H^2 = \text{Deuterium}$$
$$T = {}_1H^3 = \text{Tritium}$$
$$n = {}_0n^1 = \text{Neutron}$$

$$p = {}_1p^1{}_{=+1}\, e^{\circ}$$
$$= \text{proton.}$$
$$1\text{MeV} = \text{one million electron-volts}$$
$$= 1.602 \times 10^{-3}\ J.$$

Table 16.1. Fusion Reactions

No.	Reactants 1	Products	Energy per reaction, MeV	Energy per gram kWh
1.	$D + D$	$T + p$	4.0	22,000
2.	$D + D$	$H_e^3 + n$	3.2	27,000
3.	$T + D$	$H_e^4 + n$	17.6	94,000
4.	$H_e^3 + D$	$H_e^4 + p$	18.3	94,000

Reaction 3 is the most favourable reaction which is self-sustaining at a *temperature of* 50×10^6 *K* releasing 17.6 MeV per reaction. The first two $D - D$ reactions occur at 500×10^6 K and release less energy (3.2 MeV and 4.0 MeV). The fourth reaction releases very high energy (18.3 MeV), but requires very high temperature of 1000×10^6 K for fusion reaction.

The basic raw materials for the $D - T$ reaction is water for deuterium and tritium. Deuterium in the form of heavy water (D_2O) is extracted from water on a large scale at a moderate cost. Tritium does not occur abundantly in nature. It can, however, be produced in a lithium "breeding blanket" that surrounds the plasma core of the fusion reactor. Lithium can be extracted from sea water at low cost.

16.3. TOKAMAK FUSION REACTOR

There are two processes used for plasma confinement

1. Magnetic confinement
2. Inertial confinement

The fusion reactor with magnetic cofinment is called Tokamak reactor which in Russian means toroidal magnetic chamber. A schematic diagram of a Tokamak fusion reactor is shown in Fig. 16.1.

The plasma is contained inside an evacuated tube of about 4m. The surrounding vacuum wall through which 14 MeV neutrons from the plasma pass, is maintained at about 750°C. Outside this wall are two concentric regions, *i.e.*, the lithium breeder modulator and the magnetic shield. Tritium is manufactured in the lithium blanket. Large cryogenic superconducting magnets of 7 to 8 m diameter maintain the magnetic field.

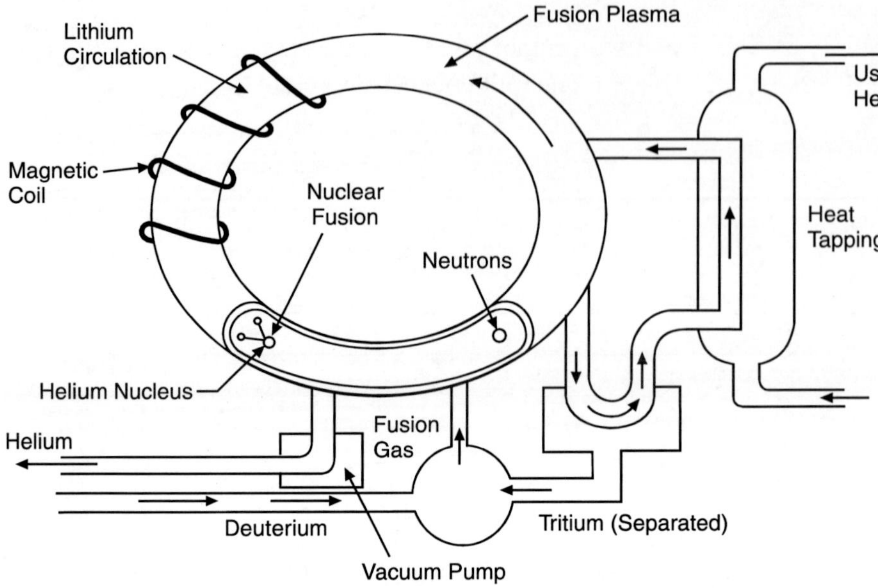

Fig. 16.1. Tokamak Fusion Reactor.

The breeding of tritium takes place as follows

$$^{6}Li + n \rightarrow H_e + T + 4.78 \text{ MeV}.$$

The heavy lithium isotope ^{6}Li acts as breeding material and moderator.

The design parameters of a projected fusion reactor are given below:

1. Thermal fusion Power = 800 MW
2. Larger plasma radius = 6.3 m
3. Smaller plasma radius = 2.05 m
4. Toroidal field strength on axis = 6 Wb/m^2
5. Plasma current = 25 – 30 MA.
6. Neutron wall loading = 0.7 MW/m^2.
7. Triple product (ntT) = 7×10^{18} MeV S/m^2

The triple product is the index for positive energy balance on net release of energy. A sufficient number of fusion reactions per unit time must occur in the reactor.

Here T is the plasma temperature of the order of 10^8 K, t is the confinement time of 0.1s and n is the plasma density of the order of $10^{20} l/\text{m}^3$.

16.4. INERTIAL CONFINEMENT REACTOR

A small pellet or sphere of a deuterium tritium mixture (solid or liquid) is heated by a very short burst of energy from either laser beams or beams of high energy charged particles. The pellet material is compressed to a high density and temperature also increases. The fusion reaction has to be completed in one trillionth (10^{-12}) second so that inertia would prevent the pellet from flying apart while fusion is in progress. No magnetic confinement is needed. The development of this type of reactor is stopped due to unavailability of high power laser.

16.5. FUTURE NUCLEAR FUSION POWER PLANT

The fusion reactor can be coupled to a triple Rankine cycle using potassium, diphenyl and steam as working fluid. A binary vapour power cycle consisting of a potassium topping cycle and a conventional steam cycle is shown in Fig. 16.2. The tritium recovery system is also shown.

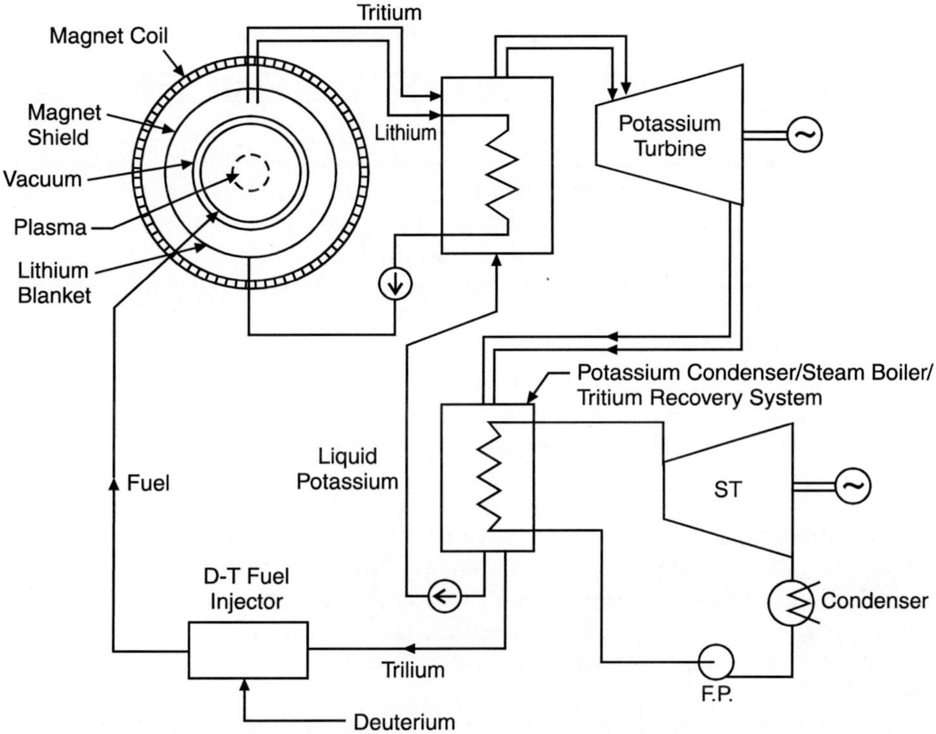

Fig. 16.2. D – T Fusion Power Plant.

16.6. ADVANTAGES OF FUSION ENERGY

1. The supply of deuterium is inexhaustible.
2. No radioactive waste is produced.
3. It is very safe to operate.
4. High conversion efficiency (60%) can be achieved.
5. Low heat rejection per kW of electricity generated to atmosphere.

TUTORIAL SHEET NO. 16.1.

1. What is fusion reaction ? What are the main fusion reactions ? Which one is the most favourable reaction and why ?
2. Describe Tokamak reactor system for power generation. What are the main design parameters of an experimental fusion reactor ?
3. Prove that nuclear fusion energy is the future energy.
4. Describe a *D–T* Fusion Reactor Power Plant with the help of a flow diagram.
5. What are the advantages of nuclear fusion energy ? What are the main problems hindering further progress ?

Chapter 17

HYDROGEN ENERGY

Introduction; Utilization of Hydrogen Energy; Hydrogen Energy for Air and Surface Transport; Power Generation; Miscellaneous Uses; Distribution of Hydrogen; Production; Limitations.

17.1. INTRODUCTION

Electrical energy is the most convenient form of energy because it can be easily controlled, transported and converted into heat and work at very high efficiencies. The only shortcoming of electrical energy is that it cannot be stored in large quantities. Alternative energy of future is hydrogen energy which can also be easily stored in addition to other qualities of electrical energy. But is highly inflamable and special handling precautions are needed during its production, transportation, storage and utilization. Hydrogen is a secondary fuel that is produced by utilizing energy from a primary source. Water and solar energy are freely and abundantly available in nature on earth. Hydrogen can be produced from water by using solar energy. All plants and hydrocarbons (fossil fuels) are sources of hydrogen.

Hydrogen can be stored as gas underground or in high pressure cylinders. Also it can be stored in liquid form at low temperatures. There are a number of metals and alloys which form solid hydrides with hydrogen. The hydrogen can be easily recovered by heating metal hydrides. Thus metal hydrides provide a possible means for hydrogen storage in solid form. Hydrogen can be similarly transported as gas through pipes or storage cylinders or as liquid hydrogen or as metal hydrides.

Hydrogen has the highest energy content per unit mass. Its burning process is non-polluting and it can be used in fuel cells to produce both electricity and useful heat. It can be used as a fuel directly in gas turbines or spark-ignition engines. It can be used as motor vehicle fuel in urban transportation where air pollution problems are critical. Its specific energy content is almost three times that of hydrocarbon fuels. Therefore it can be directly used as aircraft fuel for air transport. Hydrogen has been used as a fuel for space crafts. A $H_2 - O_2$ fuel cell liberates energy and also water as sole material product for the use of space craft passengers.

The simplest practical way to obtain hydrogen from water is its electrolysis using electricity. The latter can be generated from removable energy sources like solar energy, wind energy, geothermal energy.

Hydrogen has huge market. However, enormous capital investment is required for its production, distribution and storage. Special design precautions are needed for the safe operation of equipment and systems. This is presenting the essential obstacle in the quick introduction of hydrogen.

17.2. UTILIZATION OF HYDROGEN ENERGY

Hydrogen has the following properties which make it an attractive alternative energy source.

1. At room temperature and pressure, hydrogen is a light gas. Its density is only 1/14th of that of air and 1/9th that of natural gas.
2. At atmospheric pressure, hydrogen can be liquefied at – 253°C. The liquid hydrogen has a specific gravity of 0.07 which is 1/10th that of gasoline.
3. The standard heating value of hydrogen gas is 12.1 MJ/m^3 compared with 38.3 MJ/m^3 for natural gas.
4. The heating value of liquid hydrogen is 120 MJ/kg or 8400 MJ/m^3 as compared to 44MJ/kg or 32000 MJ/m^3 of aviation petrol. The specific energy of hydrogen liquid is superior to gasoline on mass basis but inferior on volume basis.
5. The flame speed of hydrogen when burning in air is much greater than for natural gas.
6. The ignition energy to initiate combustion is less for hydrogen than for natural gas.
7. Detonation can occur between hydrogen-air mixture between 18 and 59 percent. The internal combustion engine on hydrogen fuel can work from very rich (excess fuel) to very lean (excess air) mixture. The adjustment of air fuel ratio is less critical than for gasoline engine.
8. Mixture of hydrogen and air are combustible over a wide range of composition. The flammability limits are from 4 to 74 percent by volume of hydrogen in air at ordinary temperatures.
9. The combuston of hydrogen with oxygen from air results in release of energy and water as by-product.
10. The burning process of hydrogen is pollution free.

The possible areas of use of hydrogen in the near future are as follows :

1. Production of useful heat

(*i*) In a high-temperature combustion of hydrogen with oxygen or air.

(*ii*) In a low-temperature flameless catalytic combustion with extermely low NO_x emission.

2. Power Generation

(*i*) In reactors with direct steam generation.

(*ii*) In high-temperature and membrane fuel cells.

3. Cogeneration of heat and electricity.

(*i*) In internal combustion engine based cogeneration plants.

(*ii*) In combined-cycle power plants.

4. Automotive and aircraft fuel.

(*i*) Environmentally friendly fuel for motor vehicles.

(*ii*) Aircraft fuel.

5. Energy storage

(*i*) Compressed hydrogen gas storage under high pressure.

(*ii*) Metal hydrides.

6. Synthesis of Fuels

(*i*) Raw material to produce methanol, ammonia or hydrocarbon using carbon dioxide or nitrogen from air.

(*ii*) Raw material for manufacture of gaseous fuels.

17.3. HYDROGEN ENERGY FOR AIR AND SURFACE TRANSPORT

17.3.1. Jet Fuel

The high energy density 33.3 kWh/kg of liquid hydrogen against 12.7 kWh/kg of conventional jet fuel is the main advantage in air transportation where hydrogen energy can be used. Although volume of liquid hydrogen would be greater than regular fuel but this could be accommodated on a large aircraft. The cold liquid hydrogen can also be used directly or indirectly for cooling engine and airframe surfaces of high speed aircrafts. Liquid hydrogen may be the only practical fuel for hypersonic aircraft when developed. Because of smaller total weight, it may be possible to achieve shorter take-off runs, steeper climbing path, smaller engine thrust and less noise production.

The favorable diffusion properties and high thermal conductivity of hydrogen help to use shorter combustion chambers. Wide range of ignition for $H_2 - O_2$ mixtures (5% to 75%) by volume of hydrogen helps better control of engine operation especially under part load conditions and reduction in NO_x emission.

The heat required to vaporize hydrogen for the engines can be obtained from certain outer skin of wings and fuselage. This helps to cool the boundary layer. The laminar boundary layer so developed helps to reduce drag and fuel consumption.

The main problems are economic production of hydrogen, infrastructure for fueling of aircrafts and siting of bulky hydrogen tanks in the aircraft body.

17.3.2. Road Vehicles

The use of hydrogen fuel in engines of automobiles, buses, trucks and farm machinery can help conserve petroleum products and reduce atmospheric pollution. A mixture of hydrogen gas and air of constant ratio is introduced into the manifold. The engine speed and power are controlled by varying the quality of mixture entering the cylinder with the help of a throttle valve. In another design, hydrogen gas under pressure can be directly injected into the engine cylinder through a valve and air is admitted through another intake valve. The spark advance has to be retarded because of higher speed of flame of hydrogen in air. The engine emission will not contain carbon monoxide and hydrocarbon because the only product of combustion is H_2O. In order to control NO_x emission, the cylinder exhaust containing H_2O is injected to reduce combustion temperature.

Storage of hydrogen as compressed gas or metal hydrides for the vehicles is a disadvance against hydrogen fuel because of lesser energy density per unit volume and the weight of metal hydrides is also excessive.

The best way to use hydrogen as vehicle fuel is the use of fuel cells. The electricity generated in the fuel cells operating with hydrogen fuel can be utilized to operate electric motors for propelling the vehicle.

17.4. POWER GENERATION

17.4.1. Central Power Plants

The plants working on natural gas can be changed to hydrogen-fired plants without significant technological changes. These plants may be combined-cycle power plants with gas and steam turbines or cogeneration plants with gas turbine and internal combustion engines. Hydrogen-fired plants compared to gas-fired plants have lower capital cost and higher efficiency.

Hydrogen fuel has higher combustion velocity and flame temperature as compared to natural gas. Higher combustion velocity results in unstable combustion and higher flame temperature leads to higher No_x emissions.

17.4.2. Autonomous Power Plants

Hydrogen fuelled fuel cells can be used for domestic power generation as well as industrial power generation. Alkaline fuel cells can be used for producing electricity from pure hydrogen and oxygen. Fuel cells with acid electrolytes can be operated with impure hydrogen and hydrocarbons. High temperature fuel cells with molten carbonate electrolytes operating at 600°C are advanced generation fuel cells for electricity and heat production in cogeneration plants.

17.5. MISCELLANEOUS APPLICATIONS

Hydrogen can be used for domestic cooking replacing LPG. The burner design has to be changed with bigger holes and air supply system to take care of greaterflame speed and low specific energy per unit volume when hydrogen is used as fuel.

Hydrogen can be usefully used in radiant space heating with flameless combustion on a catalytic surface. In this case combustion temperature is low with negligible NO_x formation.

Hydrogen has many advantages over industrial gases for production of heat and other uses.

17.6. HYDROGEN STORAGE AND DISTRIBUTION

There are five principle methods that have been considered for hydrogen storage.

1. Compressed gas storage.
   Hydrogen is conveniently stored for many applications in higher pressure cylinders. The method is rather expensive and bulky.
2. Liquid storage as cryogenic storage in vacuum insulated or super insulted storage tanks. The liquid hydrogen fuel used as rocket propellant in the space programme is stored in large tanks.
3. Line pack system where it is allowed to vary the pressure in the transmission and distribution system.
4. Underground storage of hydrogen gas in depleted oil and gas fields or in aquifer systems. This is the cheapest way to store large amounts of hydrogen for subsequent distribution.
5. Storage as metal hydrides in chemically bound form. A number of metals and alloys form solid compounds by direct reaction with hydrogen gas. The metal hydrides can be transported in solid form. When the hydride is heated, hydrogen is released for use.

Hydrogen Transportation

1. Long distance hydrogen gas transmission pipelines of lengths greater than 90 km must be supplied with booster compressors. Therefore, the cost of transmitting hydrogen by pipelines must include the cost of piping, compressor and power consumption by compressors. Another problem of hydrogen transmission is hydrogen embrittlement of the pipeline materials.
2. Hydrogen in bulk can be transported and distributed as the liquid in double-walled insulated tanks. Distribution of liquid hydrogen by pipelines, jacketed with liquid nitrozen can also be considered.
3. Hydrogen can also be transported as a solid metal hydride. The main drawback is the heavy weight of hydride relative to its hydrogen yield.

Safety Precautions

Hydrogen is highly inflammable and explosive and can lead to fire and serious accidents. The production, storage and distribution of hydrogen require special precautions.

1. The system should be designed to withstand the explosion pressures.
2. The system should be designed to withstand pressure surges.
3. Proper exposition-relief system must be provided.
4. Flame traps, flame suppressors, explosion-relief devices and rapid-closing devices must be used.
5. The design, manufacture, storage should follow Petroleum Act.

17.7. HYDROGEN PRODUCTION

17.7.1. From Fossil Fuels

The conventional hydrogen production processes are shown in Fig. 17.1.

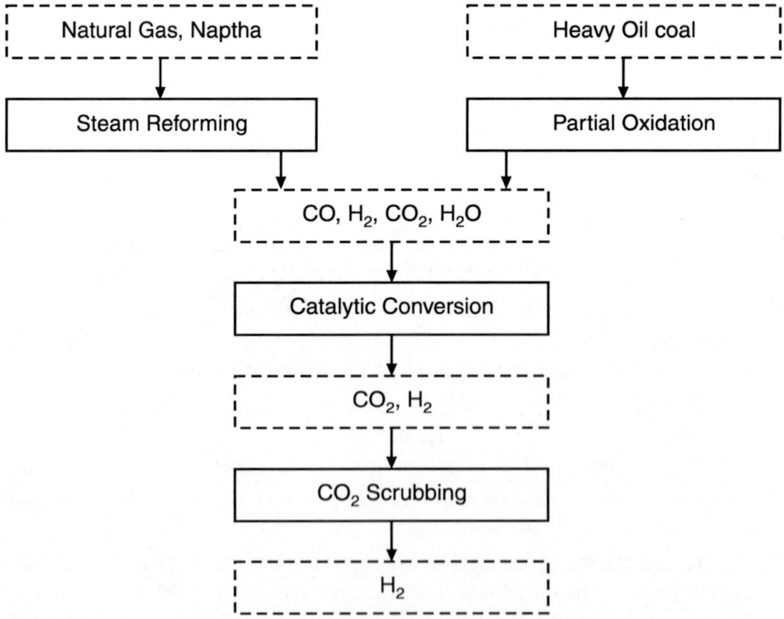

Fig. 17.1. Flow diagram of Hydrogen Gas Production from Fossil Fuels

(*i*) Natural Gas/Naptha

1. The natural gas or naptha is reformed with steam at 900°C to produce a mixture of gases by the following reaction

$$C_m H_n + m\, H_2O \longrightarrow m\, CO + \left(m + \frac{n}{2}\right) H_2$$

The gas produced has the following composition

	Component	Composition in %age by volume
1.	H_2	68
2.	CO	10
3.	CH_4	10
4.	CO_2	10
5.	N_2	2

2. The CO present is converted into CO_2 by two-stage catalytic conversion process.

   1st stage process at 300-500°C

   2nd stage process at 200°C

$$CO + H_2O \text{ (steam)} = CO_2 + H_2$$

   CH_4 is converted into CO_2 by the following reactions

$$CH_4 + H_2O = CO + 3H_2 - 205 \text{ kJ/mol}$$

$$CH_4 + 2H_2O = CO_2 + 4H_2 - 164 \text{ kJ/mol}$$

3. Finally CO_2 is removed by scrubbing process. H_2 is obtained as product gas.
4. The energy content of H_2 product gas is 75-80% of natural gas. The cost of H_2 is comparable with that of electricity on similar basis.

(*ii*) Coal and heavy Oil

1. The coal or heavy oil undergo partial oxidation or gasification at 1400°C with oxygen as gasifying medium.
2. The CO present is converted into CO_2 by two-stage catalytic conversion process.
3. CO_2 is removed by scrubbing.
4. The energy content of H_2 product gas is 55 to 60% of raw coal. The cost of H_2 produced is higher than produced from natural gas if lignite or hard coal is used as feedstock.

17.7.2. By Water Electrolysis

The principle of operation of an alkaline water electrolysis cell is shown in figure 17.2.

Hydrogen is produced by electrolytic dissociation of water

Cathode reaction :

$$4H_2O + 4\bar{e} \longrightarrow 2H_2\uparrow + 4OH^-$$

Anode reaction

$$4OH^- \longrightarrow O_2\uparrow + 2H_2O + 4\bar{e}$$

Total reaction

$$2H_2O + \text{Energy} \longrightarrow 2H_2 + O_2$$

The energy requirement is 3.5 kWh/m^3 of hydrogen produced.

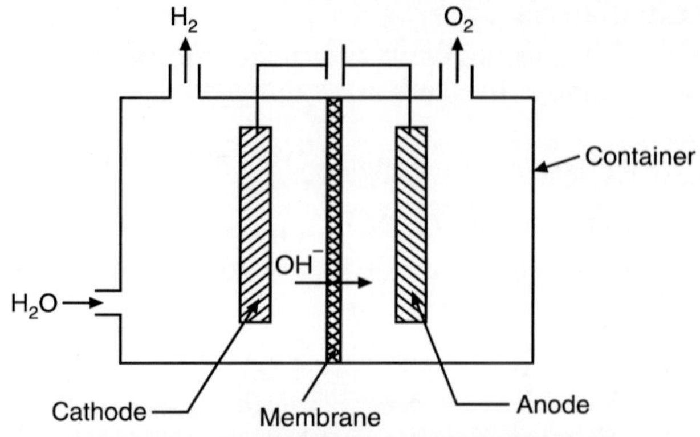

Fig. 17.2. Alkaline Cell Water Electrolysis.

The following electrolysis processes are available

1. Alkaline electrolysis
2. Membrane electrolysis
3. High temperature steam electrolysis

17.7.3. Solar Energy Methods

Hydrogen can be produced using solar radiation by the following processes :

1. Bio photolysis
2. Photo electrolysis

1. Biophotolysis

Hydrogen can be produced very cheaply by using green algae's ability to generate hydrogen gas from water and sunlight. The green algae is a simple plant. It should be kept in an anaerobic (oxygen-free) environment free from sulphur.

2. Photo Electrolysis

Hydrogen and oxygen gases are liberated at the cathode and anode respectively by decomposition of water when both the electrodes are subject to sunlight. It is a very attractive method of producing hydrogen as the process is intermittent. There is no need for storage of gas.

The solar electricity can be produced by photo-electrically or thermochemically.

17.8. LIMITATIONS OF HYDROGEN ENERGY

Hydrogen has very attractive properties and can be a future source of energy :

1. Availability
2. Manufacture without harm to environment
3. Good storage properties
4. Readily convertible to other forms of energy.

It can easily replace electricity, natural gas, synthesis gas, etc. There is still a need for considerable research and development in the following areas of hydrogen technology.

1. Production of hydrogen of photoelectrolysis of water using solar energy.

2. Production of hydrogen by blue green algae and by certain bacterial species.
3. Storage of hydrogen through metal hydrides and non-metal hydrides (rice husk and misch-metal)
4. Liquid hydrogen production, storage and utilization.
5. Development of suitable engines and burners to utilize hydrogen as a fuel.

TUTORIAL SHEET NO. 17.1.

1. Write briefly about utilization of hydrogen energy.
2. How can hydrogen energy be used for air and surface transport ?
3. Discuss briefly the role of hydrogen fuel for power generation.
4. Discuss the main problems and technologies available for storage and distribution of hydrogen
5. What are the different methods for hydrogen production ? Explain in brief.
6. What are the areas where research and development efforts are needed to make sure that hydrogen becomes future source of energy ?

❑❑❑

Chapter 18

EXPLORING NEW ENERGY SOURCES

18.1. INTRODUCTION

Modern societies require increasing amounts of energy for domestic, industrial, commercial, agricultural and transport uses. These energy needs are being met by a combination of short-term, depletable fossil fuel supplies and long-run, renewable energy sources. Economic growth will require massive infusions of energy in the foreseeable future. Urbanization and industrialization are both energy intensive. Even rural development and agricultural productivity and production depend on larger and larger amounts of power. There is an urgent need for the development of new and more efficient energy sources.

18.2. REQUIREMENTS OF NEW ENERGY SOURCE

Heavy demand for energy poses the problem of developing new, efficient methods of energy production based on latest achievements in science and technology. At the present time, the existing methods of converting different forms of energy into electricity are limited because these are mostly based on depleting chemical fuel reserves. The conversion technologies are inefficient based on large material consumption, Large amounts of heat and by-products are released heating the nearby water sources and contaminating the environment with release of harmful waste matter into the atmosphere. This may be regarded as one of the most important social problems.

Research of new energy sources and development of efficient conversion technologies have to meet the following requirements.

1. The energy source and conversion technology to usable form should be harmonious to the five bodies of nature, *i.e.*, the earth, the water, the wind, the fire and the sky. These natural sources should be used as minimum as possible and the harmful waste should not be rejected to these bodies by the energy system.
2. The source of energy should be an indigenous resource. It should be available in considerable quantities.
3. The local source of energy should be used to conserve foreign exchange and generate local employment.

4. The conversion technologies should be based on available scientific and technological knowledge. It should be possible to design, manufacture, assemble, install, operate and maintain the energy system by local people using local skills and materials.
5. The conversion technology should be flexible and modular and rapidly deployed.
6. There should be ease in adding new capacity, less risk in investment, lower interest on borrowed capital because of shorter lead times and reduced transmission and distribution costs.
7. It should be financially and economically competitive with current systems.
8. There should be no technical and economic uncertainties in the new energy systems.
9. It should have high energy efficiency and low economic and financial costs.
10. The new energy source should have flexible applications for lighting, heating, cooling, mechanical work.

18.3. DESIGN REQUIREMENTS

The design and analysis of a new energy system have to be approached from a system analysis view point combining technical design with economic analysis. There is no single solution to a given task in energy utilization, and each problem has to be analysed separately from fundamental principles. It is necessary to match the available energy source to the task at hand and there is no general solution. Proper design and optimization of new energy systems may require a high level of engineering analysis. The system engineer has to decide on the basis of the task at hand, the system available to achieve a technical solution and economies involved.

The design and analysis of solar energy systems is given below as an example of illustration.

18.3.1. Collection System

Energy delivery of a collector depends on its physical configuration, its operating temperature, and the climatic parameters of the solar radiation level, ambient temperatures and wind speed. Collector performance is also determined by secondary variables, including fluid flow rates, collector orientation, geographic location and system control strategy. Table 18.1. provides a collector-task classification showing the best combination of collectors and tasks.

The detailed design of a collector will depend upon the selection of the type of collector, type of focussing and concentration ratio and also type of selective coating for the absorber. All these are functions of fluid temperature required as shown in Fig. 18.1.

18.3.2. Solar System Model

The rational design of a solar-thermal system requires the knowledge of the dynamic interaction of all system components for solar collection, thermal storage, fluid circulation and energy distribution, control and non-solar auxiliary energy source.

Table 18.1. Collector-Task Classification

Collector Type	Space heating small Bldg.	Space heating large Bldg.	Water heating	Space cooling	Process heat	Crop drying	Dis-tillation
Flat Plate (Liquid)	S	S	S	S/U	S	S	S
Flat Plate (Air)	S	U	U	U	S	S	U
Evacuated Tube	S	S	—	S	S	—	—
Intermediate Concentration	—	—	—	S	S	—	—
High Concentration	—	—	—	—	S	—	—

S = Satisfactory
U = Unsatisfactory
— = Uneconomic or task mismatch

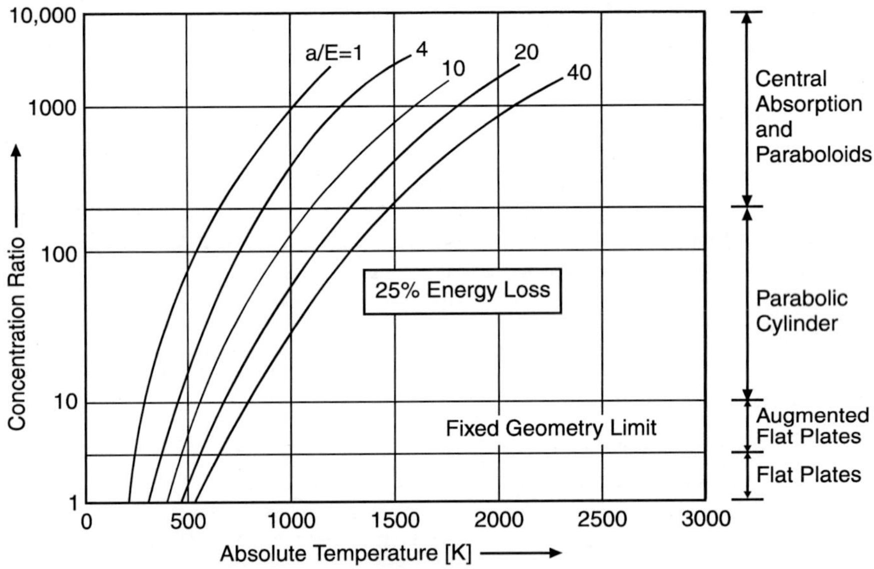

Fig. 18.1. Effect of Concentration Ratio on Absorber Temperature

Computers are generally required for calculation of numerical solar models because many component models are non-linear and a closed-form simultaneous solutions of the equations describing a system is not possible in all but the simplest cases. In addition, most solar systems operate in a continuously changing, transient manner subject to short-time scale changes in all forcing functions. Since meaningful modeling results require delivery totals for fairly large periods (months or years), and since solar systems respond on fairly short time scale (minutes or hours), a great many calculations are required. For these reasons computerised models are the method of choice for analysis of most solar-thermal systems.

A simple block diagram of the computer model of a solar system using hourly weather data is shown in Fig. 18.2.

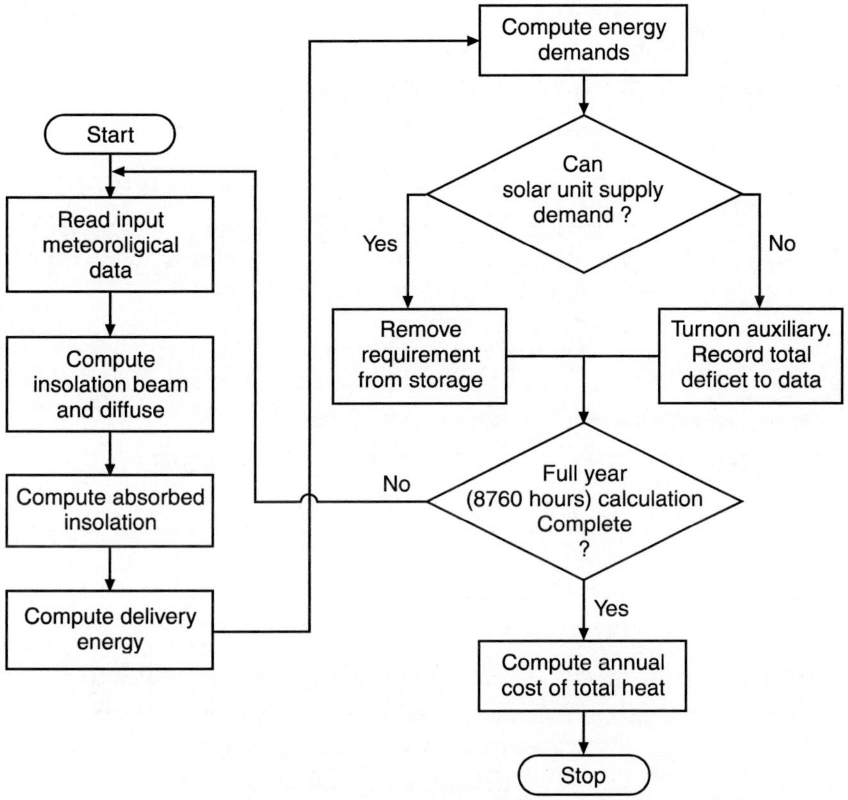

Fig. 18.2. Simplified block diagram of Computer-Simulation Model of a Solar thermal System

18.3.3. Optimum Task to Energy Level Match

Solar energy is the only major source where entropy level of the form in which it is collected can be manipulated to provide an optimum task-to-entropy-level match. This match can be accomplished by varying degrees of concentration of solar energy or by improved receiver design, for example, with an evacuated cover. For high-entropy uses such as space heating, water heating and crop drying, low temperature, high-entropy solar energy collected by flat-plate collectors provides the best match between the energy source and task. The operation of a solar-thermal generating station requires low-entropy (high-temperature) thermal energy to match the task of producing low-entropy shaft work.

18.4. EXERGY ANALYSIS OF ENERGY SYSTEMS

The exergy analysis of air energy system enables us to identify the sources of irreversibilities and inefficiencies with the aim of reducing the losses and achieving the maximum resource and capital savings. This can be achieved by a careful selection of the technology and optimization of design of the system and components. Exergy analysis involves the determination of exergy efficiency or second-low efficiency.

18.4.1. Exergy Efficiency

The analysis of energy systems based on second law of thermodynamics calculates the changes in the quality of energy or entropy. The exergy of a system decreases as a process loses its quality.

An exergy balance of an energy system gives

$$E_{in} = E_u + E_d \text{ [kW]}$$

where E_{in} = Rate of exergy input [kW]

E_u = Rate of useful product exergy [kW]

E_d = Rate of destruction or loss of exergy [kW]

Exergy efficiency, ε of an energy conversion system is the ratio of actual performance of system and ideal performance of the system.

$$\varepsilon = \frac{E_u}{E_{in}} = 1 - \frac{E_d}{E_{in}}$$

18.4.2. Exergy Efficiency of Solar Collectors

The exergy efficiency of solar collectors can be given as :

$$\varepsilon_{sc} = \frac{\dot{m}_{sc}\,\Delta e}{E_{SR}}$$

$$= \frac{\eta_{sc}\,E\,\Delta e}{E_{SR}\,\Delta h}$$

where $\dot{m}_{sc}$ = mass flow rate of collector fluid [kg/s]

Δe = exergy increase of collector [kJ/kg-K]

Δh = enthalpy increase of collector [kJ/kg-K]

E = Incident solar radiation [kW]

E_{SR} = exergy flow rate of solar radiation [kW]

η_{sc} = efficiency of solar collector

The efficiency of solar collector,

$$\eta_{sc} = \frac{\dot{m}_{sc}\,\Delta e}{E}$$

$$\frac{E_{SR}}{E} = 0.933$$

$$\therefore \qquad e_{SC} = 0.933\,\eta_{sc}\left[1 - \frac{T_0}{\Delta T_{sc}}\right] ln\,\frac{T_2}{T_1}$$

where T_0 = Ambient temperature [K]

T_1 = Inlet temperature of collector fluid [K]

T_2 = Outlet temperature of collector fluid [K]

ΔT_{sc} = Temperature increase in collector [K]

The exergy efficiency of a solar collector will increase with the increase of solar collector efficiency, η_{sc}.

The exergy efficiency of flat plate solar collector is low, because this type of solar collector transforms low-entropy (high-temperature) solar radiations into high-entropy (low-temperature) energy of air or water.

The exergy efficiency of a concentrating type solar collector is high as it produces low-entropy (high-temperature) fluids at 700 to 1400 K.

Example 18.1. *A flat plate collector heats water from 305K to 335K. Calculate the exergy efficiency of the collector if its efficiency is 48% and ambient temperature is 293K.*

Solution :

$$T_0 = 293\ \text{K}$$
$$T_1 = 305\ \text{K}$$
$$T_2 = 335\ \text{K}$$
$$\Delta T_{sc} = T_2 - T_1 = 30\ \text{K}$$
$$\eta_{sc} = 48\%$$

The exergy efficiency of solar collector,

$$\varepsilon_{sc} = 0.933\ \eta_{sc}\left[1 - \frac{T_0}{\Delta T_{sc}}\right] ln\frac{T_2}{T_1}$$

$$= 0.933 \times 0.48\left[1 - \frac{293}{30}\right] ln\frac{335}{305}$$

$$= 0.037 = 3.7\%$$

The ratio,

$$\frac{\varepsilon_{sc}}{\eta_{sc}} = \frac{0.037}{0.48} = 8\%$$

Example 18.2. *Air is heated in a central receiver solar power plant from 693K to 1073K. Calculate the exergy efficiency if ambient temperature is 303K and collector efficiency is 70%.*

Solution :

$$T_0 = 303\text{K}$$
$$T_1 = 693\ \text{K}$$
$$T_2 = 1073\ \text{K}$$
$$\Delta T_{sc} = 1073 - 693 = 380\ \text{K}$$
$$\eta_{sc} = 0.70$$

The exergy efficiency of solar collector,

$$\varepsilon_{sc} = 0.933\ \eta_{sc}\left[1 - \frac{T_0}{\Delta T_{sc}}\right] ln\frac{T_2}{T_1}$$

$$= 0.933 \times 0.7\left[1 - \frac{303}{380}\right] ln\frac{1073}{693}$$

$$= 0.425 = 42.5\%$$

The ratio,

$$\frac{\varepsilon_{sc}}{\eta_{sc}} = \frac{0.425}{0.7} = 60.7\%$$

18.5. ECONOMIC EVALUATION OF ENERGY SYSTEMS

The following methods may be used for economic evaluation of energy systems :

1. Life cycle costing method,
2. Net benefit (cost saving) method, and
3. Net benefit/cost ratio method.

Life Cycle Costing Method

The total present value of total annual value of the alternative energy systems are calculated. The system with the lowest cost is the best economic alternative.

The present value of a system,

$$PV = 1 - \frac{RV}{(1+i)^n} + \frac{\sum OM_j}{(1+i)^j} + \sum p_{con}\, Q_{con}\left[\frac{1+e}{(1+i)^j}\right]$$

where
I = Capital cost or investment cost.
RV = Residual value at nth year
n = no. of years of economic evaluation
i = interest rate
OM_j = Annual cost of operation, maintenance, repair and replacement in the year j.
p_{con} = initial price of conventional energy
Q_{con} = energy requirement (conventional)
e = escalation rate of energy cost

Net Cost Savings Method

The present value of cost savings,

$$PS_{net} = \sum\left\{-p_{com}\,\Delta Q_{con}\left[\left(\frac{1+e}{(1+j)^j}\right)\right]\right\} - \left[\Delta I - \frac{\Delta V}{(1+i)^n}\right] + \frac{\sum OM_j}{(1+j)^i}$$

Δ = Change due to new energy

Net Benefit/Cost Ratio Method

$$S/I = \frac{(PV_{net} + I)}{I}$$

$$(S/I)_{\text{net}} = \frac{PV_{net}}{I}$$

$(PV)_{\text{net}}$ = Present value of cost savings
I = Capital investment

18.6. INTERNAL RATE OF RETURN FOR INVESTMENT IN NEW ENERGY SYSTEM

A variety of technical, economic and financial factors determine the financial feasibility and economic attractiveness of investments in new technologies. Investment decisions are made by comparing the cost of a project for new energy system, which depends upon the price of capital with expected returns (yearly cash flow). The price of capital is either the interest rate of loans or the rate of return on equity or investment in the new system. The degree of risk associated with a new system affects the price of capital ; high risk ventures cost more.

Figures that are widely used to evaluate the financial and economic aspects of new projects are the internal rate of return (IRR), the gross payback period (GPB) and net present value (NPV), all of which are related.

The internal rate of return (IRR), which measures how profitable a project is, is basically the discount rate (r) at which NPV is zero. For a project with an expected equipment life of N years and expected net cash flow of Q_t in year t, the $IRR(r)$ is defined by :

$$o = \sum_{t=1}^{N} \frac{Q_t}{(1+r)^t}$$

where the net cash flow (Q_t) is given by

$$Q_t = (1 - T_t)(S_t - OM_t - ALP_t)$$

where for the year t,

S_t = savings on fuel,

OM_t = operation and maintenance cost.

ALP_t = annual loan payment

T_t = corporate income tax

Given the capital cost, equipment lifetime and cash flow sequence, it is possible to calculate the *IRR* as shown in Fig. 18.3. The IRR should be greater than the cost of capital.

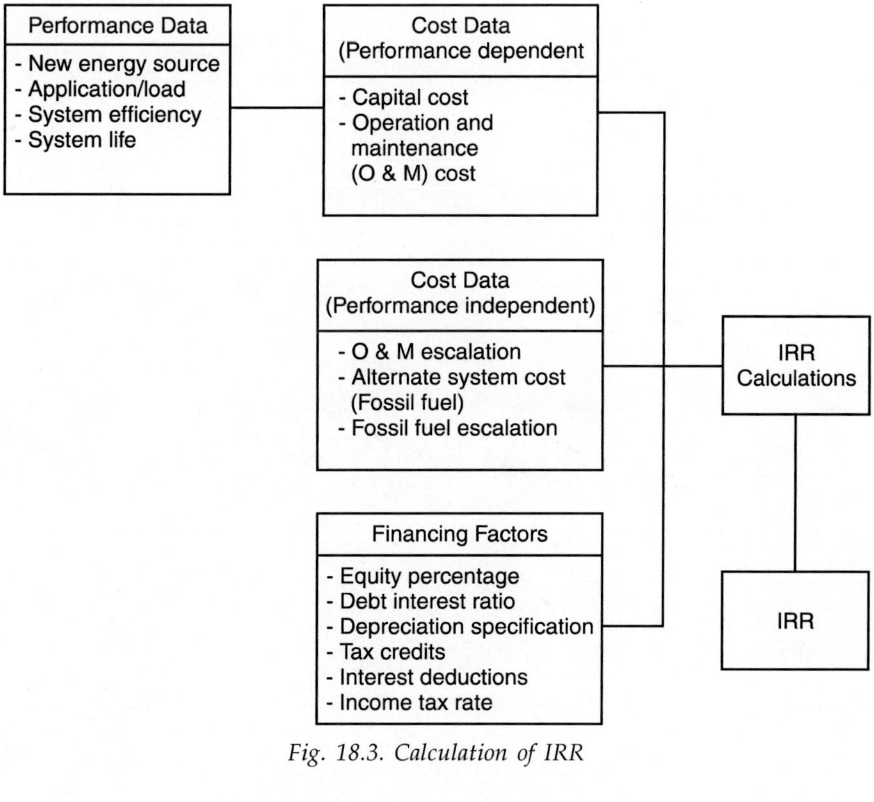

Fig. 18.3. Calculation of IRR

TUTORIAL SHEET NO. 18.1.

1. List the requirements of a new energy source.
2. What are the design requirements of a new energy source ? Illustrate with an example of solar collectors.
3. What is exergy efficiency ? How will you determine the exergy efficiency of a solar collector ?
4. What methods can be used for economic evaluation of a new energy system ?
5. What is the significance of Internal Rate of Return ? How is IRR used for investment decisions for a project on new energy system development ?

PART V

ALLIED ENERGY SYSTEMS

Chapter 19

ENERGY RESOURCE CONSERVATION

Introduction; Transport sector; Rural Energy; Urban energy; Industrial Energy Consumption, Power generation and distribution; Energy Efficient Building; Heating Load of a Building; Thermal Insulation of a Building;

19.1. INTRODUCTION

Energy saved is energy produced. Therefore, substantial energy savings can be achieved through energy conservation measures. Energy savings can be considered as additional source of energy. This will also help in reduction of environmental pollution. Energy savings can be achieved by :

1. Modifying the consumer behavior and avoiding unnecessary energy consumption.
2. Employing energy efficient and high performance equipment and plant. The specific energy consumption and energy intensity should be reduced.
3. Using waste heat recovery plant.
4. Using energy management system and reducing energy losses.

Plenty of opportunities exist for energy conservation at the level of end use, supply of energy and structural improvements. The problem of energy conservation can be best tackled by study of suitable measures in various sectors of energy consumption.

19.2. TRANSPORT SECTOR

Transport swallows major part of all imported oil causing great problems in the balance of payments. Lack of planning has resulted in the congestion of cities causing extremely low energy efficiencies in urban transport. There is a large potential for energy savings in transport sector of the order of 20–25%. The opportunities for conservation differ per means of transport. The driver's style and conduct can save much energy in trucks and buses. In case of passenger cars more energy can be conserved by improving the performance of these cars. The possibility of fuel conservation potential has been projected in Table 19.1.

Table 19.1. Fuel Conservation Potential (% age)

Mode of Transport	Fuel Consumption share	Conservation Potential				
		Efficiency Improvement	Traffic Regulation	Improved operation	Demand Reduction	Total
Trucks	35 – 45	6	2	10	5	23
Buses	8 – 12	2	< 0.5	5	< 0.5	7
Private cars	25 – 35	20	2	2	5	29
Ships	5 – 10	2	5	5	< 0.5	12
Air planes	5 – 10	20	5	5	< 0.5	30
Trains	2 – 5	1	< 0.5	5	< 0.5	6

19.2.1. Infrastructure Improvements

1. **The energy conservation should be part of planning**
   Physical distances and choice of mode of transport determine energy consumption. The location of facilities in a city for housing, working, shoping, entertainment, etc. has great effects on energy use.

2. **Traffic regulation**
   Decrease of traffic congestion by planning of traffic lights, one - way traffic, limited parking, etc. can reduce energy consumption.

3. **Telecommunication**
   Efficient telecommunication can reduce travelling and freight transport.

4. **Construction and maintenance of roads, bridges and tunnels**
   Proper roads can shorten travelling distances, faster speeds and reduced tyre friction.

5. **Construction and maintenance of waterways and ports**
   Transport by ship is very efficient as compared to surface transport and can be an important energy conservation measure.

6. **Construction, renovation and maintenance of railways**
   Railways are capital-intensive but considerably more energy efficient than roadways. It is important to rehabilitate, maintain and upgrade railways.

19.2.2. Improvement of Road Transport

1. **Promoting non-motorized transport**
   Promotion of use of bicycles is possible through public education, good and cheap bicycles availability. The animal cart design should be improved.

2. **Discouraging private cars**
   This can be achieved through efficient public transport, car pooling and restricted parking.

3. **Improved buses, trucks and cars**
   Preventive maintenance, tuning of ignition and carburetor, right tyre pressure, etc. can reduce fuel consumption.

4. **Availability of new vehicle models**
   The new models of vehicles of streamlined bodies, air shields, nose cones

cousume less fuel (15% or more). The commercial vehicles should run on diesel rather than gasoline.

5. **Public and Freight transport**
   The energy efficiency can be improved through optimum routes, less trip frequency, location and distribution of freight distribution centres.

19.2.3. Sea Transport

1. **Promotion of sea transport**
   Ships are 5 to 10 times more energy efficient than trucks.
2. **Navigation speeds**
   The optimum speeds of ships and firing of engines are important for fuel saving.
3. **Proper maintenance of vessels**
   Energy savings can be affected through proper maintenance of engines and clean and smooth hulls.
4. **Modifications in vessels**
   The old engines can be replaced by more efficient engines.

19.2.4. Railways

1. **Promotion of rail transport**
   Transport by rail is more energy efficient than by road.
2. **Rationalization of fares and tariffs**
   Lower fares and tariffs can shift the load to railways.
3. **Proper maintenance of trains**
   Maintenance of locomotives is very important for energy savings.
4. **Electrification**
   The intensely used lines should be electrified.

19.3. RURAL ENERGY

The major energy consumption in villages is by households followed by agriculture and cottage industries. The main energy consuming activities are given in Table 19.2. The main energy sources are wood, dung and agricultural residues. The energy balance of a village Pura in India is given in Table 19.3.

Table 19.2. Energy Consuming Activities

Domestic	Agriculture	Cottage Industry
– Cooking	– Soil preparation	– Machines and looms
– Space heating	– Irrigation, drainage	– Furnaces, foundries
– Hot water	– Fertilizing	– Hot water
– Lighting	– Harvesting	– Lighting
– (Cooling)	– Threshing, milling	– Process heat
	– Drying	– Space heating

Table 19.3. Energy Balance of Pura Village (GJ/Year)

Energy Source	Energy consumption				
	Agriculture	Domestic	Lighting	Industry	Total
Fuel wood	–	3308.6	–	142.2	3450.8
Kerosene	–	–	72.9	5.9	78.8
Electricity	26.2	–	11.1	3.0	40.3
Total (Fuels)	26.2 (0.7%)	3308.6 (92.7%)	84.0 2.4%)	151.1 (4.2%)	3569.9 (100%)
Men	33.4	212.8	–	20.8	267.0
Animals	52.0	–	–	–	52.0
Total (Muscle power)	85.4 (26.8%	212.8 (66.7%)	– (0%)	20.8 (6.5%)	319.0 (100%)
Grand total	111.6 (2.8%)	3521.4 (90.6%)	84.0 (2.2%)	171.9 (4.4%)	3888.9 (100%)

The use of commercial fuels is very low due to high cost and non-availability. Cooking accounts for major rural energy consumption (70 to 8%). Therefore, conservation of cooking energy is very important. The other areas of interest are lighting, space heating and cottage industries.

19.3.1. Cooking

1. **Improving Combustion efficiency**

   The oxygen supply for hearths should be controlled. The size and shape of wood logs, briquetting of agricultural residues and moisture content of fuels influence the combustion efficiency.

2. **Stoves**

   Improved cooking stoves can save much energy.

3. **Cooking method**

   Depending upon the type of food, the method of cooking has influence on energy requirements. The food can be boiled, fried, stewed or baked.

4. **Cooking efficiency**

   Use of pressure cooker can save the energy three fold. Similarly use of pots and pans of high thermal mass and conductivity will save energy.

5. **Food quantity**

   Cooking in large quantities offers more opportunities for energy saving.

6. **Type of food**

   Use of millet in place of rice can save energy.

19.3.2. Food and Fuel Production

1. **Optimum use of land**

   The land should be suitably divided for agriculture, animal grazing and fuel production depending upon the energy requirements.

2. **Cultivation methods**

   The use of seeds, fertilizers, water and pesticides should be reduced by better timing as per soil conditions and type of crop.

3. **Energy Crops**

   Cross-breeding and choice of plant requiring minimum inputs can save energy indirectly.

4. **Agriculture machines and implements**

   Better design and cooperative use of machines and implements can save energy and costs.

5. **Draught animals**

   The efficiency of draught animals should be improved by crossbreeding and better feeding. Replacement of tractors by draught animals can save energy.

19.3.3. Processing, Storage and Transport of Food

1. **Post harvest losses**

   Reduction of food losses between harvest and consumption is an indirect method of saving energy.

2. **Processing and distribution**

   Reduction in requirement of transport, storage, processing of food by optimization can save considerable energy.

3. **Energy efficient techniques**

   Energy can be saved through use of alternative techniques of drying, presalting before cooking, recycling of packing material.

4. **Equipment design**

   Improvement in design of equipment for processing, carts, cooking can save energy.

19.3.4. Cottage Industry and Handicraft

1. **Energy efficient kilns**

   Energy requirements for brick making and charcoal production can be reduced by use of modern steel kilns.

2. **Tools and implements**

   Simple improvements of tools and implements used for making of handicrafts can save energy and work.

3. **Appropriate technology**

   Appropriate technology should be used to save energy in village industries.

19.3.5. Lighting

1. **Kerosene lamps**

   Improved kerosene lamps help to save fuel.

2. **Alternative energy sources**

   Vegetable oils and animal fats can be used for lighting to save kerosene.

19.4. URBAN ENERGY

Urban energy includes hot water, cooking, lighting, space conditioning and electrical appliances. Use of electricity and LPG with higher efficiency of cooking ranges and hot plates help save energy. The largest savings are possible in restaurants, hospitals and school hostels where large scale cooking is employed. Hot water is less of a basic need than cooking. If hot water is used only in winter, integration with space heating becomes rational. In poor households same fire is used for cooking and water heating. Fluorescent lights in place of incandescent lights should be used where possible. In large buildings, optimal use of day light, changes in working hours and cogeneration of heat and electricity can result in large energy savings.

Space conditioning including cooling, heating, ventilation and air purification plays an important role in middle and high income group households. The energy consumption for air-conditioning may be ten times that for lighting. Cleaning and maintenance of air-conditioners can save large energy quantities.

The introduction and use of household appliances like TV's, refrigerators and washing machines is becoming more and more widespread. Large amounts of energy can be saved by controlling the product specifications, quality, taxes and education at Government level.

19.4.1. Cooking

1. **Improved stoves**
   Shift from wood stoves to kerosene stoves saves energy. Similarly LPG hot plates are efficient.
2. **Food preparation**
   Pressure cooking is energy efficient technique.
3. **Pans**
   More efficient cooking pans should be used suited to the type of fuel used.
4. **Cooking customs**
   Cooking customs change when a family moves from the village to a city.

19.4.2. Hot water

1. **Waste prevention**
   The water temperature must be adjusted as per its use. Leaking of pipes and taps of hot water should be avoided.
2. **Geysers**
   Solar water heaters should be used in conjuction with geysers.
3. **Insulation**
   Insulation of boilers and pipes are very important for energy saving.
4. **Cogeneration**
   In large buildings, cogeneration of heat and electricity and recycling of waste heat from air-conditioning and space heating plants should be used for water heating.

19.4.3. Lighting

1. **Use of day light**
   Properly located and sized windows and cleaning of glass panes can reduce need for electrical lighting.
2. **Efficient lamps**
   Use of fluorescent lamps is an energy and money saving measure.
3. **Electricity**
   Electric lights are more efficient than kerosene lamps, however electrification leads to use of more lights.

19.4.4. Space conditioning

1. **Building construction**

   Proper design and construction of buildings and use of vegetation have **favourable climatic effects.**

2. **Air-conditioning equipment**

   Proper setting and control of temperatures in air-conditioning and space heating equipment save energy.

3. **Replacement of old equipment**

   Replacement of old and obsolete equipment by new equipment can save energy.

4. **Insulation of buildings**

   Insulation of buildings can reduce heat losses in cold countries and inlet of heat in warm countries.

19.4.5. Electrical Applicances

1. **Change of habits**

   Washing of clothes at lowar temperature, use of less detergents, proper food preservation outside refrigerator can save energy.

2. **Appliances**

   Refrigerators should be placed in cool rooms and washing machines and refrigerators should be set at optimal rates.

3. **Models**

   Different models and makes of washing machines, refrigerators and TV sets may consume different amounts of electricity

19.4.6. Municipal Supplies

The local authorities can plan and save large amounts of energy in water supply, sewerage, waste disposal, street lighting and public transport services.

19.5. INDUSTRIAL ENERGY CONSUMPTION

The large and medium scale industries are more energy intensive and labour extensive whereas small scale industries are relatively labour intensive. Process industries are more energy intensive than engineering industries. Metallurgical plants are very sensitive to quality of energy. The energy conservation measures will depend upon the technology level of individual plants. However some measures are common to all industries. Most energy is consumed by large industries and that is where most energy can be saved.

19.5.1. Industrial Development

1. **Policy matters**

   The project reports for new industries should also be based on energy policy. Recycling of waste heat, energy intensities and conservation should be included.

2. **New technologies**

   The procurement of latest technology should be encouraged by tax rebates, subsidies, import permits. The new technologies are most efficient technologies.

3. **Plant location**

   Proper plant location can help a cooperative cogeneration plant for different industries. It can also make the use of waste heat from the neighbour.

4. **Small scale industry**

   High unemployment makes labour intensive and energy and capital extensive small scale industries very attractive.

19.5.2. Operation Management

1. **Reduction of process time**

   By increasing the productivity, the saving in operation time is always associated with saving in energy consumption.

2. **Training of operators**

   Training of machine operators can make them energy conscious.

3. **Maintenance management**

   Preventive maintenance of machines and energy equipment is necessary for energy efficient operations.

4. **Energy management**

   Proper load management, reduction of start-up losses and efficient process heat utilization can help in energy savings.

5. **Thermal insulation**

   Insulation of furnaces, boilers and piping can save large amount of energy.

6. **Continuous operations**

   Continuous operations are more energy efficient as compared to batch operations.

7. **Waste heat recovery**

   Waste heat can be used to heat air, water by heat exchangers.

8. **Energy Cascading**

   The waste heat discharges from one equipment can be used as input to another lower temperature process.

9. **Co-generation**

   Cogeneration of heat and electricity from the same power plant is a very powerful energy conservation tool.

10. **Electrical equipment**

    Replacement of obsolete electrical equipment can improve energy efficiency and load factors.

11. **Recycling of materials**

    Reduction and recycling of waste materials save energy directly or indirectly.

12. **Internal transport system**

    Internal transport system of an industry can consume large amounts of energy and optimization can help conserve energy.

13. **Lighting**

    Factory lighting should be designed with automatic switching, optimal use of daylight, proper lamps, regular cleaning of lights. These measures can save more than 50% energy.

14. Water heating

Hot water is needed in factory canteen and energy conservation measures should be used.

15. Space conditioning

It is necessary to ensure proper working conditions for workers and products and processes. Waste heat can by used for space heating.

19.6. POWER GENERATION AND DISTRIBUTION

The maximum efficiency of thermal power plants is 35 – 40% and 60% of energy is lost as waste heat. The average generation efficiency in India is 27%. By superimposing 23% transmission and distribution losses, the overall efficiency is only 21%. The decentralized power generation by antonomous power plants working on commercial fuels as well as local resources of solar, wind, hydro, biomass, etc. can help to save lot of energy.

19.6.1. Power Generation

1. Preventive maintenance

Absence of preventive maintenance at power plants due to lack of trained workers, spare parts and finances results in heavy energy losses. Better training of personnel and detailed operating instructions can contribute to energy efficiency.

2. Operating parameters

Better control of pressures and temperatures in the steam turbine cycle are needed to achieve higher efficiencies.

3. Unit size

Installation of a number of smaller units in place of large unit can save energy by operating at full capacity.

4. Prime movers

Replacement of gas turbines by diesel engines and use of cogeneration can help save energy.

5. Energy source

Replacement of oil fired plant by hydropower, wind turbines and plants using local fuels can save lot of energy.

19.6.2. Power Supply Management

1. Peak shaving

The peaks in demand can be shaved by energy storage system and electricity pricing.

2. Autonomous power generation

Use of autonomous small capacity generators having low efficiencies should be discouraged by ensuring high quality public supply of electricity.

3. Captive power plants

These plants can become technically fesasible and economically viable if excess electricity is returned to the power grid.

4. Types of generators

Use of different types and sizes of generating units for base and peak loads will increase overall efficiency.

5. Grid size

The grid size should be increased by interconnection of power plants for better utilization of base load units. Antonomous power generators is a better alternative when demand is low on the grid.

6. Load management

Use of computer is necessary to avoid overloading.

7. Incentives

Sufficient incentives should be provided to the consumer by differential pricing to encourage energy saving.

19.6.3. Power Transmission and Distribution

1. Grid maintenance

Regular maintenance of power lines, transformers and capacitors can save energy.

2. Capacitors

High quality capacitors can be installed to improve power factor above 95%. These should be located near to generating units.

3. Transformers

The no-load losses in the old transformer can be prevented by installing transformers with 15% impedence.

4. Size of distribution system

The length of distribution lines should be less than 500m. Increase number of transformers and split the distribution lines to achieve the above target.

5. Conductor size

The diameter of cables should be increased to reduce losses in the distribution lines.

6. Voltage

Resistance losses are lower at higher voltages.

7. Distribution losses

Twenty five to fifty percent of distribution losses are due to inaccurate metering and theft. Better management and control can reduce these losses.

19.7. ENERGY EFFICIENT BUILDING

A residential or commercial or institutional building requires electrical energy for lighting, airconditioning and gadgets (refrigerator, TV set, washing machine, computer, etc.) and thermal energy for space and water heating. The annual energy requirements in a cold place can be as high as 220 to 270 kWh per m^2 of building floor area. The annual energy load depends upon:

1. Shape and size of the building.
2. Thermal properties of building materials.
3. Energy storage capacity of the building envelope (walls, windows, celing, roof, floor, basement).
4. Efficiency of airconditioning, heating and ventilation system.
5. Use of solar energy.
6. Behaviour of the residents or users.

The following factors can reduce the energy requirement.

1. **Shape and size**

   The energy requirement for space conditioning of a building can be reduced by proper architectural design. A cube shape of a building has the lowest ratio of exposed surface to the volume of the building. Any deviation from cubic shape will lead to heat transmission through the envelope; more heat loss in winter and more heat load in summer.

2. **Multi-storey compartment**

   Flat in a multi-story building has lower overall thermal transmittance than a bungalow or a cottage.

3. **Thermal Design**

   The thermal design of a building based on use of passive solar energy and proper building materials and components (windows) can help reduce the annual energy requirements. The thermal insulation of the walls and the ceiling can further help in lowering the energy requirements.

The annual energy requirements of new houses have been reduced to 40 – 75 kWh/m^2/year by following proper design guidelines.

19.7.1. Low Energy Building

A low energy building has the following features. Although the additional building costs are higher by 3 to 8% but annual energy requirements of 50 – 70 kWh/m^2/year are achieved.

1. Compact building design.
2. Airtight and energy efficient windows.
2. Thermal insulation of the building envelope.
4. Passive solar energy utilization.
5. Use of construction materials with high thermal resistance.
6. Elimination of materials and components of low thermal resistance in any part of the building.
7. Mechanical ventilation without infiltration of outside cold air.

19.7.2. Passive Solar House

The annual energy requirements of a passive solar house has been reduced to 15 kWh/m^2/year. This has been achieved by:

1. Effective thermal insulation
2. Airtight windows
3. Controlled ventilation
4. Combination of active and passive solar energy use.

19.7.3. Zero Energy Building

A zero energy building uses only natural energy and no fossil fuels. The natural energy sources can be solar, wind, biomass, ground and water heat. It has the following features :

1. Thermal protection
2. Use of internal energy resources
3. Waste heat recovery

4. Use of solar photovoltaic and captive thermal plants for water heating and space conditioning
5. Seasonal heat storage.

19.8. HEATING LOAD OF A BUILDING

The heating load of a building is the difference between building heat loss and energy gain from solar energy and internal energy sources. The building heat loss is the sum of heat loss by transmission through the envelope and heat loss due to infiltration of cold air.

The heating energy of a building is given by :

$$Q_h = (U_b\, A_b + C_p\, Z\, V_p)(t_i - t_o) \quad [W]$$

where U_b = Overall heat transfer coefficient of the building [W/m^2–K]

A_b = Total exposed area of the building envelope [m^2]

C_p = 0.34 Wh/m^3 – K = Specific heat of air.

Z = Air exchange rate (l/h)

V_b = Building space [m^3]

t_i = Indoor temperature [°C]

t_o = Outside air temperature (°C)

The transmission heat loss can be calculated as :

$$Q_{tr} = 24 \times 3600\; U_b\, A_b\, DD\ (J)$$

where 24 × 3600 = the time factor [s/day]

DD = Heating degree-day for a month [K-days]

The U_b of a building can be calculated as.

$$U_b = \frac{\sum U_i\, A_i}{A_b} \quad [\text{W/m}^2\text{–K}]$$

Example 19.1. *A home has total space volume of 350m^3, envelope area is 378 m^2, design indoor temperature 18°C and outside temperature – 3°C. The annual heating hours are 1570 hour. Calculate the design heat load and annual energy requirements of the building. Take U_b = 0.65 W/m^2 – K and Z = 0.5 l/h.*

Solution : 1. Design heat load

$$Q_l = (U_b\, A_b + C_p\, Z\, V_b)\,(t_i - t_o)$$
$$= (0.65 \times 378 + 0.34 \times 0.5 \times 350)\ (18 + 3)$$
$$= 6409.2 \text{ W.}$$

2. Annual energy requirement

$$Q_e = Q_l \times N = 6409.2 \times 1570$$
$$= 10.062 \text{ MWh/a.}$$

19.9. THERMAL INSULATION OF A BUILDING

The thermal resistance of a component of a building can be calculated as :

$$R = \frac{1}{U} = \frac{1}{h_i} + \sum\left(\frac{\delta}{k}\right) + \left(\frac{\delta}{k_i}\right) + \frac{1}{h_o} \quad [m^2K/W]$$

where U = Overall heat transfer coefficient [W/m^2–K]

$\left(\dfrac{\delta}{k_i}\right)$ = thermal resistance of insulation [m^2 K/W]

$\sum\left(\dfrac{\delta}{k}\right)$ = thermal resistance of wall (m^{2}K/W)

h_i = Coefficient of convective heat transfer of inside film [m^2 K/W]

h_o = Coefficient of convective heat transfer of outside film [m^2 K/W]

δ = thickness of component [m]

l_h = thermal conductivity of material [W/mK]

The required thickness of thermal insulation

$$\delta_i = k_i\,(R_i - R) = k_i\left(\frac{1}{U_i} - \frac{1}{U}\right)$$

Example 19.2. *Calculate the thickness of thermal insulation for reducing transmission losses by 80%. The value of U = 1.4 W/m^2 – K. The thermal conductivity of insulation is 0.04 W/m–K.*

Solution :

$$U_i = 0.2U = 0.2 \times 1.4$$
$$= 0.28 \text{ W/m}^2 - \text{K.}$$

The thickness of insulation required,

$$\delta_i = k_i\left(\frac{1}{U_i} - \frac{1}{U}\right)$$

$$= 0.04\left[\frac{1}{0.28} - \frac{1}{1.4}\right] = 0.114 \text{ m}$$

TUTORIAL SHEET NO. 19.1.

1. Recommend energy saving measures for transport sector.
2. Discuss the energy conservation measures to be adopted for villages.
3. Write brief note on the energy savings possible for urban sector.
4. Explain the following :
   (a) Energy conservation in industries
   (*b*) Energy conservation in power generation and distribution.
5. Explain the following :
   (*a*) Low Energy Building
   (*b*) Passive Solar House
   (*c*) Zero Energy Building
6. What do you understand by 'Energy Efficient' Building? Explain the various design features of such a building.
7. How is heating load calculated for a building?
8. Explain with an example the thermal insulation design of a building.

Chapter 20

ENERGY STORAGE

Introduction; Characteristics of Energy Storage System ; Storage of Mechanical Energy ; Electrochemical Energy Storage (Battery) ; Thermal Energy Storage ; Hot Water Storage System ; Packed Bed Storage System ; Latent Heat Storage System ; Medium and High Temperature Storage ; Storage System for Solar Plants ; Thermal Chemical Energy Storage System.

20.1. INTRODUCTION

An energy storage system is required to meet the difference between the energy demand by the customer and energy supply by the power plant. It is very important for solar, wind energy and other renewable forms of energy because these are intermittent in nature. The energy storage systems can be classified as follows :

1. Mechanical energy storage
   Flywheel energy storage, pumped hydro power storage and compressed air energy storage.
2. Thermal and thermochemical energy storage,
3. Electrical energy storage
   Battery storage and magnetic field energy storage.

20.2. CHARACTERISTICS OF ENERGY STORAGE SYSTEM

An energy storage system is characterised by the following performance characteristics :

1. Storage capacity
2. Energy density
3. Charging and discharging rate
4. Storage duration
5. Storage efficiency
6. Economics, initial cost, life span and payback period.

20.2.1. Storage Capacity

An energy storage system consists of a reservoir which contains a storage medium. Storage capacity is the maximum energy quantity stored in one cycle. A hot water storage system has the following capacity.

$$E = m\, C_p\, (t_i - t_f) \text{ [kJ]}$$

where m = mass of water [kg]

C_p = 4.186 kJ/kg-K
= Specific heat of water
t_i = initial temperature [°C]
t_f = final temperature [°C]

20.2.2. Energy Density

Energy density of a storage system is the energy stored per kg or m^3 of storage medium

$$e = \frac{E}{m} \text{ [kJ/kg]}$$

The energy density of various storage media and systems are given in Table 20.1

Table 20.1. Energy Density of Storage Systems

Storage System/Medium	Energy Density [Wh/kg]
1. Sensible Heat Storage [ΔT = 50K]	
Water	58
Iron	8
Coarse gravel	25
Concrete & rock	13
2. Latent Heat Storage	
Ice (MP = 0°C)	93
Paraffin (MP = 42-67°C)	47-52
Inorganic salt (MP 29-70°C)	48-70
Lithium Fluoride (LiF, MP = 848°C)	283
3. Mechanical Energy Storage	
Flywheel	20-30
Pumped hydro (Δh = 300m)	0.82
4. Hydrogen and Synthetic Fuels	
Liquid hydrogen	33000
Metal hydrides	600-2500
Methanol	7390
Ethanol	7695
5. Batteries	
Lead-acid battery (20-30°C)	40
Iron-nickel battery	60
Ni-Cd battery	100
Sodium sulphur and Lithium sulphur batteries (300-375°C)	150

20.2.3. Charging and Discharging Rate

Charging rate is the amount of energy supplied to the storage system per unit time. The discharging rate is the amount of energy withdrawn from the storage system per unit time.

20.2.4. Storage Duration

The operating cycle of an energy storage system consists of :

1. Charging from an energy source,
2. Energy storage in the system,
3. Withdrawal of energy from the system and supply to the consumer.

The total storage cycle time,

$$\tau = \tau_c + \tau_s + \tau_d$$

where τ_c = charging duration

τ_s = storage duration

τ_d = discharging duration

A hot water storage system may have storage duration for 8 to 24 hours. In other solar heating plants, storage duration may be for a few months.

20.2.5. Storage Efficiency

The storage efficiency may be defined as the ratio of discharged energy to charged energy

$$\eta = \frac{E_d}{E_c} = 1 - \frac{\Delta E_l}{E_c}$$

where E_d = Discharged energy [J]

E_c = Charged energy [J]

ΔE_l = Loss of storage energy [J]

20.3. STORAGE OF MECHANICAL ENERGY

20.3.1. Flywheel Energy Storage

It consists of a rotating mass (flywheel), a motor generator set and an enclosure. The rotating mass acts as storage medium and energy is stored as kinetic energy of flywheel. During charging the motor adds energy to the flywheel and during discharging, the energy is generated to cover the peak load. Various shapes of flywheel include a circular disc or a circular ring or a thin long shaft.

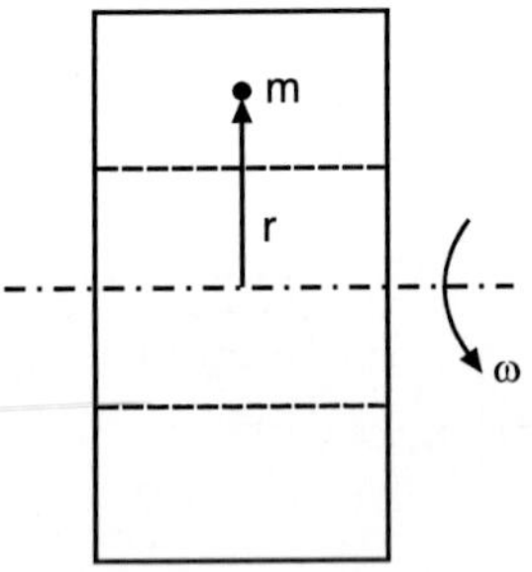

Fig. 20.1. Flywheel

The flywheel energy,

$$E = \frac{1}{2} I\omega^2 [j]$$

where I = Moment of inertia of flywheel [kg m^2]

ω = Angular velocity [s^{-1}]

$$I = \frac{1}{2} mr^2$$

where m = mass [kg]

r = radius [m]

The maximum energy in fully charged state,

$$E_{max} = \frac{1}{2} M \omega_{\max}^2 \quad [J]$$

The energy which can be withdrawn,

$$E_d = \frac{1}{2} I(\omega_{\max}^2 - \omega_{\min}^2) \quad [J]$$

where ω_{min} = Minimum angular velocity $[s^{-1}]$

The energy density per kg of flywheel mass,

$$e = \frac{E}{m} = \frac{1}{4} r^2 \left(\omega_{\max}^2 - \omega_{\min}^2\right)$$

$$= k\left(\frac{\sigma}{\rho}\right) \quad [J/kg]$$

where k = shape factor from 0 to 1

σ = allowable yield stress of flywheel $[N/m^2]$

ρ = density of material $[kg/m^3]$

The value of energy density depends upon the size, shape and material of flywheel. A disc of uniform strength has the shape factor equal to one. Titanium gives the highest energy density of the value 220 kJ/kg. Composite materials are used for higher energy densities. A flywheel made of circular rings of glass/epoxy composite can give an energy density of 450 kJ/kg and graphite/epoxy composite ring a value of 540 kJ/kg.

20.3.2. Pumped Hydro Power Storage

A pumped hydro power plant as shown in Fig. 20.2 consists of an upper reservoir and a lower reservoir connected by a pressure pipe line through a pump-motor-turbine-generator set. When electrical power demands are low, the water is pumped from lower reservoir to higher reservoir by spare power. During peak load, the potential energy of stored water is converted into its kinetic energy and is used to generate extra power by turbine-generator set.

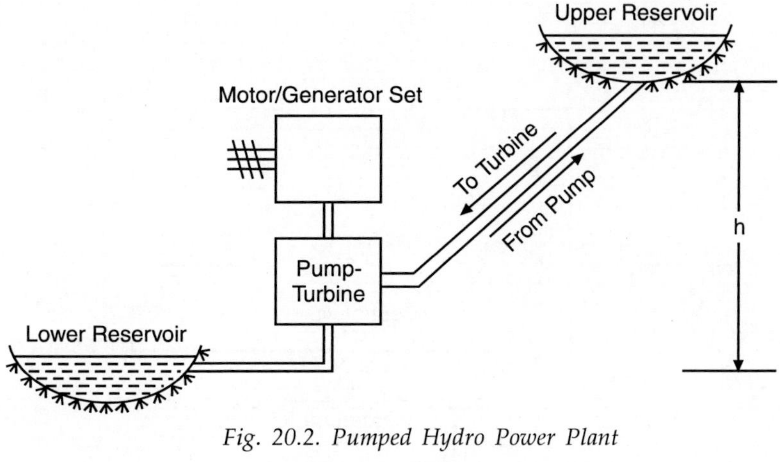

Fig. 20.2. Pumped Hydro Power Plant

The pumping power required,

$$P_p = \frac{\rho g Q_p h_p}{\eta_p} \quad [W]$$

where $g = 9.81\ m/s^2$
= acceleration due to gravity
ρ = Water density [≈ 1000 kg/m³]
Q_p = Pump water discharge [m³/s]
h_p = Pump head [m]
η_p = Overall pump-motor efficiency

The turbine power available,

$$P_T = g \rho Q_T h_T \eta_T \eta_G \quad [W]$$

where Q_T = Turbine water discharge [m³/s]
h_T = Turbine head [m]
η_T = Turbine efficiency
η_G = Generator efficiency

The overall efficiency of pumped hydro storage power plant,

$$\eta_o = \frac{P_e}{P_p} = \eta_p \eta_T \eta_G.$$

20.3.3. Compressed Air Energy Storage

A compressed air energy storage system as shown in Fig. 20.3. consists of the following components :

1. An air compressor (C)
2. Motor-generator set, (M/G set)
3. Air cooler,
4. Underground air reservoir
5. Gas turbine (GT)
6. Combustion Chamber (CC)

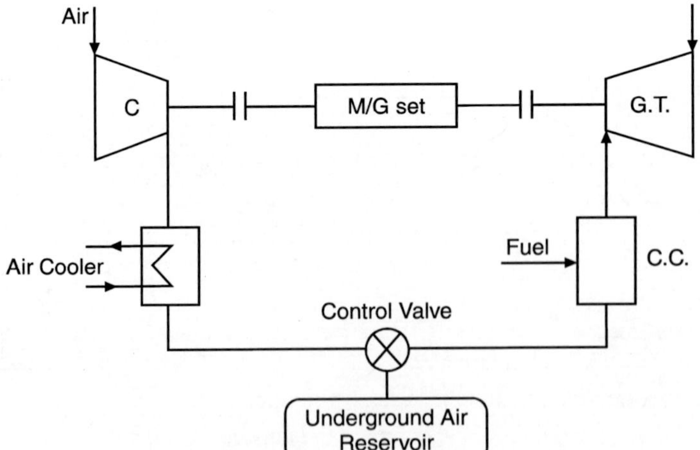

Fig. 20.3. Compressed Air Storage System

When electrical demand is low, the spare power is used to drive air compressor with the help of motor. The cooled air is stored in the underground reservoir. When demand is high, the compressed air from the underground is used to drive gas turbine to meet the peak loads.

Presently, there are only two compressed air storage systems operating in the world. The specifications of Huntorf of (Germany) storage system are :

Volume of underground reservoir	= 300,000 m^3
Depth of reservoir	= 650-800 m
Pressure of stored air	= 72 bar
Compressor power	= 60 MW
Air flow rate from compressor	= 108 kg/s
Turbine power	= 290 MW
Air flow rate to turbine	= 425 kg/s
Energy density	= 4 kWh/m^3
Overall efficiency of plant	= 42%
Storage duration	= 4 hours

The turbine power,

$$P_T = m_T \, \Delta h_T \, \eta_T \quad [\text{kW}]$$

where m_T = mass flow rate through gas turbine [kg/s]

Δh_T = isentropic enthalpy drop through gas turbine [kJ/kg]

η_T = isentropic efficiency of turbine.

The power required to drive the air compressor,

$$P_c = \frac{m_c \, \Delta h_G}{\eta_c} \quad [\text{kW}]$$

where m_c = mass flow rate in the compressor [kg/s]

Δh_c = isentropic enthalpy drop in the compressor [kJ/kg]

η_c = isentropic efficiency of compressor

The enthalpy drop through compressor or turbine can be calculated as :

$\Delta h = C_p \, \Delta T$ [kJ/kg]

C_p = Specific heat of air [kJ/kg-K]

ΔT = Change of temperature in compressor or turbine [K]

The overall efficiency of compressed air storage plant,

$$\eta = \frac{P_T}{P_c}$$

20.4. ELECTROCHEMICAL ENERGY STORAGE (BATTERY)

An electrochemical cell or a battery consists of two electrodes and electrolyte in a box. When the cathode and anode are interconnected externally through a resistance, electrochemical reaction takes place in the cell. Lead-acid batteries, nickel-cadmium batteries and other batteries are used to store electrical energy.

A lead-acid battery has anode of porous lead, cathode of lead oxide and acqueous solution of sulphuric acid is used as electrolyte.

The overall reaction is

$$2H_2SO_4 + Pb^- + Pb^+O_2 \underset{\text{Charge}}{\overset{\text{Discharge}}{\rightleftharpoons}} PbSO_4 + 2H_2O + PbSO_4$$

The batteries have the following specifications

Power density = 50 W/kg
Energy density = 200 Wh/kg
No. of cycles = 1000
Life span = 4 to 6 years

The principle of lead-sulphuric acid battery is shown in Fig. 20.4.

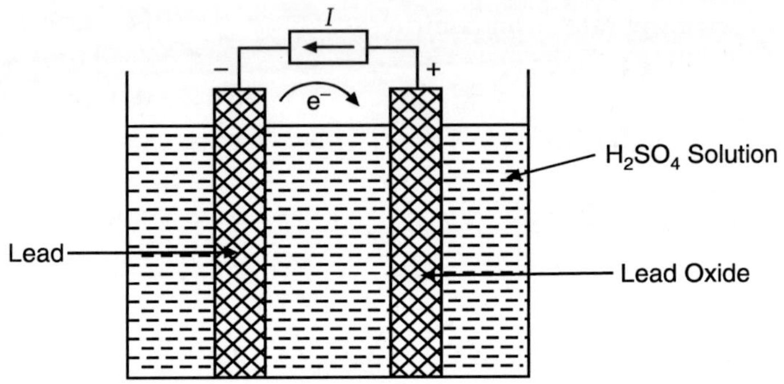

Fig. 20.4. Lead-acid battery

The voltage drop of internal resistance,

$$V = E - IR_i \quad [V]$$

where E = emf [V]
I = current [A]
R_i = internal resistance [Ω]

The current in the external circuit,

$$I = \frac{E}{R_i + R_e} \quad [A]$$

where R_e = external resistance [Ω]

Also, $I = nF$ [A]

where n = No. of electrons/mole of H_2SO_4
= 2
F = Faraday's constant = 96487 C/mole

The power output of the battery

$$P = VI \, [W]$$

Example 20.1. *A 12-V lead-acid battery supplies a current of 60A. Calculate the power output and consumption of H_2SO_4.*

Solution : $P = VI = 12 \times 60 = 720$ W

The number of acid moles,

$$N = \frac{2I}{nF} \quad [\text{moles/s}]$$

The consumption of sulphuric acid,

$$m = NM$$

The molecular mass of H_2SO_4 = 98 g/mole

$$N = \frac{2 \times 60}{2 \times 94687} = 6.34 \times 10^{-4} \text{ moles/s}$$

$$m = 6.34 \times 10^{-4} \times 98 = 0.0621 \text{ g/s.}$$

20.5. THERMAL ENERGY STORAGE

There are three types of systems :

1. Sensible heat storage
2. Latent heat storage
3. Thermal chemical energy storage

The thermal energy storage systems can also be classified as :

1. Low temperature, below 100°C
2. Medium temperature, 100-500°C
3. High temperature, above 500°C.

These systems can use :

1. Liquid as storage medium,
2. Solid as storage medium

The main parameter is specific heat of storage medium and change in temperature.

The latent heat storage system uses sensible heat as well as latent heat. There is a phase change, *i.e.*, melting and freezing. The heat is supplied from a solar collector which melts the storage medium. When heat is withdrawn, the storage medium is frozen again. The storage material should be physically and chemically stable in the operating range of temperature and should possess high heat of fusion. The latent heat storage system has higher energy density than sensible heat storage system.

20.6. HOT WATER STORAGE SYSTEM

A hot water storage system is shown in Fig. 20.5. Water is used for storage below 100°C. For high temperatures steam, organic fluid, molten salts or molten metals can be used as storage materials.

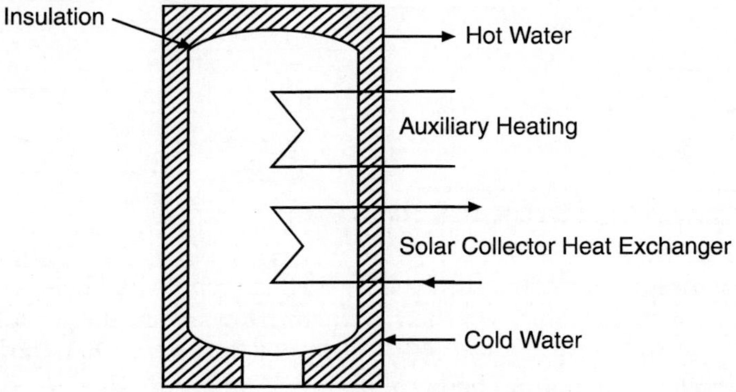

Fig. 20.5. Hot Water Storage Tank

The storage capacity is

$$Q_s = mC(t_{max} - t_{min})$$
$$= \rho QC\,(t_{max} - t_{min})\ [J]$$

where m = mass of water [kg]

C = 4.187 kJ/kg-K = Specific heat of water.

Q = Water volume [m^3]

ρ = 1000 kg/m^3 = Density of water

t_{max} = maximum storage temperature [°C]

t_{min} = minimum storage temperature [°C]

The energy density,

$$q = \frac{Q_s}{m} = C\,(t_{max} - t_{min})\quad [J/kg]$$

Charging heat rate,

$$Q_c = m_c C\,\Delta T_c \quad [W]$$

Discharging heat rate

$$Q_d = m_d\, C\,\Delta t_d \quad [W]$$

Example 20.2. *Determine the mass and volume of water in hot water storage system for a capacity of 300 kWh.*

The maximum and minimum temperatures of water are 80°C and 20°C.

Solution : 1. Mean temperature of water

$$= \frac{80 + 20}{2} = 50°C$$

$$C = 4181\ J/kg\text{-}k$$
$$\rho = 988\ kg/m^3$$

2.
$$m = \frac{Q}{C(t_{max} - t_{min})}$$

or
$$m = \frac{300 \times 3600}{4.181 \times (80 - 20)} = 4305.2\ kg$$

3.
$$Q = \frac{m}{\rho} = \frac{4305.2}{988} = 4.357\ m^3$$

4. Energy density,

$$q = \frac{Q}{m} = \frac{300}{4305.2} = 0.07\ kWh/kg.$$

20.7. PACKED BED STORAGE SYSTEM

Concrete, earth, rock, gravel and granite are suitable materials for low temperature heat storage. In a solar air heater a vertical pebble bed heat storage is used. During charging hot air from solar collector flows from the top of pebble bed. There is thermal stratification and the upper and lower beds are at different temperatures, *i.e.*, the upper bed is at higher temperature than lower bed. During discharging the cold air enters at the bottom of the bed and hot air leaves at the top. The thermal stratification helps to get air at higher temperature from the bed top. The temperature difference between air and pebbles is small due to large surface area of the bed and high air-to-pebble heat transfer coefficient.

The storage capacity is given by

$$Q_s = Q(1 - \varepsilon)\,\rho_p\,C(t_{max} - t_{min}) \quad [\text{J}]$$

where Q = Storage volume [m^3]

ε = Pebble bed porosity

C = Specific heat of pebbles [J/kg-K]

ρ = Density of pebble [kg/m^3]

t_{max} = max. temperature of storage [°C]

t_{min} = min. temperature of storage [°C]

20.8. LATENT HEAT STORAGE SYSTEM

The storage capacity can be calculated as follows :

$$Q_s = m\left[C_s(t_m - t_{\min}) + h_m + C_l(t_{\max} - t_m)\right] \quad [\text{J}]$$

where m = mass of storage material [kg]

C_s = Specific heat of storage material in liquid state [J/kg-K]

C_l = Specific heat of storage material in liquid state [J/kg-K]

h_m = heat of fusion of storage material [J/kg]

t_{max} = maximum storage temperature [°C]

t_{min} = minimum storage temperature [°C]

t_m = fusion temperature of storage material [°C]

The storage capacity includes major part of latent heat and some part of sensible heat.

The charging process consists of :

1. heating of storage material in solid state from initial temperature (t_{min}) to melting point (t_m). This is sensible heat of storage.
2. melting of storage material at t_m. This is latent heat of storage.
3. heating of liquid storage material to maximum temperature ($t_{\max}$). This is also sensible heat of storage.

The discharging process follows the reverse process and consists of :

1. the cooling of liquid from t_{max} to t_m.
2. solidification (crystallization) at t_m.
3. cooling of material in solid state from t_m to t_{min}.

Storage Material

The efficient operation of latent heat storage system depends upon proper selection of storage material.

1. It should have large heat of fusion.
2. It should have large specific heats.
3. The melting point of material should be within the working range of storage system.
4. The material should have physical and chemical stability.

5. The volume change during melting and fusion should be minimal.
6. The thermal conductivity should not be very low.

The main advantage of latent heat storage system is reduction in the mass and volume of the material and the system.

The comparative indices of latent heat and sensible storage system are given in Table 20.2.

Table 20.2. Comparative Study of Latent Heat and Sensible Heat Storage System.

Parameter	Sensible heat storage		Latent heat storage	
	Water	Pebble bed	Glauber salt (Na_2SO_4, $10H_2O$)	Paraffin
1. Mass (kg)	7933	39667	2200	2500
2. Relative mass	1	5	0.28	0.32
3. Volume (m^3)	7.93	38.1*	1.5	3.18
4. Relative volume	1	4.8	0.19	0.40
* Pebble bed porosity = 0.35				

The following problems are faced in the low-temperature latent heat storage system.

1. There can be subcooling of liquid below the melting point during discharging process.
2. There are substantial changes in the volume during phase change.
3. Heat transfer rate can be very low between the working fluid and storage material.

These difficulties can be overcome by the use of hybrid system.

There can be a hybrid heat storage system. The sensible heat storage has water as storage material and latent heat storage system has paraffin.

The total storage capacity of a hybrid plant can be calculated as follows :

$$Q_{hs} = \left[m_s\, c_s\,(t_{\max} - t_{\min}) + m_l \left[c'_s (t_s - t_{\min}) \right] + h_m + C_l\,(t_{\max} - t_m) \right] \quad [\text{J}]$$

where m_s = mass of sensible heat storage material [kg]

m_l = mass of liquid [kg]

c_s = Specific heat of sensible heat storage material [J/kg-k]

c'_s = Specific heat of latent heat storage material in solid phase [J/kg-k]

c_l = Specific heat of latent heat storage material in liquid phase [J/kg-k]

t_{max} = maximum storage temperature [°C]

t_{min} = minimum storage temperature [°C]

t_m = melting point of latent heat storage material [°C]

h_m = heat of fusion of latent heat storage material [J/kg]

20.9. MEDIUM AND HIGH TEMPERATURE STORAGE

Thermal energy storage system for medium and high temperature applications are very important for solar plants used for power generation, process heat, refrigeration and space cooling. The high temperature storage plant (above 500°C) can enhance the performance and availability of a solar power plant. The storage capacity should be sufficient to ensure full-load operation of solar power plant for $\frac{1}{2}$ to 3 hours.

20.9.1. Medium Temperature Storage Systems

The following materials and systems can be used for medium temperature storage systems.

Table 20.3. Medium Temperature Storage Systems

Temperature Range	Storage System
1. upto 200°C	Hot water/Steam Storage
2. 200-400°C	Oil Storage
3. 300--500°C	Sodium (molten salt sensible heat storage system)
4. 280-500°C	Latent heat storage system with Nitrates of alkali metals and earth-alkali metals.
5. 500-1300°C	Sensible heat storage with Silica (SiO_2), Alumina (Al_2O_3) and Magnesia (MgO)

The thermo-physical properties of storage materials used for medium temperatures are given in Table 20.4.

Table 20.4.Properties of Storage Materials for Medium Temperatures

Storage Material	M.P. t_m (°C)	Fusion heat h_m (kJ/kg)	Con-ductivity k (N/mk)	Specific heat C (J/kg-K)	Density ρ (kg/m^3)
$LiNO_3$	252	530	1.33	2030	2130
KNO_3	337	150	0.43	1750	1860
NaCl	800	520	4.00	950	2160
$NaNO_3$ + NaCl (93.6% + 6.4%)	294	171	0.61	1800	1880
KNO_3 + KCl (94.5% + 4.5%)	320	150	0.48	1210	1890
KNO_3 + KBr + KCl (88% + 4.7% + 7.3%)	342	140	1.00	1000	1887
Steel	—	—	35	550	7850
Cast iron	—	—	45	500	7800
Concrete	—	—	2.1	1000	2400

20.9.2. Steam Storage System

Hot water is used as storage material. Steam is supplied through nozzles and pressure of storage tank increases from p_1 to p_2. The steam gets condensed due to mixing with water. There is a temperature rise from t_1 to t_2 and enthalpy rise from h_1 to h_2.

The mass of steam supplied can be determined from the equation of balance of storage energy.

$$m_s = \frac{m_{w_1}(h_1 - h_2)}{(h_s - h_2)} \quad [\text{kg}]$$

where m_{w_1} = Initial mass of water in the storage tank at p_1 and t_1.

h_1 = initial enthalpy of water [kJ/kg]

h_2 = final enthalpy of water [kJ/kg]

h_s = enthalpy of steam supplied [kJ/kg]

The mass of boiling water after charging,

$$m_{w_2} = m_{w_1} + m_s \ [\text{kg}]$$

It is available at p_2 and t_2.

During discharging, steam is withdrawn from the storage tank. There is an adiabatic pressure drop Δp. A portion of water is vaporised and temperature is changed to saturation temperature. The saturated steam is available for further processing.

A steam storage system used for a nuclear power plant of 400 MW capacity has the following specifications.

Storage capacity : 64 units of 580m^3 capacity each

Steam superheaters : 8 units of 180m^3 capacity each

Charging pressure : 20 bar and 48 bar

Charging temperature : 212°C and 260°C

Storage duration : 2 hours

20.10. STORAGE SYSTEMS FOR SOLAR PLANTS

The heat storage system of a parabolic trough solar power plant as shown in Fig. 20.6 consists of twin tanks. In one tank, hot oil is supplied from solar collector which is fed to second tank. The cooled oil is returned to the solar collector again. The maximum allowable oil temperature is limited to 400°C to avoid its decomposition.

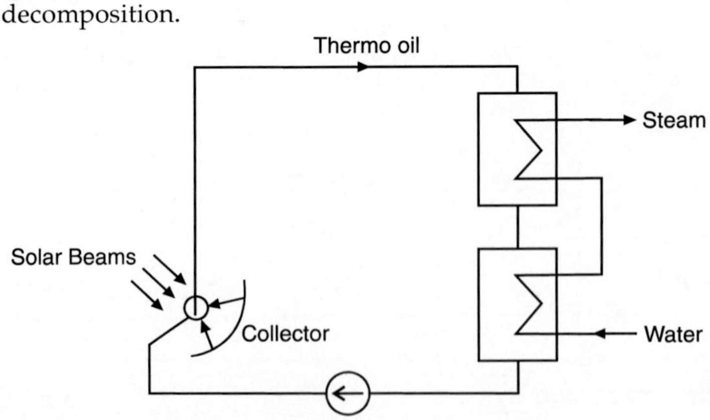

Fig. 20.6. Twin Tank Heat Storage System

Example 20.3. *The storage duration for a solar power plant should be 3 hours full load operation. The following are the specifications*

Power plant rating = *50 MW*
Plant thermal efficiency = *38%*
Storage efficiency = *85%*
Storage material = KNO_3
Storage temperature = *±30K*

Calculate the amount of potassium nitrate salt.

Solution : 1. Storage capacity,

$$Q_s = \frac{P\tau_s}{\eta_{th}\,\eta_s} = \frac{50 \times 3600 \times 3}{0.38 \times 0.85}$$

$$= 1671750 \text{ MJ}$$

$$= 1.672 \times 10^9 \text{ kJ}$$

2. From data book,

Fusion temperature, t_m = 337°C
Heat of fusion, h_m = 150 kJ/kg
Specific heat, C = 1.75 kJ/kg-K
Δενσιτψ, ρ = 1860 kg/m^3

3. The amount of storage material,

$$m = \frac{Q_s}{[h_m + C(t_{\max} - t_{\min})}$$

If $\Delta t_s = \pm 30\text{K}$

$$t_{max} - t_{min} = 60\text{K}$$

$$\therefore \qquad m = \frac{1.672 \times 10^9}{(150 + 1.75 \times 60)}$$

or $\qquad m = 6556863 \text{ kg} = 6557 \text{ tonnes}$

4. Volume capacity

$$Q = \frac{m}{\rho} = \frac{6556863}{1860} = 3525.2 \text{ m}^3$$

Example 20.4. *A latent heat storage system is to be designed for a solar plant. The storage capacity should be 2 MWh and highest temperature is 260°C. The temperature variation should be less than 20 K. Select a suitable storage material and its quantity required.*

Solution : From Table 20.4, $LiNO_3$ is most suitable material

$$t_m = 252°\text{C}$$

$$h_m = 530 \text{ kJ/kg}$$

$$C = 203 \text{ kg/kg-K}$$

$\therefore$ mass of storage material,

$$m_s = \frac{Q_s}{(h_m + C\Delta T_s)}$$

$$= \frac{15 \times 3600}{(530 + 203 \times 20)} = 94.4 \text{ kg}$$

20.11. THERMOCHEMICAL ENERGY STORAGE SYSTEM

Heat of reaction of reversible chemical reactions is used to store energy. The system is suitable for storing solar energy. An endothermic reaction is used for energy storage. Steam reforming of methane or dissociation of sulphur trixoide, methanol or ammonia is used and reaction takes place at 500°C and more in the presence of a catalyst. The reaction products can be stored locally or transported in a pipeline to the energy consumer location. At the consumer point, an exothermic reaction takes place and heat is released due to recombination of reaction products.

This system has the following advantages :

1. It has high energy density
2. The energy losses during storage and transport are negligible.

Examples of reactions

$$Na_2S \cdot 5H_2O \rightleftharpoons Na_2S + 5H_2O \; (e_V = 500 \text{ Wh/m}^3)$$
$$H_2SO_4 \cdot H_2O \rightleftharpoons H_2SO_4 + H_2O \; (e_V = 300 \text{ Wh/m}^3)$$
$$Ca(OH)_2 \rightleftharpoons CaO + H_2O \; (e_V = 250 \text{ Wh/m}^3]$$

20.11.1. Chemical Heat Pipe

The following endothermic reaction takes place at 960°C.

$$CH_4 + H_2O = CO + 3H_2 - 6020 \text{ kJ/kg of } CH_4$$

The reaction products ($CO + H_2$) carriers the energy through a pipe over long distances, from North Africa to Europe. The exothermic reaction at consumer point.

$$CO + 3H_2 = CH_4 + 6020 \text{ kJ/kg of } CH_4$$

The catalysts used are rhodium or nickel.

This system is very useful for transportation of solar energy over long distances.

TUTORIAL SHEET NO. 20.1.

1. How is an energy storage system characteristised ?
2. Describe the flywheel mechanical energy storage system
3. Differentiate between pumped hydro and compressed air storage systems
4. How is a battery used to store electrical energy ?
5. Compare the performance of the following systems.
   (a) Water storage system
   (b) Packed bed storage system
6. How storage capacity of a latent heat storage system is determined ? Compare the performance of various materials ?
7. What is the significance of medium and high temperature storage for solar energy ?
8. Write notes on :
   (a) Steam storage system
   (b) Twin tank heat storage system
9. How can solar energy be stored and transported over long distances ?

□□□

PART VI

PROJECTS AND PAPERS (BY THE AUTHOR)

PROJECTS AND PAPERS

Paper-1

PROSPECTS FOR WIND ENERGY UTILIZATION IN HARYANA*

The importance of wind as an alternative source of energy will increase with the depletion of fossil fuels and the need to maintain our ecosystem. In many countries as well as in some parts of India, electric power is being commercially produced from windmills. There is a special need for lift irrigation in many parts of Haryana where surface water is scarce and topography does not permit effective flood irrigation by the canal system. Therefore, there is a large potential in these areas to exploit wind energy for pumping of surface and ground waters, especially to operate water sprinkler systems for irrigation.

The paper is based on the analysis of wind data for various parts of Haryana and study of technological feasibility of wind energy utilization for water pumping, taking into consideration the requirements of crops and availability of groundwater. The paper also describes special research and development projects so that wind energy can be used in the near future in the plains of Haryana and other similar areas.

1. INTRODUCTION

The winds account for a power of 2×10^{13} W. Though this is only a small portion of the incident solar energy when compared, but it is more than the total energy consumption rate of the world. It is estimated that just one per cent of potential wind power is equivalent to 2×10^{11} W or approximately three per cent of the current world energy consumption rate. If wind power were used for the generation of electricity, its thermal equivalent would be as high as 8–9 per cent of the total (Daniel M. Simmons 1975). Technical and economic problems related to the location and variability of wind intensity tend to limit the attractiveness of wind energy conversion systems, but the potential extractable wind power is not insignificant.

Following a better understanding of aerodynamics of rotating aerolifts, there is a revival of interest in harvesting wind energy. The importance of wind as alternative source of energy is increasing with the depletion of fossil fuels and the need to maintain our ecosystem. The Government of India has formulated

The United Nations Library, Tokyo

and launched a special Wind Energy Programme. At a conservative assessment, wind power potential in India is around 20,000 MW (*The Hindustan Times* New Delhi, 29 October 1992). At present, an aggregate wind power capacity of about 43 MW has been established in the country and it is proposed to establish an additional 100 MW capacity in the Eighth Plan. Although in many countries and some parts of India, electric power is being commercially produced from wind-mills, wind energy can be more appropriately utilized for pumping water for irrigation due to high variation of power density of wind over time and location (Srivastava, Girija Sharma 1987). In India, windmills have been tried for pumping water for drinking., irrigation for fodder crops, forest nurseries, and agro-forestry plantations. But, so far, windmills have not been tried among individual farmers and in the irrigation of annually cultivated crops.

There is a special need for lift irrigation in many parts of Haryana where surface water is scarce and topography does not permit effective flood irrigation by the canal system. Therefore, there is a large potential in these areas to exploit wind energy for pumping of ground water, especially to operate sprinkler systems.

2 WIND DATA

The average output of water from windmill pumping system must be matched with the requirements of various crop combinations and availability of groundwater. Haryana is a small state (44 212 sq km) with almost uniform topography (97% alluvial plains and three per cent hilly tracts). Therefore wind data have been compared for two typical tracking stations and is given in Table P.1.1.

Table P.1.1. Mean Daily Wind Speed Data for Haryana

Month	Mean daily wind speed (kmph)	
	Hisar*	Karnal**
January	5.3	4.8
February	6.3	6.9
March	7.5	6.2
April	7.9	6.5
May	8.9	7.4
June	10.7	7.8
July	9.9	4.8
August	7.7	4.7
September	6.7	4.9
October	5.5	4.5
November	4.7	4.3
December	5.4	5.4
Average	7.2	5.7

* Lat. 29° 10'N; Long 75° 44'E; Elev. 221 masl

** Lat. 29° 42'N; Long. 77° 02'E; Elev. 249 masl

Source : Annamani, DA Modey. 1983. Wind Energy Data for India. Indian Institute of Tropical Meteorology. New Delhi : Allied Publishers Pvt. Ltd.

The examination of the wind speeds indicates that Haryana is a region of light breeze with wind speeds 6 to 11 kmph (1.6 to 3.3 m/s). The southern part of Haryana is slightly more favourable than the northern part.

The mean annual energy content of wind can be calculated as equal to 0.00001 ρV_i^3; where V, is the hourly wind speed in kmph for hour i and ρ is the mean air density equal to 2.8703 kg/m^3. The mean annual energy content comes out to approximately 250kWh/m^2.

3. GROUNDWATER SITUATION IN HARYANA

The entire Haryana region has been surveyed for groundwater availability by Groundwater Directorate. Haryana State: and Minor Irrigation and Tubewell Corporation, Karnal. District wise groundwater data are given in Table P.1.2. The broad features of substantial parts of Haryana are promising from the point of used groundwater potential. The irrigation requirements for Haryana are summarized as below (Singal 1985).

Table P.1.2. District-wise Groundwater Data for Haryana

District	Utilizable recharge (ha-m)	Net draft (ha-m)	Pre monsoon water level June, 1989 (m)	Post-monsoon water level October, 1989 (m)
Ambala	35 054	30 276	9.27	7.96
Yamunanagar	44 089	41 013	7.52	6.58
Kurukshetra	63 648	51 546	13.63	13.36
Karnal	121 634	107 426	8.24	8.00
Panipat	62 162	47 448	7.61	7.65
Sonipat	59 855	24 041	5.63	5.63
Faridabad	54 605	20 501	6.81	6.45
Kaithal	89 768	61 298	6.78	7.02
Jind	106 689	20 212	7.66	7.43
Rohtak	86 905	16 391	4.91	4.74
Gurgaon	46 633	35 440	9.15	8.91
Rewari	22 591	15 584	13.98	14.42
Sirsa	94 110	17 161	10.02	9.98
Hisar	120 990	34 557	9.28	9.35
Bhiwani	47 900	13 325	19.66	19.45
Mehendragarh	31 902	22 230	26.89	26.66
Average	—	—	10.44	10.22

Source : Groundwater Directorate, 1990 Groundwater monitoring in Haryana based on observations during June, 1990–*a Technical report*. Haryana: Groundwater Directorate.

Drip irrigation for corn, wheat, cotton, millet = 25 m^3/ha-day Flood irrigation for rice, sugarcane = 50 – 60 m^3/ha-day.

Live stock = 40 – 50 liters/head-day
Lifting height = a few meters to 20 m.
Power required = 135 W/ha for surface water;
400W/ha for groundwater

4. CHARACTERISTICS OF WIND TURBINES

Specific speed or tip-speed ratio of a windmill rotor is the ratio of its peripheral speed to wind speed. Rotors of low tip-speed ratio are preferred for driving piston pumps, which require a high starting torque and low rotational speed. The rotor of low tip-speed ratio also gives high efficiency and low cost. The other important characteristics of windmill rotors are cut-in speed and cut-out speed.

The main characteristics of windmill suitable for Haryana are given below:

1. It should be water pumping, low tip-speed ratio windmill preferably of vertical axis design.
2. It should be suitable for installation on open wells, canals, river beds, ponds, lakes, etc.
3. The total head over which water has to be lifted should be 15–20 m with suction head 10.5 m, if installed on ground level.
4. It should be suitable for rated wind speeds of 2 m/s with cut-in speed of 1.5 m/s.
5. Discharge at 2 m/s of wind speed and total head of 15 m should be 0.5 to 1 litre/s.
6. The system design should be suitable for operating a sprinkler system for irrigation.

The Windmill WP-2, developed at National Aeronautical Laboratory, Bangalore can pump 2.9 m^3/hour against a head of 10 m at a wind speed of 3.5m/s. A Sail Wind developed at the same laboratory uses six triangular canvas sails on a rotor of 7 m diameter and pumps 5.4 m^3/hour against a head of 10 m at wind speed of 3.5 m/s (Singal 1985).

The specifications of a propeller type windmill pump-set, which has been standardized for installation in India, are listed below :

Tower height = 7 m
No. of propellers = 12
Rotor diameter = 5 m
Machine power ratio = 0.36
Water depth = 20 m
Maximum daily output = 23 m^3 at a wind speed of 10 m/s.

The total cost of the machine including transportation and erection is Rs 11, 750 and a similar expenditure is to be incurred on well-boring and construction of storage tank. The government provides a subsidy of Rs 7500.

The windmills installed so far are not only of small size (about 5 m diameter), but also require high maintenance and support. Research and Development (R & D) efforts are now needed so as to reduce the maintenance expenditure of windmills to occasional oiling and changing the pump washer only. Somewhat larger windmills have to be developed than the presently available.

5. R & D EFFORTS NEEDED

The present windmill technology is inadequate for the low wind speed regions of Haryana. Special R & D projects in the following areas must be taken up so that wind energy can be used in near future in the plains of Haryana and other similar regions.

5.1. Artificial Winds

Generation of artificial winds to drive windmills by heating large surfaces with favourable thermodynamic properties is technically feasible. A project proposal has been prepared to heat a large surface in which case the resulting current (artificial wind) will drive turbines (Venikov. Putyatin 1981). The efforts needed to pursue the project in the form of money, manpower, and time are huge.

5.2. Aeroelectric Plant

The low wind velocity in Haryana can be augmented by the use of diffusers at intake to windmills. Besides the propellers Madaras and Darrieus, there has been a plethora of designs for wind machines. One intriguing power plant design, called the aeroelectric plant, uses the flow up a tower that looks like a cooling tower (El-Wakil 1985). Its walls are heated by solar radiation. Since the walls are circular, the sun's rays need not be tracked as it changes position in the sky during the day. The heated walls, in turn, heat the inside air and a flow up the tower is established. This air flow is made to drive a number of air turbines located near the top of the tower. The turbines, in turn, drive electric generators. The driving pressure causing air flow is given by the well-known chimney effect.

P. Carlson of California has proposed a slightly modified form in which, the interior air in a very tall tower would be cooled by pumping water to the top. The water evaporates in the low pressure air there, causing a downward flow of cooled air. The driving pressure can be calculated in a manner similar to that for wet cooling towers. A conceptual design of such a plant called for a 1.5-mi-high, 900 ft-diameter tower located in a hot desert and 10 wind turbines surrounding the tower periphery at the bottom producing 2500 MW. The above design can be scaled down for producing water pumping wind turbines in Haryana.

5.3. Low Wind Turbines

The turbines available in India and abroad are suitable for a rated wind speed of 3.5 m/s or more whereas low wind speed turbines for rated values of 1.5–2 m/s are needed for Haryana. Special efforts are, therefore, needed to develop cheap and simple rotors, which can cut-in at low wind speeds available in Haryana.

The Savonius rotor and American multi-blade type windmills have optimum power coefficients at a very low tip-speed and can therefore be used as starting point to develop windmills suitable for low wind speed. However, M/s H-Energiesysteman B.V. of Holland have developed twin blade propeller type rotors for low wind areas (wind speed less than 3 m/s) (The Netherlands Agency for Energy and Environment).

5.4. *Energy Storage Schemes*

A combination of a windmill and a micro-hydro turbine can be used for storage of energy as potential energy of water (Fig. P.1.1). Water could be pumped to high reservoir during periods of high winds and the potential energy could then be used during periods of low winds, assuming a constant output from the wind water generating system as the basis for sizing of the water generator and the storage system.

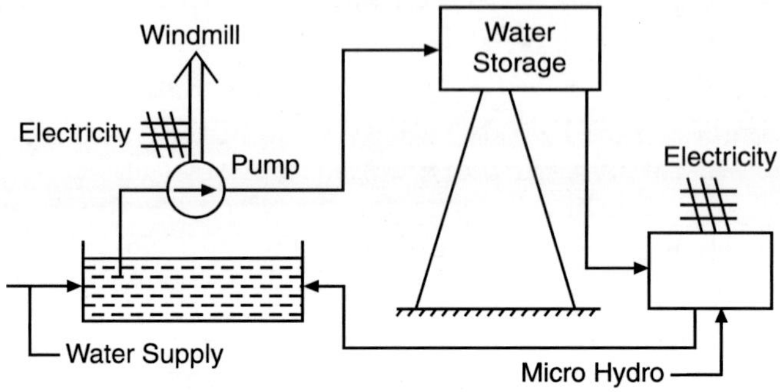

Fig. P.1.1 Integrated Energy System.

5.5. *Windmill with Hydraulic Transmission*

Hydraulic transmission consisting of a split fluid coupling/torque convertor can be used to run a centrifugal pump to lift deep ground water for the operation of a sprinkler system for irrigation (Fig. P.1.2). The windmill is directly coupled to hydraulic pump part to coupling/torque convertor while the hydraulic turbine part, which is directly coupled to the centrifugal water pump, is lowered deep in the open well. The variable speed/torque hydraulic pump sends pressure fluid through a pipe to drive the hydraulic turbine to operate the centrifugal pump.

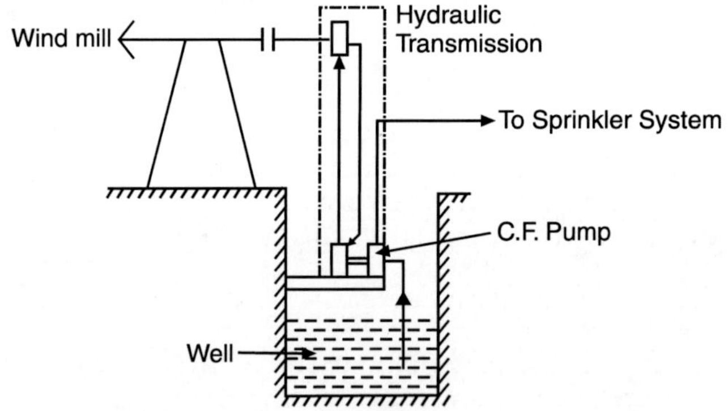

Fig. P.1.2. Windmill with Hydraulic Transmission.

The use of hydraulic transmission has many advantages such as optimum adaptation of rotor speed to guarantee a maximum output; no gear box; no controlling and switching equipment, as no electrical generator is involved; simple maintenance, as no reciprocating pump and allied crank mechanism; and the centrifugal pump can be installed on ground level or lowered into the deep open well to suit the suction lift available (The Ministry of Economic Affairs 1986).

CONCLUSIONS

Haryana has ample underground water to be tapped for irrigation. However, the wind characteristics are not favourable for the existing windmill technology. Special efforts are needed to develop suitable turbines and systems for adaptation in Haryana and other similar regions, so that abundant, reliable, clean wind can be harnessed cost-effectively as alternative and renewable source of energy to meet the irrigation and other requirements of plains of Haryana. It is recommended that a special research institute may be set up for carrying out the development projects.

REFERENCES

1. Daniel M. Simmons, *Wind Power*, Noyes Data Corporation, Park Ridge, New Jersy (1975).
2. *The Hindustan Times*, New Delhi, 29 October, 1992
3. R.K. Singal, *Non-conventional Energy for Lift Irrigation,* Electrical India, vol. XXV : 8 (1985).
4. V.A. Venikov and E.V. Putyatin, Introduction to Energy Technology, Mir Publishers, Moscow (1981).
5. M.M. El-Wakil, *Power Plant Technology*, McGraw-Hill Book Company, New York, 1985, 624 pp.
6. The Netherlands Agency for Energy and Environment, *Holland Wind Energy*, Export Catalogue, 3503 RE Utrecht, 18pp.
7. The Ministry of Economic Affairs, *Holland Energy*, Export Catalogue, The Hague (1986).
8. U.K. Srivastava and Girija Sharma, *Wind Energy for Water Pumping*, Oxford and IBH Publishing Co. Pvt. Ltd., New Delhi (1987).

❑❑❑

Paper-2

AN ENERGY EFFICIENT CITY

ABSTRACT

The fuel cell has all the virtues of high efficiency, low pollution and silent operation. It can meet the entire energy/power requirements of a city, may it be individual apartments, commercial buildings and other complexes by on-site power units or street lighting and other municipal consumption by scattered substations, or industrial estates by co-generation plants based on fuel fells. The local trains and buses, trucks and personal cars can be propelled by methanol fuel cells.

The paper describes a conceptual design of a city where local transport, personal cars, domestic, commercial, municipal and industrial power supply have been planned using different types of fuel cells. A continuous and dependable power supply can be ensured by the proposed system for comfortable living, entertainment, efficient commutation, communications, commercial and industrial activities in a pollution- and noise-free environment.

1. INTRODUCTION

India is essentially an agricultural country with bulk of its population living in 5,64, 276 small and big villages scattered all over the land. The post independence uneven development and industrilisation has created moral, mental and spiritual degradation and frustration among rural population in India, who have remained isolated from the economic planning and growth. Transportation of commercial fuels like coal, diesel and petrol over long distances is very difficult and transmission and distribution of electricity through power grids to far flung areas are prohitively expensive. The farmers and craftsman are not in a position to pay for high costs of energy; thus small scale industries in rural areas are reduced and there are limited irrigation possibilities in agriculture.

There is a continuous flux of population to new urban economic centres which has created the problem of mushrooming of slums and destruction of social life. The urban infrastructures have been overstretched and chaotic conditions in the cities can explode with serious consequences. The problem must be tackled by checking the migration of rural population to cities by creating opportunities in the villages. At the same time, new cities with fully planned, adequate infrastructure based on modern technologies must be created. A continuous and dependable power supply must be ensured for comfortable living, entertainment, efficient commutation, communications, commercial and indus-

Electrical India, Mumbai

trial activities in a pollution and noise free environment. The fuel cell has all the virtues of high efficiency, low pollution and silent operation. The paper describes a conceptual design of a city where local transport, personal cars, domestic, commercial, municipal and industrial power supply have been planned using different types of fuel cells.

2. NEW DEMANDS

Indian societies are rapidly urbanising. The growth of urbanisation is projected in Table 1. While per capita energy consumption in India as a whole is abysmally low, the growing rate of urbanisation will put pressure on the energy sector. The cities accommodate an energy intensive urban society with per capita energy consumption five times or more than the national average energy consumption. The energy intensity of urban society will increase multifold as there will be demand for multi-dimensional space conditioning of houses and offices. The efficiency and creativity of the blue-collared as well as white-collared manpower in industries and commercial houses can be lifted by ENVIRONMENTAL CONDITIONING of working and living space. The audio-visual system and air-conditioning system would be integrated to artificially create natural situations of hill stations, water falls, forests, sea coasts or other work-friendly environment in the houses, offices and factories. Specially prepared videotapes would be projected on the walls and ceilings of the living and work space; audiotapes would be played to create complimentary music and sounds of trees, water falls sea waves, breeze, etc. The temperature, humidity, wind velocity would be controlled by air-conditioning system which would also be integrated to a fume spray system to create matching odours for the selected situations. The integrated environmental conditioning system could be programmed for creating different natural sitauations on the same day or different days.

Table P.2.1. Growth of Ubranization

Year	1971	1981	1991	1996	2001	2006	2011
Urban as %age of total population.	19.9	23.3	25.7	27.8	30.5	33.5	36.6

3. PRESENT SCENARIO

Near chaotic conditions prevail in all the metropolis and big cities of the country due to population explosion. All the infrastructural systems including energy and power supply systems for domestic, commercial, transport and industrial needs, commutation, water supply, health services, communications, civic life, etc. have been overstretched and broken down. The power generated in 35 percent inefficient thermal power plants have to be transported over long and tedious transmission and distribution lines to the cities and only 75 percent of the power generated reaching the consumers. The low voltage and frequency is available only for a fraction of a day. The power supply is therefore supplemented by small, noisy and highly polluting gensets even installed inside the apartments, offices and on the crowded streets, roads and industries. The roads are in bad shape, flooded with stinking water with no drainage whatsoever, overcrowded with all types of vehicles, trucks, buses, cars, scooters fuming and contributing maximum to the chemical and thermal pollution of the city life.

The problem can be relieved by planning new cities and economic centres with proper and sufficient infrastructural systems. One such city planned with fuel cell technology is described in the following paragraphs.

4. FUEL CELL TECHNOLOGY

If an electric current is passed through a dilute solution of an acid or an alkali by means of two platinum electrodes, hydrogen is produced at the cathode and oxygen is evolved at the anode. If this process is reversed by removing the power supply and connecting the two electrodes through a suitable resistance, the presence of hydrogen at one electrode and oxygen at the other will produce a small current in the external circuit, water being produced as a by-product. This reverse process of electrolysis is the essence of the fuel cell technology as chemical energy stored in hydrogen and oxygen have been combined to produce electricity.

Unlike other heat engines, fuel cells donot have moving components. Therefore, the fuel cell will be quieter and should require less maintenance and attention in operation. The manufacturing costs should also be less, as close tolerances as in pistons are not required. Because the fuel cells convert chemical energy directly to electrical energy, the conversion efficiency can be much higher; they may be cheaper to operate, and there will be less heat produced, resulting in small radiator and exhaust systems.

Although much of the interest in the fuel cells is due to their efficient use of fuel, there are considerable pollution control advantages to be gained as well. Because the fuel reacts electro-chemically rather than by combustion with air, no nitrogen oxides are formed. For the same reason, emissions of unburned and partly burned gaseous and particulate products are essentially nil. There is relatively little thermal pollution because less heat is lost as waste. Energy conversion via fuel cells, therefore, represents one of the best ways to achieve this goal, because it is possible, simultaneously, to obtain more work and less pollution from a rupee's worth of fuel with a cell than any other energy conversion device.

With this impressive list of advantages, it is perhaps surprising that fuel cell has not been more readily accepted. Cheap, practical and long-lasting devices are still not available in the market. Fuel cells have not been widely used or developed as terrestrial energy sources because of their high costs (partly due to expensive catalysts) and short operating lifetime (due primarily to degradation of electrode materials). The early development of fuel cells for electrical power generation was overshadowed by the steam turbine generator and for mobile applications, it could not compete with internal combustion engines. The biggest boost to fuel cell development has been space power applications where high power density and low weight are important. Fuel cells have been an integral part of the Gemini and Apollo manned space flight systems. It is only recently that some of the problems of high cost and short lifetime may be overcome and fuel cell batteries may be commercially viable within short period.

Hydrogen and oxygen were selected as reactants for the fuel cells used for spacecraft power supplies because of relatively high reactivity of hydrogen. Presently hydrogen is a relatively expensive fuel. The reactants for commercial fuel cells should be as cheap and readily available as possible, for example, air as oxidant and natural gas or petroleum derivatives as fuel. As coal gasification technology matures, very satisfactory feed streams for fuel power plants will be available.

The output of fuel cell is low voltage dc power, cells may be connected in various series and parallel arrangements to get whatever voltage and power are desired, and highly efficient inverters are available for conversion to ac.

Fuel cells have been proven as practical power sources in certain specific applications, such as space missions and remote site operations. The development of fuel cells for widespread commercial applications is now underway. Although advantages of fuel cells over other heat engines could swing the balance in favour of fuel cells, the relatively high development costs involved means that substantial markets must be sought by finding new applications.

City life depends upon continuous and reliable power supply for comfortable living, entertainment, efficient commutations, communications, commercial and industrial activities in a pollution and noise free environment. An attempt has been made in this paper to discuss the characteristics and requirements of different types of fuel cells for their applications in individual houses, apartments, commercial buildings, street lighting, industrial sites, communication systems, electric vehicles for rail and road transport so that total power requirements of a city can be planned on fuel cell batteries only.

5. FUEL CELL AS POWER GENERATION UNIT

The fuel cell converts chemical energy directly to electrical energy, electrical power generation is the most natural application of fuel cells. For this application, fuel cells must compete with steam turbine, which are remarkably efficient devices with efficiency approaching 40 percent at rated load. However, demand for electrical energy is far from constant. Over the course of a year, the actual output of a power plant may very by a factor of four, and the daily variation in load can be almost a factor of three. To adjust to this changing demand, either the large base load plants must sometimes operate at part load, or smaller cycling or peaker units must be used during periods of high demand. Either way, efficiency suffers and pollution increases. On the other hand, the fuel cell system not only has a greater efficiency at full load, but this efficiency is retained and even increases as load diminishes, so that inefficient peaking/standby genesets may not be needed.

A fuel cell system, unlike a steam turbine, need not be big to be efficient. This characteristic, taken together with two others, low emissions and capability of operation on a variety of fuels, allows fuel cell system to be operated almost anywhere. A small power plant for a community can be operated on the optimum fuel available locally with nearly the same efficiency achieved by a large central power plant. An electric supply company of a large metropolis can disperse a number of generators throughout its area and match capacity to local demand, substantially reducing the expense and other problems associated with transmission and distribution of electricity. Costs and other problems involved with local distribution of electrical energy are likely to be greater with the adoption of underground lines in urban areas.

5.1. Efficiency

Fuel cells have high conversion efficiency especially at low power levels as they are not subject to Carnot limitation on efficiency. It is a one-step process without moving parts and both mechanical and heat losses are absent. This indicates that these systems have the potential for very high reliability and silent and unattended operation.

The hydrocarbon fuel cells operate at much the same efficiency in the 100 kW range is large multi-megawatt units. Efficiency may be further increased to about 55 percent if pure hydrogen is used in place of processed fossil fuels, and again to about 60 percent when oxygen is additionally substituted for air. While the efficiency of fuel cell increases as the power level is decreased to about 40 percent load.

5.2. Pollution

In conventional power plants, a considerable amount of NO_x' SO_2' H–C and particulates are emitted due to combustion of fuel. Fuel cells emit mostly non-toxic and harmless air, CO_2' water vapour and small heat as exhaust. Noise is low because of absence of moving parts. Thermal pollution of waterways is not a problem because fuel cells are air cooled and are few restrictions on site locations. Air pollutants are reduced by a factor of more than 10 over conventional systems. The environmental impact of fuel cells is compared with code requirements of conventional systems in Table P.2.2.

Table P.2.2. Environmental Impact of Fuel Cells Kilogram of Pollutants per Million kJ Heat Input USA Federal Standards

	Gas-fired Central Station	Oil-fired Central Station	Coal-fired Central Station	Fuel Cell
Particulates	0.05	0.05	0.05	0.0000015
NO_x	0.10	0.15	0.35	0.006 – 0.009
SO_2	Not required	0.40	0.60	0.000012
Smoke	20% Opacity	20% Opacity	20% Opacity	Negligible

5.3. Scale

Fuel cell efficiency is insensitive to size. Scaleup in power output is accompanied by interconnecting of fuel stacks (modules) due to the limit to the size of thin electrodes. There is a saving in cost due to mass production of identical components for scale-up.

The fuel cell is a very simple device with no moving parts except a few fuel and coolant pumps. Controls are simple and automatic.

5.4. Modularity

Fuel cell units can be added to a power plant system incrementally over a period of time and built rapidly. There is no need to tie up considerable capital in unused initial capacity which is a very serious drawback of large power plants with long construction periods.

5.5. Versatility

Fuel cells have inherent adaptability towards a broad range of applications. The direct nature of electro-chemical process minimizes the effects of both scale and load level on operating characteristics. The power plant modular construction and ease with which units can be linked permits matching of capacity to demand. Increased capacity can be added quickly and simply. Because the power plant is air cooled it is not dependent on cooling water availability and thus has few location restrictions. It has the capacity to operate on a variety of fuels as given in Table P.2.3.

Table P.2.3. Fuels And Fuel Cells

Fuel	Fuel Cell	Fuel to ac Power Heat Rate (kJ/kWh)
Hydrogen	Alkaline	6500
Impure Hydrogen	Acid	7750
Syngas	Molten carbonate	8200
Methanol	Acid	6500
SNG	Molten carbonate/Acid	7500
Naptha	Molten carbonate/Acid	7500

Small-scale fuel cells power units can be used with advantage for planning the complete power requirements of a city. For individual houses, apartments, commercial buildings and industrial sites, independent on-site power units could meet power demands from a kW to several kWs. They can be integrated into total energy packages that not only generate electricity, but provide control of temperature, humidity and cleanliness. They could also decentralize the generation of electricity and trim the expenses of transmission over networks by serving as substations. Fuels can be processed on site or supplied to individual power units through underground pipelines.

6. CO-GENERATION UNITS

Co-generation presents an efficient way of utilising our limited energy resources because the same fuel source is used similtaneously to produce two forms of useful energy including electricity and heat. Co-generation applications are process and site specific for a given industry or a commercial building. The phosphoric acid fuel cell can be used for a co-generation system. It is capable of operating on a variety of fuels, including natural gas, light distilates, propane, and coal-derived synthetic fuels. Synthetic fuels include hydrogen/carbon monoxide/methane mixtures and methanol that can be refined, reformed and/or shifted to produce hydrogen.

The co-generation system incorporates fuel cell stacks, turbo-compressor for supply of pressurised air to the cell cathodes, and for recovery of pressure and heat energy from the cell exhaust streams, a desulphurizer for removal of sulphur from the raw fuel, a steam reformer for conversion of hydrocarbon fuels to hydrogen and carbon monoxide, shift convertors for reaction of carbon monoxide and water to produce hydrogen and carbon dioside, and a heat recovery system for generation of process steam for the industry and reformer steam from waste heat transferred from the fuel cell.

The overall process in the fuel cells consists of the continuous electrochemical reaction of hydrogen in the fuel stream and oxygen from air to produce electric power and by-product hot water and heat. Electric efficiency of the system, after parasitic power, is in the range of 0.38 to 0.41. If process steam generated at 6 bar is not sufficient for the industry, supplementary firing may be used.

7. MOBILE UNITS

Fuel cells may provide practical, low-pollution vehicles with useful performance and range. Batteries, fuel cells and combination of both have been tried for electrically powered vehicles. The most practical systems will probably be hybrid power plants in which batteries provide peak power and the fuel cells act as charging units during low power periods.

In transport industry, high efficiency and low pollution make the fuel cell attractive. Fuel cells are expected to meet the requirements of high energy density (available energy/weight ratio) handily since the amount of energy available is determined by the size of fuel tank. A fuel cell powered vehicle can have a good long range without refuelling, and can be refuelled rapidly, just as can present day internal combustion vehicles.

Fuel cell systems of adequate performance can be built to propel local trains for the city. It would be very smooth, and quiet, virtually pollution-free, and could operate on conventional fuels. The fuel cell could replace the conventional diesel engine/generator set directly or small fuel cell with a motor could be fitted to each wheel.

It is possible to meet the high power density (power/weight ratio) criteria for large buses and trucks. To propel a vehicle of the weight comparable to an intermediate car with speeds and accelerations usable in present traffic conditions (almost instant starting characteristics and very intermittent use), it is probably necessary to achieve a power density of about 220 W/kg, which is equivalent to 6kg/hp. It may be possible to meet that goal by hybridising a fuel battery with one of several high goal by hybridising a fuel battery with one of several high power storage devices such as one of new generation flywheels.

The criteria of high power density is more difficult to achieve for small personal cars. Changed driving patterns of decreased speed and acceleration requirements can help to introduce low power vehicles. In case of a new city, roads can be planned with multi-lane and unlevel crossing to mitigate the necessity for high power densities.

8. RECENT DEVELOPMENTS

It was reported in the Times of India dated 24th October, 1997 that a fuel cell stack producing 50 kW (power needed to operate a car) has been successfully developed by Plug Power LLC, latham, New York. It produces electricity from many fuels such as gasoline, ethanol, methanol and natural gas. A fuel processor developed by Arthur D Little, Cambridge, converts the raw fuel into hydrogen, which is run through a carbon monoxide removal system before it is fed into a fuel cell to produce electricity. The carbon monoxide removal system was developed by Los Alamos National Laboratory. The technology can be used for powering vehicles and in homes and apartments to provide heat and electricity. To match the internal combustion engines now in use, the fuel cell system must be reduced in size, weight and cost. A six fold reduction in cost was needed which is possible if there were high-volume production.

Technorama of October, 1997 describes a fuel cell of 1.5 kW capacity developed recently by an Australian Company for meeting the power requirements of a house. The power/efficiency ratio shows a significant advantage for the Ceramic Fuel Cell-upto 200 percent as compared with traditional technologies.

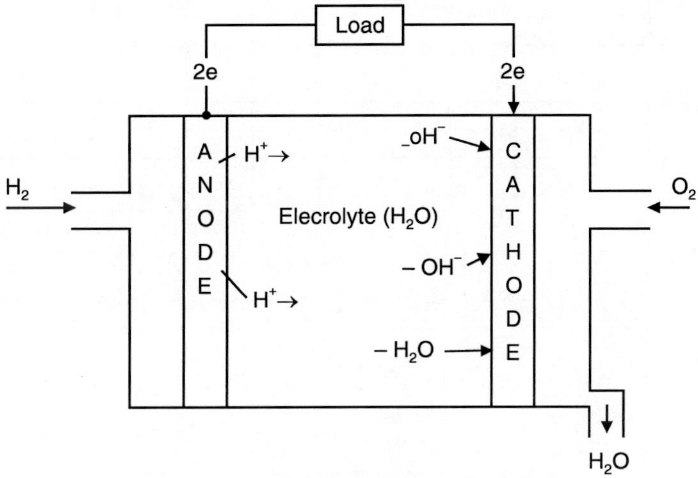

Fig. P.2.1. Hydrogen-Oxygen Fuel Cell

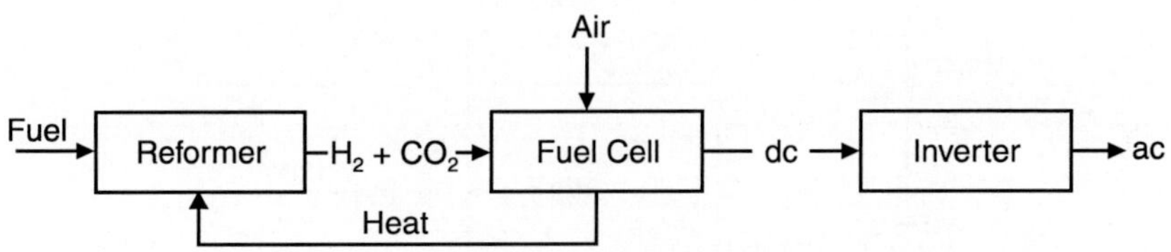

Fig. P.2.2. Hydrogen Fuel Cell Power Unit

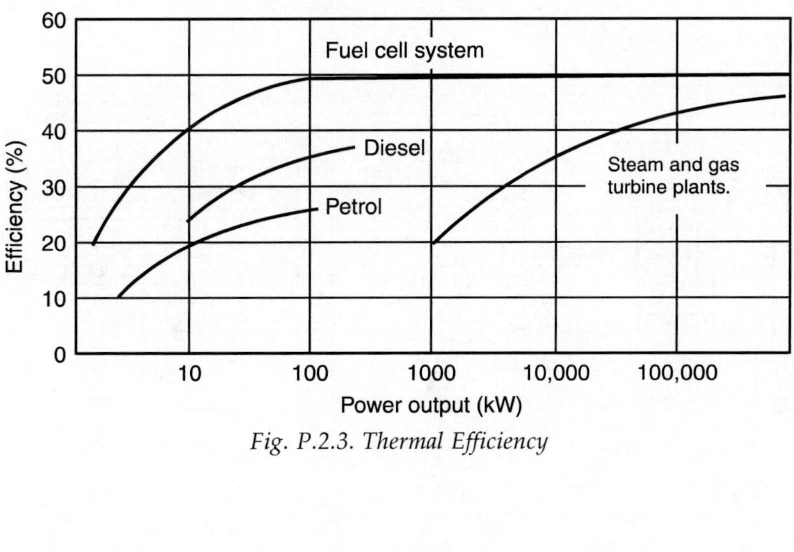

Fig. P.2.3. Thermal Efficiency

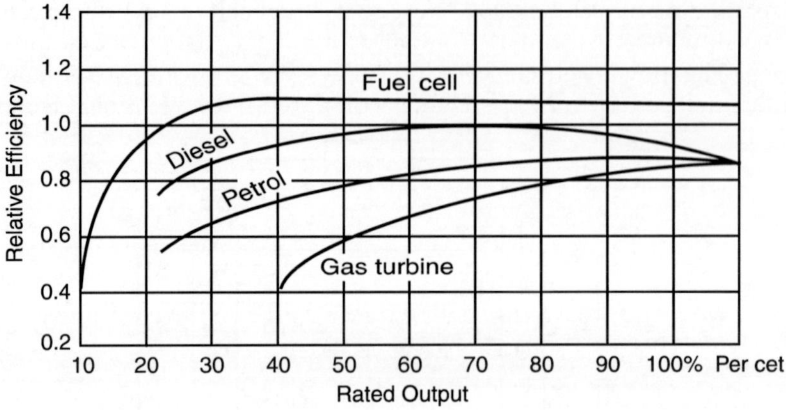

Fig. P.2.4. Part Load Efficiency

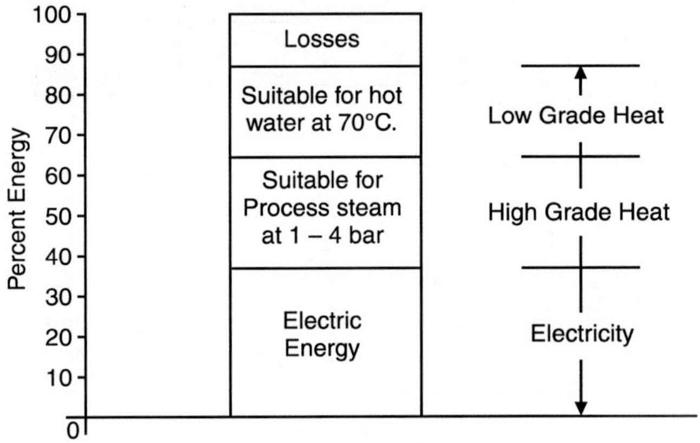

Fig. P.2.5. Co-generation Potential

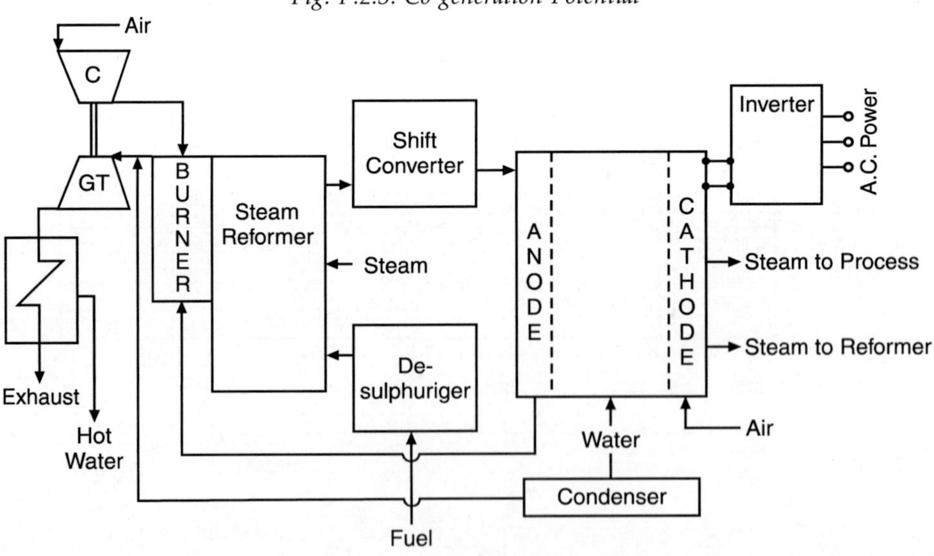

Fig. P.2.6. Phosphoric Acid fuel Cell Co-Generation Plant

9. CONCLUSIONS

It may be concluded that fuel cells have all the virtues of efficiency, low pollution and silent operation for meeting all the power/energy requirements of a city, may it be individual apartments, commercial buildings and other complexes by on-site power units or street lighting and other municipal consumptions by scattered substations, or industrial estates by co-generation plants based on fuel cells, or local trains and buses, trucks or personal cars propelled by methanol fuel cells. Future cities can be planned on fuel cell systems for their power and energy requirements.

The above discussions prove the existence of huge markets for fuel cells. These should be seriously considered for our future power requirements. The commercial power units are technically feasible. Research and development should be aimed at reducing costs and increasing life so that its full potential for power generation as well as engines for all types of vehicles may be exploited.

10. REFERENCES

1. K.H. Spring, "*Direct Generation of Electricity*", Academic Press, London, 1965.
2. Robert Noyes, "Fuel Cells for Public Utility and Industrial Power", Noyes Data Corporation, New Jexrsey, 1977.
3. S.David Hu, '*Cogeneration*", Reston Publishing Company, Virginia, 1985.
4. Rechnorama, the Institution of Engineers (I), October, 1977.
5. "*Now Gasoline can Power Low-Pollution Electric Cars*", The Times of India, 24th October, 1977.

❑❑❑

Paper-3

AN OLD MODEL AND NEW TECHNOLOGY FOR AN ENERGY SUFFICIENT VILLAGE

ABSTRACT

Cow has been used as a unit around which sound economy of a village had been built and sustained in ancient India. This paper proposes to remodel the system and cow has been evaluated as a unique energy conversion unit for an energy sufficient village. It has been proved with the help of suitable data that with certain new technological inputs, the old model can be used to meet all the energy requirements of a village.

1. BACKGROUND

Lord Krishna had been a colourful and many facet personality. With rural brought-up, he showed himself as possessor or great physical prowess and went on, later in life, to demonstrate that he could win great battle by shear planning and diplomacy without active involvement otherwise. As a preacher of Bhagwat Gita, his practical philosophy of Karma Yoga has overshadowed his many other original and unconventional trends in thinking. One of his great contributions to suffer was in the field of rural economy. He visualised cow as an economic unit around which sound economy of a village could be built and sustained. More interesring was his commitment to this idea and aggresiveness he used for the implementation of his programme.

2. INTRODUCTION

The post independence uneven development and industrialisation has created moral, mental & spiritual degradation and frustration among rural population in India, who have remained isolated from the economic planning and growth. It has now become compelling to think about the 80 percent of rural population for the satisfaction of their basic needs, their self-reliance based on social participation and harmony with the environment to ensure the sustainabiity of their development.

The solution of the problem lies in making available cheap and abundant supply of energy for cooking, heating, lighting, community development, pumping of water for drinking and irrigation, powered agriculture and small scale industries. The energy conversion techniques s0hould be pollution free, based

Commonwealth, Conference on Energy, Bangalore

on renewable and locally available sources and appropriate technology involving local manpower, local materials and local resources which can be locally managed and amalgamated with the rural surroundings.

In this context, the wisdom of Lord Krishna can be tested. His idea of cow as a unit for rural economy can be remodelled to evaluate it as energy conversion unit. The cow eats biomass (fodder) and gives out milk and cowdung which can be properly processed to yield useful energy. Cow is symbolic and includes buffalo and other milching and associated animals.

3. RESOURCES FROM A COW

Cow is a very rewered animal in Hinduism because of plentiness of resources obtained from her. The daily income resources are milk and cowdung. Milk is one of the most wonderful and almost complete food available in nature. Not only it has high nutrient constituents but also it is used for a large variety of domestic and industrial food products in the form of curds and kafeers, butter and butterghee, cream, cheese, sweets, milk-powder, ice milk only and can meet their full energy (food) requirements from it.

Cowdung is another wonderful waste to meet the fuel and fertilizer requirements in rural areas. A major portion of collected cowdung is used as a domestic fuel after drying the cowdung cakes. The above together with some supplements of agrowastes and firewood meet the fuel requirements of the rural population in India.

The cow periodically gives birth to draught animals which contribute substantially to the energy demands of agriculture, irrigation and transport sectors.

Useful food and industrial raw materials are obtained after the death of cow, i.e., meat, leather, glues, blood, bones, etc. Therefore, cow is very useful as a source of raw materials for Dairy and Milkfood industries, Meat industry, Leather, Chemicals and Glue industries, etc.

3.1. Milk and Milk Products

The milk yield of cow and buffaloes varies widely from 5 litres to 30 litres per day; the total milk production in India may be of the order of 700 million litres per day. The daily calories need of man varies from 2500 (light work) to 5000 (heavy work). Although, man does not depend wholly on milk and milk products for his calories intake, however, a model can be conceived to evaluate the contribution of cow to meet the total energy intake of man.

The calories content of some milk product are: cow milk (670 calories per litre), buffalo milk (950 calories per litre), curd (600 calories per litre), cheese (3480 calories per kg), butter (7290 calories per kg) and butter ghee (8280 calories per kg). It may, therefore, be assumed that cattle milk can give on an average 900 calories per litre of milk. Thus the present cattle population in India can fully meet the total energy requirements of 180 million people of average size.

The energy equivalents of man are : 1050 kJ per manhour; 840 kJ per woman hour and 500 kJ per childhour. Therefore, the total energy equivalent of milk production in India in 416,000 MWh.

3.2. Draught Animals

Draught animal is an excellent energy device readily available in the rural scene and within the means and technical competence of the farmer for ploughing and water lifting and small scale cart operators for transportation (1) In India, there may be about 130 million draught animals including young stock; and 80 million work animals ploughing two-thirds of land and carry two-thirds of rural transportation. These are more economical than tractors for ploughing farms of sizes less than four hectares and more competent than trucks for handling vehicles for small scale transportation (about one tonne) over short distances (below 20 km). Trucks become economical only for heavier loads of 3 - 5 tonnes and when distances moved per day are more than 200 km.

The animal power is 9650 kJ/Bullock-hour and total draught animal energy of working animals in India is 215,000 MWh. Five pairs of bullocks can replace one tractor and therefore, eight million tractors can be replaced by 80 million work animals, saving huge amount of capital investment in addition to avoidance of dependence on scarce diesel/kerosene oil, mostly imported by India. On an average a buffalo gives 6 to 7 births and a cow 10 to 12 births per life.Therefore, the cow contributes substantially to this form of renewable energy source.

3.3. Cowdung

The cattle population in India is 235 million heads with a dry dung yield of about 170 million tonnes per year. The composted manure prepared from cowdung has 0.75 to 1 per cent Nitrogen after losses and is used as a fertilizer for agriculture. The collection efficiency of cowdung is 70 per cent and loss constituent of 30 per cent pollutes the environment.

The prevalent practice of burning cowdung cakes, firewood and farm wastes in rural India is not an attractive one. A major portion of the collected cowdung is used as domestic fuel after drying the cowdung cakes. The practice of making cowdung cakes is unhygienic and low combustion efficiency of 11 per cent is wasteful. This can rapidly result in the problem of deforestation on one hand and loss of soil nutrients on the other.

3.3.1. Biogas

The livestock and even human wastes have much potential as an energy source in developing countries via an earobic digestion; the wastes are treated with simultaneous evolution of methane fuel and retention of nitrogen and other vital plant nutrients in the remaining sludge, which can be recycled as a fertilizer. 'Biogas' generation is a versatile process ranging from small domestic 'Gobar' gas plants catered by a few cows to much bigger community type plants where relatively large volume of cowdung is continuously available. These depend greatly on hot climate and manual labour for its functioning. These can go a long way to providing much of the energy required by a large population.

The technology for conversion of animal wastes into biogas via anaerobic digestion is well established, and quite a large number of biogas plants (about one lakh) are already in operation in Indian villages (2). Our country has a potential of producing about 22,000 million cubic meters of biogas with a heating value of 21,800 kJ per cubic meter.

The Gobar gas plants are mainly developed by Khadi and Village Industries Commission New Delhi. In every state, Khadi and village Industries Boards look after the erection work of the Gobar gas Plants. The Ministry of Agriculture, Government of India is also helping for the construction of Gobar Gas Plants.

For a typical Indian village of population 500, the number of cattle would probably be around 250. The amount of dry dung collected from them would be approximately 180 tonnes per year calculated on the basis of 13.5 kg of cattle dung as daily produce of one cattle. This would be more than sufficient to meet the complete fuel and other energy requirements of the village population on conversion to methane.

The human wastes can also be used for production of biogas. Community latrines can be planned in the villages for collection of night soil for feeding to biogas plants. Wastes of 200 persons can be used to produce about 5 cubic meter of gas per day to extract 12 kWh of equivalent energy.

3.3.2. Biogas Engines

The requirement of Indian farmers for engines varies from 3.7 kW to 7.5 kW in vertical high speed and vertical low sped types all over the country. Normally, the engine runs on an average for 5 hours per day. Biogas produced by a digester that is functioning well is approximately 65 per cent methane and has a heating value of nearly 21000kJ per cubic meter. The biogas engines for pumpsets produced in India are essentially diesel engines where air intake is connected to biogs supply. The diesel is required for starting the engine and about 20 percent diesel and 80 per cent biogas are mixed automatically to produce power for running the pumpset. The biogas consumption is about 0.46 cubic meter per kWh which can be obtained from 13.5 kg of cattle dung, i.e., daily produce of one cattle. The price of a typical 3.7 kW, 1500 rpm biogas engine with centrifugal pump of 100 × 100 mm including accessories and trolley is about Rs. 15000. It is financed by the State Agro Industries and Land development Banks and other nationalised banks and subsidised by Government. The cost of a typical gobar gas plant of 85 cubic meters capacity is abut Rs. 1.5 lakhs.

3.3.3. Spent Slurry

The disgestion process in the Gobar gas Plant enhances the fertilizer quality of cowdung. The nitrogen content of the spent slurry from the plant is very high, 1.5 to 2 per cent which is double the nitrogen content of original cowdung.

4. CONCLUSIONS

The cow is a very useful animal converting solar energy through biomass into useful products and services. It is an unique energy conversion unit. At the same time it should be remembered that cow lives largely on crop residues, grass and other such products which are not readily useful to man. Also they graze on marginal lands where edible crops cannot be raised.

Lord Krishna's wisdom is fully vindicated. It is proposed that rural economy should be planned with cow as energy conversion and economic unit. More important is that the committment and aggressiveness of Lord Krishna must be used in the implementation of the plans. The cattle breeding, health care, research and development inanaerobic digestion, farm and transport implements should be taken up more vigorously.

REFERENCES

1. Ramaswamy, N.S., Appropriate Technology-Draught animal Systems", Seminar on Appropriate Rural Technology Strategies for Development and Extension, Lucknow, April 16-17, 1984
2. Singla, R.K., "*Non-Conventional Energy for Lift Irrigation*", Electrical India, March 1985.

Paper-4

AN ALPINE HYDRO VILLAGE

1. INTRODUCTION

India is a country of villages. These villages vary in size and population and are scattered all over the country. Some of these are located on the hills and in dense forests and can be reached only on foot or by pony. They have no road connections to other villages, towns and cities. The population of these villages may be one hundred to two hundred people. Most of them earn their living by cultivating the land, growing fruits and some of them may be craftsmen who earn their living by weaving woolen carpets, by making earthen pots, by weaving baskets and even by making beautiful carved articles from wood, brass or other local materials. Some of them keep cattles like cows, sheep and goats and make their living on animal yields. In a nutshell, life is hard and hilly villages are very low income areas.

The main wealth of hilly areas is vegetation varying according to temperature, soil and rain and mainly depends upon the altitude. Near the base of the mountains the vegetation will be thick evergreen forests while the top of the mountains will have no vegetation, but will be covered with eternal snows. The vegetation belts of the Himalayas are given in Table 1.

Table P.4.1. Vegetation Belts of the Himalayas

Altitude (m)	Description	Trees	Fruits	Crops
0-750	Tarai	Sal, Bamboo	Tropical Fruits	Barley, rice, Tea Potatoes
750-1000	Deciduous Forests	Oak, Horse-chestnut	Temperate Fruits	Rice
1000-1600	Mixed Forests	Oak, Beech	Walnuts	Tea
1600-3300	Coniferous Forest	Pine, Chir, Fir, Deodar, Cedar, Spruce	—	—
3300	TIMBER	LINE		
3300-4200	Alpine Grasslands	—	—	—
4200	SNOW	LINE		
4200-8500	Eternal Snow	—	—	—

Electrical India,. Mumbai

The valleys are surrounded by the mountains and are protected from the cold winds blowing on the mountains. The climate is warm and the soil is more fertile than that of the mountains. In almost all valleys, there re rivers and sometime lakes. Good crops especially fruits like apples, pears, cherries, etc., can be grown in the valleys.

India has made great strides in industrialisation during post independence year. This has brought prosperity to a selected few and created concentrated centres of employment near the industrial complexes and big cities where economic activities have flourished. The material and energy resources have been diverted to feed these industrial and commercial 'gluttons'. The high technology industry could not penetrate the hilly areas. The population of these areas felt neglected and many of them are flocking to the new economy centres in search of their livelihood. The continuous migration of population and uneven distribution of wealth, material and energy resources have created many social and political problems which can lead to very serious consequences, if not further checked.

2. NEED FOR CHANGE

The continuous flux of population to new economy centres has created the problems of mushrooming of slums and destruction of social life; the introduction of high technology will give birth to widespread unemployment. These problems can explode with very serious consequences. These can be checked only by adoption of Gandhian model of socio-economic development with village as the unit. Each unit must be self-sufficient economically and the ratio of exports to imports of goods into a village must be more or equal to one. The development, welfare and prosperity of hilly village depend upon the availability of cheap an abundant energy for domestic cooking, heating and lighting, community development, irrigation water, mechanised agriculture and small industrial parks based on local materials.

Transportation of commercial fuels like coal, diesel and petrol over long distances to hilly areas is very difficult and transmission and distribution of electricity through power grids to far flung areas are prohibitively expensive. The farmers and craftsmen are not in a position to pay for high costs of energy, thus small scale industries in these areas are reduced and there are limited irrigation possibilities in agriculture.

3. APPROPRIATE TECHNOLOGY

The modern high technologies donot adopt to rural ecology. Only appropriate technology involving local resources, manpower and raw materials can ensure environmental harmony and sustainability of the development. However, there should be provision for upgrading of technology in the future. The wealth of the hills can be increased by inputting ample energy and technology to increase agricultural, forest and animal yields. Appropriate technology can be worked out by a technical institute specially established for the development of hilly areas. An energy production technique based on local resources and appropriate technology which can be taken care by trained local people for its design, manufacture, installation, operation and maintenance is ideally suited. The utilisation of energy flowing water through small hydel schemes is a good example of appropriate technology for the development of hilly areas. An alpine hydro village should meet most of its energy requirements from small hydel schemes. One such model is described in the following paragraphs.

The utilisation of hydroenergy to operate agricultural and industrial devices is one of the oldest and widespread techniques of human achievements. Hydraulic rams have been used to raise water for lift irrigation and drinking purposes. Water mills of various types have been used to generate mechanical power to drive irrigation pumps, machine tools and small electric generators. The Himalaya Mill is well known in the mountain areas from Afganistan to Burma which is mostly used for grinding grains. The present model of alpine hydro village is based on these devices and are briefly described before discussiong the hydro model proper.

4. HYDRAULIC RAM

Hydraulic ram is a contrivance to raise a part of large amount of water available at some height, to a greater height. This can be employed in hilly areas where some natural source of water like a spring or a stream is available at some altitude. Work done by a large quantity of water in falling through a small height is used to raise a small part of it to a greater height. Action of water hammer makes it feasible.

No external power is, therefore, required to work this machine. Other attractions are: negligible amount of maintenance and supervision costs, continues operation, high efficiency, quiet operation and possibility of automatic adjustment of water supply. The liftable volume of water diminishes asymptotically with lifting height. In case of medium lifting heights, the hydraulic ram operateswith an efficiency absolutely comparable to a piston pump of the same performance.

As the hydraulic ram does not need a driving unit, it is to be considered as ideal for lift irrigation and supply of drinking water for hilly areas, where the supply in fossil energy carriers or electricity is problematic. The hydraulic ram operates with water streams of 1 to 40m^3/s and fall heads of 1.5 to 30m and with lifting heights upto 300m. These are a number of different types which can be distinguished by the operation of oscillating valves.

Hydraulic ram is a simple and rugged device for operating irrigation schemes. This can be easily assembled from local materials. The proposed technical institute for the development of hilly areas has to put in some development work to improve the performance and adoption of hydraulic ram to irrigation.

5. HIMALAYA MILL

It is a vertical shaft, horizontal runner hydraulic turbine used to drive grain mills. The wooden radial blades are attached to a boss fitted into a wooden or steel shaft. The water is fed through open wooden channel. It produces about 0.5hp with a water head of 1m. There are about 10,000 such mills scattered in the hilly areas from Afganistan to Burma producing power equivalent of about 45000 kWh daily.

The horizontal water wheel can be improved in performance to raise its efficiency substantially so that it can be adopted in a small-scale industry park as a prime-mover for flour mills, thrashers, ricehusking machines, vertical lathe, circular saw, trip hammer, small electrical generator, carpet loom, etc. The vertical lathe can be used by the craftsmen to produce wooden vessels and trip hammers for embossed copper vessels. The saw mill can be used by the carpenters. All these machines can be employed together to meet the local needs to produce domestic and agricultural implements and packing boxes for the export of fruits and other products.

The Himalaya Mill can be developed into an efficient and modern Pelton turbine using metallic buckets fitted to a vertical shaft supported on roller bearings. The water can be directed onto the buckets, through adjustable nozzle from a piping system collecting water from natural streams existing abundantly in the hilly areas. The stopping of turbine can be achieved through manual brakes and load speed regulation by the nozzle. The design documentation can be provided by the proposed technical institute for the development of hilly areas and machine can be manufactured in the small scale industry park. The output of the machines can be standardised at 5 to 10 kW.

The tail race water can be recirculated by lifting it with a hydraulic ram for irrigation purposes.

6. FLOATING MILL

Poncelet water wheel or a Darrius Turbine rotor can be submerged in a river or water stream available in the hilly areas to tap current energy from water of high speeds. The mechanical energy thus obtained is used for driving irrigation pumps, small electric generators and small compressors for cold storages and ice mils. The small cold storages are necessary in the hilly areas to store potatoes and other perishable goods to get better returns. Ice mill can be used to chill and store milk before it is sent to bigger collection centres.

The electricity generated is required to light the houses, schools, community centres and to run TV sets and other domestic devices. All these measures can improve the wealth of the village multifold and raise the standard of living of the population which will arrest their migration to big cities.

The blades of the water mills can be made from fabricated steel, ferrocement, glass fibre or timber depending upon the availability of the local materials and skills. The power output depends upon the swept area and cube of water velocity and very large energy can be tapped from hill streams flowing at high speeds.

The turbine rotor can be mounted on a pontoon or may be hung from a bridge or a beam over the stream. The pontoon can float on two rows of barges or empty oil drums under which wooden keels are fixed. The mooring cable can be adjusted with a winch to ensure sufficient angle of the keels to the current direction so that water side thrust on the keel keeps the pontoon at a distance from the bank. A turbine with swept area of $4m^2$ and rotor efficiency of 25% can give an output of 8kW at a water current speed of only 2 m/s. These turbines can be adopted to run small cold storages and ice mills, irrigation pumps and to produce all the electricity requirements of the village.

7. HYDRO VILLAGE

A typical layout of an alpine hydro village is shown in fig 1. The capacity and number of each type of hydro machines will depend upon the size of the village and characteristics of available water stream. The following units have been included in the scheme.

1. A small cold storage and ice mill run by a floating water mill. The capacity of the plant can be around 10 kW. Such cold storages can be built along the flow of water stream to store potatoes, fruits and other perishable products.

This will help to avoid spoilage of such products and realize better returns by disposing off uniformly over the year. Similarly the ice mills attached to the cold storage can be used to chill and store the milk cooperatively and can be drawn for disposal to collection centres and self use efficiently.

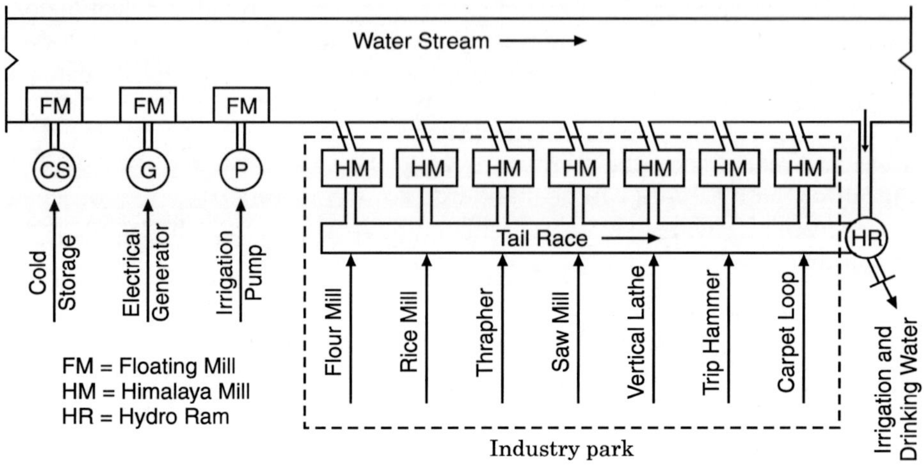

Fig. P.4.1. Hydro Village

2. Electrical generator run by a floating water mill. The capacity will match the village requirement of electricity for lighting, heating, cooking and running of radios, TV sets and other small electrical gadgets in the houses, schools, dispensaries and community centres.
3. Irrigation pump run by a floating water mill. The capacity will suit the irrigation water requirements.
4. Industry Park consisting of flour mills, rice mills, thrashers, saw mills, vertical lathes, trip hammers, carpet looms, etc., run by improved Himalaya mills individualy. The capacities can be standardised at 5 to 10 kW each. The industry park will meet all the domestic and agricultural requirements and make the village independent of imports of implements and spare parts. This will help to produce the crafts goods like wooden, brass and other vessels, toys, beauty carvings woolen carpets for exports with mechanical aids. All packing boxes for export of forest, agricultural and animal yields will be prepared in the industry park.
5. Hydraulic ram to lift the used water from tail race for irrigation and domestic water consumption.

8. CONCLUSIONS

The proposed alpine hydro village can meet the productivity requirements for self-sufficiency as per Gandhian Model. A technical institute for the development of hilly areas should be set up who can undertake research and development for the improvement and adoption of hydraulic machines for a hydro village. It should also plan and implement such schemes and monitor the results. such dydro village can go a long way in bringing prosperity and self-sufficiency to these villages. The population migration can be arrested and country can be prepared to absorb high technology in the required sectors.

9. REFERENCES

1. Singal, R.K., "*Floating type Micro hydro Power Plants*', Electrical India, Bombay, 30th Septemeber, 1984.
2. Singal, R.K., '*Non-Conventional Energy for Lift Irrigation*' Electrical India, Bombay, March 1985.
3. '*Upgrading and Further Development of Traditional Devices for the Utilisation of* Hydro-energy', Bremen Overseas Research and Development Association, Bremen.

❑❑❑

Paper-5

NON-CONVENTIONAL ENERGY FOR LIFT IRRIGATION

1. INTRODUCTION

India is essentially an agricultural country with 70 per cent of its population living in rural villages and 70 per cent of its gross domestic product emanating from the rural agricultural sector. There are nearly 350,000 villages with a population less than 500. These villages are remotely scattered and widely dispersed from the industrial belt. Their energy needs are meagre. On the average, these villages need 150 kWh of energy per day for energizing irrigation pumps, for lighting, for cooking, etc. Drip irrigation of corn, wheat, cotton, millets requires about 25 m^3 of water per hectare-day whereas food irrigation of rice, sugarcane, etc, would require 50 to 60 m^3 of water per hectare-day. In addition, water requirement for live stock is 40 to 50 liters per head-day. The lifting height of water varies from a few meters to 30m. The equivalent energy requirement would be 135 w per hectare for surface water and 400 w per hectare for ground water.

As the load factor is very low and because these villages are widely dispersed, it is most uneconomical to electrify them. Only 11 per cent of these villages have been electrified so far. It is equally difficult to reach them with fossil fuels. Consequently, they have to find an energy source which is locally available. If locally available energy sources could be converted to mechanical energy to drive irrigation pumps, the rural scene in India would change appreciably, and it would help to boost agricultural production which is vitally important in view of growing population.

Out of the total 143 million hectares of land under cultivation in the country, hardly 43 million hectares — a barely 25 per cent of the land under cultivation is irrigated. Out of this irrigated area, 23 million hectares are served by tube-wells, ponds or such similar minor irrigation schemes. There are about 2-3 millions of open wells in our country which are not yet motorized.

The non-conventional and renewable energy sources have assumed considerable importance and urgency after oil crisis. Various modes of irrigation schemes based on non-conventional energy sources have been discussed in the following paragraphs which are immediately available.

2. SOLAR ENERGY

India is bestowed with plentiful of sunshine for 250-300 days per year. In

Industrial India, Mumbai

south and central India, the annual average daily incidence of solar energy is about 4.5 kWh/m^2 with the maximum exceeding even 7.0 kWh/m^2. Therefore solar energy offers a viable solution if it can be converted to mechanical energy to drive irrigation pumps. Requirements of small-scale irrigation and domestic plants are well within the reach of solar energy pumps. A pair of bullocks working continuously can raise one m^3 of water per hour from a depth of 30 m. The equivalent of this work can be done by an electric motor pump rated at 0.8 kW for which a solar pump can be designed.

2.1. Solar Water Pumps

Organic vapour Rankine Cycle systems with rotary machines or screw expender or reciprocating engine or spiral expanders can be used to drive water pumps, Sofretes of France have developed simple, rugged, reliable solar water pumps which work on low temperature thermodynamic cycle. In consists of solar flat plate collectors with methyl chloride as heating fluid, a heat exchanger to transfer heat to working fluid (Butane or Freon), expansion motor working at 150/200 rpm and water pump with hydraulic transmission. A collector area of 60 m^2, for example, can lift 10m^3 | hr of water over a height of 12m. A collector surface area of 1500 m^2 can be used to lift 150 m^3/hr of water over a height of 40m to irrigate 20 to 30 hectare of land.

The National Physical Laboratory, New Delhi has developed Abhimanyu Solar Pump of 1 kW rating using R114. An area of 10m^2 is used to operate the pump for four hours a day. When not pumping, it can be used to drive a generator or thresher or a lathe. The total cost of the components (condenser, boiler expansion motor, storage tank, collector assembly and pump) is Rs. 12,500 and if we include the development charges, overheads and labour for assembly, the total cost will be of the order of Rs. 37,000.

2.2. Solar Photo Voltaic Water Pumping System

In a solar photovoltaic water pumping system, the sun light is directly converted into electricity and this energy is used to run an electric motor, pumpset. The Central Electronics Ltd., Sahibabad (UP) have developed the following system :

Solar Panel	:	300w (peak)
	:	2.4 × 2.5m (size)
System Voltage	:	60V DC nomianl
Motor Pump Set	:	DC motor-centrifugal pump monoblock Configuration.
Peak water output	:	1.5 to 2.0 1/s.
Average water output (at 5m total head)	:	30 to 40m^3 per day.
Max Suction Head	:	6m
Max total head	:	10m.

The solar array consists of 5 rows of 4 modules each. Each module consists of 36 cells each 76mm diameter connected in series to generate about 16 peak watts of power at incident radiation of 100mW/cm^2 at 28°C. Four such modules are connected in series and five such series connected module chains are kept in parallel. The complete array is mounted on an angle iron structure which can be oriented to face the sun. The system can be provided with storage batteries and power conditioning equipment. The total cost of the system including solar photovoltaic array, on off switch, motor pump set, accessories including suction and delivery valves and piping is about Rs. 40,000 which is highly subsidised and installed at a cost of Rs. 5000 only for a small or marginal farmer.

3. WIND ENERGY

The wind is a free, clean, inexhaustible energy source. Until this century, it has served mankind well by sailing ships, driving windmills to grind grains and pump water and even generating small amounts of electricity. The support for wind power is increasing again due to doubtful position of fossil fuels and safety of nuclear power. Following a better understanding of the aerodynamics of rotating aerolifts, there is a revival of interest in harvesting of wind energy. For a wind speed of 10m/s, the effective power obtained is 0.25kW per m^2 of swept area. As wind velocity of 10m/s occurs only for a few hours a year, the mean annual available power is very less. Depending upon local wind conditions, the annual available energy per m^2 of swept surface may vary between 100 kWh for calm areas and 500 kWh for very windy areas. Atleast 25 per cent of our land especially coastal areas are situated in reasonably windy areas. The scope for small water pumping wind mill is quite significant.

The wind energy data for Veraval situated on the windy west coast of India is as follows:

Max usable energy in wind for $30m^2$ swept area	:	20,000 kWh
Machine Power output	:	4,000 kWh
Water depth	:	10m
Average daily output	:	$273m^3$ of water
Max daily output	:	$2428m^3$ of water.

Windmill WP-2 developed at National Aeronautical Laboratory. Bangalore can pump $2.9m^3/hr$ against a head of 10m in wind speed of 3.5m/s. A sail wind developed at the above laboratory uses six triangular canvas sails on a rotor of 7m diameter and pumps $5.4m^3/hr$ against a head of 10m in wind speed of 3.5m/s.

The specification of a propeller type wind mill pump set which has been standardised for installation in India are :

Tower Height	:	7m
No. of Propellers	:	12
Rotor Diameter	:	5m
Machine Power Ratio	:	0.36
Water depth	:	20 m
Max daily output	:	$23m^3$ at a wind speed of 10m/s.

Total cost of machine including transportation and erection is Rs. 11750 and a similar expenditure is to be incurred on well/boring and construction of storage tank. The Government provides a subsidy of Rs. 7500

The current philosophy is to use the same wind rotor for different wind regimes with modified down-stream components and higher ratings. Given that the power contained in the wind goes as cube of the wind velocity, a two-fold increase in the wind velocity implies an eight fold increase in the power contained.

The windmills installed so far are not only of small sizes (about 5m diameter). But these also require high maintenance support. Research efforts are now needed so as to reduce the maintenance expenditure of windmills to minimum, i.e., to occasional oiling and changing the pump washer. Somewhat larger windmills are required to be developed than presently available. It is also needed to develop cheaper and simpler S-rotors which can cut in at low wind speeds of 1.5-2m/s to cover wider areas of regions.

4. BIOGAS ENGINES

The technology for conversion of animal wastes into biogas via anaerobic digestion is well established, and quite a large number of biogas plants (about one lakhs) are already in operation. Our country has a potential of producing about 22,000 million cubic meters of biogas. The Gobar gas plants are mainly developed by Khadi and Village Industries Commission, New Delhi and in every state, Khadi & Village Industries Boards look after the erection work of the Gobar Gas plants. The Ministry of Agriculture is also helping for the construction of Gobar Gas plants.

In consists of a well made of bricks putting 2 to 3 steel rods to make it more strong and plastered. The top of the well is covered by 1.6mm thick steel sheet floating dome. The design of the above plant is available from KIhadi & village Industries Commission. The size of thc plant will depend upon the gas requirement by the engine.

The requirement of Indian farmers for engine varies from 3.7kW to 7.5kW in vertical high speed and vertical slow speed types all over the country. Normally, the engine runs on an average for 5 hours per day. Biogas produced by a digester that is functioning well is approximately 65 per cent CH_4 and has a heating value of approximately 21000kJ/m^3. The biogas engines for pumpsets produced in India are essentially diesel engines where air intake is connected to biogas supply. The diesel is required for starting the engine and about 20 per cent diesel and 80 per cent biogas are mixed automatically to produce power for running the pumpset. The biogas consumption is about 0.46m^3/kWh which can be obtained from 13.5kg of cattle dung., i.e., daily produce of one cattle. The price of a typical 3.7 kW, 1500 rpm biogas engine with centrifugal pump of 100 × 100mm including accessories and trolley is about Rs. 6500. It is subsidised by the Government and can be financed by the State Agro Industries, Land Development Banks and all nationalised banks. The cost of a typical gobar gas plant of 85m^3 capacity is about Rs. one lakh.

The human wastes can also be used for production of biogas. Community latrines can be planned in the villages for collection of night soil for feeding to biogas plants. Wastes of 200 persons can be used to produce about 5m^3 of gas per day to extract 12 kWh of equivalent energy by running a biogas engine.

5. BIOMASS GASIFIERS

The estimated production of agricultural residues in India is 200 million tonnes per year compared to about 130 million tonnes of wood. In equivalent energy content, this amount to about 54 million tonnes of coal which is 50 per cent of our annual coal production of 108 million tonnes in the country. Of this total residue, about 40 million tonnes is used in commercial sector, the remaining is burnt in fields, used for fodder roofing material, for composting and other needs. It appears that a substantial amount of residue is available as fuel.

The direct burning of biomass in domestic hearths is grossly inefficient, giving only 8 to 11 per cent useful energy. Biomass residues because of their low bulk density and extensive smoke formation are the most inconvenient and pollutant fuel. It is therefore, desirable to use biomass more efficiently.

Gasifier is essentially a thermochemical reactor in which the biomass undergoes partial oxidation and producer gas is obtained. The agricultural and forest residues can first be drived in solar driers to restrict the moisture content below 15 per cent. These are pressed in compaction briquetting machines to prepare feed size of 50-75mm for the gasifier. The biomass briquets are ignited in the gasifier and air flow is controlled to ensure partial combustion. The producer gas is reduced to CO and H_2 by carbon present in the burnt lower layers. The main constituents of fuel gas are CO, H_2 and CH_4.

The gas produced by the gasifier is hot and contains tar, vapours and soot particles. To make it engineworthy, it is cooled to near ambient conditions and cleaned to remove tar and dust by cross-current water scrubbers. Air and clean gas are supplied to the engine in pre-specified proportions. On an average 1 kg of biomass produces 2.5nm^3 of fuel gas and 1kWh of shaft energy to drive the irrigation pump.

The diesel engine utilizes the gas as a supplementary fuel operating on duel-fuel mode. The solar driers, briquetting machines and gasifiers are now made in India. The total cost of a biomass gasifier with gas cooler and scrubber of 15m^3/hr capacity, 5kW/1500 rpm diesel engine and 100 × 100mm centrifugal pump with necessary controls and accessories is about Rs. 37500.

6. HYDRAULIC RAMS

The utilization of hydroenergy to operate agricultural and industrial devices is one of the oldest and widespread techniques of human achievements, Hydraulic rams manufactured by John Blake Ltd., UK was installed at Taj Mahal, Agra in 1900 for raising water from the Yamuna at the rate of 2700m^3 per day. It is still working satisfactorily with very low maintenance cost. Blake Hydrams installed at Risalpur (Pakistan) in 1925 raises 4860m^3 of water per day to a vertical height of 20m and to a horizontal distance of 1600m. Hydraulic rams are also used in the Alpine areas of the world for the water supply of remote farms for irrigation and drinking water purposes.

Hydraulic ram is a contrivance to raise a part of large amount of water available at some height, to a greater height. This is employed when some natural source of water like a spring or a stream is available at some altitude, e.g., in a hilly region. Work done by a large quantity of water in falling through a small height is used to raise a small part of it to a greater height. Action of water hammer makes it feasible. No external power is therefore, required to work this machine. Other attractions are: negligible amount of maintenance and supervision costs, continuous operation, high efficiency, quite operation and possibility of automatic adjustment of water supply. The liftable volume of water diminishes asymtotically with lifting height. In case of medium lifting heights, the hydraulic ram operates with an efficiency absolutely comparable to a piston pump of the same performance.

As the hydraulic ram does not need a driving unit, it is to be considered as ideal for hilly areas, where the supply in fossil energy carries or electricity is problematic. The hydraulic ram operates with water streams of 1 to $40m^3/s$ and fall heads of 1.5 to 30m and with lifting heights up to 300m, which means an efficiency in the range of about 50 per cent to 2 per cent at the minimum. There is a number of different types which can be distinguished by the operation of the oscillating valves.

Hydraulic ram is a simple and rugged device for operating irrigation schemes and deserves R&D inputs for its performance improvement and adoption to irrigation.

7. WATER MILLS

Poncelet water wheel or a Darrieus Turbine Rotor can be sumberged in a river, water stream or canal to tap current energy from water speeds as low as 1m/s. The mechanical energy thus obtained is used for driving irrigation pumps. The blends of water mills can be made from fabricated steel, ferrocement, glass fibre or timber depending upon the availability of local materials and skills. The power output depends upon the swept area and cube of water velocity. A vertical axis rotor of $3.75m^2$ swept area runs at 13.5rpm in 1m/s current velocity. From this, a belt drives a pulley on an intermediate shaft which rotates at about 110rpm and from this shaft another belt drives a centrifugal pump mounted on the pontoon at approximately 950rpm. A 50mm floating pipeline is connected to the pump to lift water from the river. In a current speed of 1.18m/s, it lifts 3.51/s of water against a vertical height of 5m and horizontal distance of 25m and can be used to irrigate a one or two hectare riverside horticultural plot.

Although the machine will work at lower speeds, it is unlikely to be economical proposition in a current speed below 1m/s. The energy available in river current increases with cube of water velocity; a two-fold increase in water velocity has an energy flux eight times greater.

8. CONCLUSIONS

The bulk of future energy needs of India are not going to be met by fossil fuels and hydroelectric plants alone and have to be met from a variety of sources. Similarly, geographical conditions dictate the exploitation of locally available energy sources for lift irrigation utilising the appropriate intermediate technologies to convert them to mechanical energy for driving irrigation pumps. Some of the sources and technologies immediately available for this purpose have been discussed. An all out effort is needed to upgrade the performance of these devices and for adopting them to irrigation schemes. Special attention must be paid to the development and adoption of hydraulic rams and water mills for irrigation schemes which have relatively lacked behind.

9. REFERENCES

1. Singal, R.K., 'Floating Type Micro Hydro Power Plants', Electrical India, Bombay, 30th September, 1984.
2. Daniel M.Simmons, 'Wind Power', Noyes Data Corporation, Park Ridge, New Jersey, 1975.
3. Nirmal K. Pande, '*Energy Constraints—Need to Tap Renewable Sources*' Economic Times, July 5, 1984.

Paper-6

INTEGRATED RURAL SERVICES

SYNOPSIS

The provision of domestic amenities in the villages and other rural services for community the irrigation agriculture and cottage industries require abundant energy. Local energy resources available in the village and appropriate technologies developed have been explored. Alternative modes of supply of rural services and amenities using an integrated approach have been recommended. The information of the above study can be used by modern village planners.

1. INTRODUCTION

In most of the rural communities, quality of life is extremely low. There are people having no asset to talk about, no land, no shelter, no food, and are forced to live a miserable life under the trees because of their financial distress. Even when there is enough food, sanitation, medical care, education, lighting, hot water and community life are severely lacking.

A minimum shelter or dwelling is a must for any human being next to food, to enable him to live a useful social life. The desire to build a reasonable good house of one's own with maximum possible amenities is cherished by one and all.

In order to improve the living standards of rural population, a number of projects like 'Low Cost Housing' or 'Houses for the Millions' have been launched by Central and state Governments to build new and low cost houses. The provision of maximum possible amenities with these houses and other rural services require cheap and abundant energy for domestic cooking, food production and processing, heating and lighting, community development, irrigation water, mechanised agriculture and cottage industries, based on local resources. Transportation of electricity through grids to far flung areas are prohibitively expensive. Supplying commercial fuels like coal, diesel and patrol over long distances is very difficult.

An attempt has been made in this paper to identify need for domestic amenities and other rural services and their present status of supply. Potential of renewable alternative energy sources locally available in the village and status of supplementary and appropriate technologies developed have been surveyed in details. Alternative modes of supply of these services based on local resources and available technologies have been explored. Under the severe constraints of resources and paucity of funds, an integrated approach to supply of rural services is recommended.

International Conference on Habital and Sustainable Development, New Delhi.

2. RURAL SERVICES

Domestic amenities for a typical village have been identified in Table P.6.1. Cooking energy is the main requirement followed by drinking water, wash water, lighting and heating. The common modes of energy inputs presently used are listed in the 2nd column of the Table. The proposed alternative sources are shown in the 3rd column. The potential of these resources and latest technology status are discussed in the following paragraphs. Typical services for community life are included in Table P.6.2. The community services as on today are absent Therefore, alternative sources are recommended in the Table. Similarly, the utilities in the form of electricity, water and steam are needed for the development of cottage industries in the villages. The potential technologies based on renewable and local energy sources are listed in Table P.6.3.

Table P.6.1. Domestic Amenities

Sr. No.	Amenities	Common mode of Energy input (present)	Potential alternate sources (Proposed)
1.	Cooking	Wood, cowdung	Biogas,solar cooking plates.
2.	Drinking water	Animal powers, Human labour	Wind water current energy of water, electricity, solar pumps
3.	Wash water	Open surface water,	Water pumps run from Wind, solar, micro hydro, electricity.
4.	Lighting	Kerosene	Biogas lanterns, solar lights, electricity
5.	Heating	wood, Cowdung	Solar, biogas heaters.

Table P.6.2. Community Services

Sr. No.	Services	Common mode of Energy input(present)	Potential alternate Source (proposed)
1.	Radio. TV Sets	Nil	Solar, Electricity
2.	Street lighting	Nil	Solar, biogas electricity
3.	Milk chilling	Nil	Solar refrigeration, electricity

Table P.6.3. Utilities for Industries

Sr. No.	Utilities	Common mode of Energy input (present)	Potential alternate Source (proposed)
1.	Electricity	Nil	Micro Hydro, Solar, wind, biogas generators
2.	Water	Open ponds, canal system	Wind, Solar, Micro, Hydro, Biogas water pumps
3.	Hot water/ steam	Wood, Cowdung	Solar heaters/boilers on biogas, biomass

The main human efforts in rural areas are directed towards securing sufficient food. Nearly 65-80% of energy consumed is in the food chain systems. The typical food supply systems and traditional energy inputs are illustrated in Figs. P.6.1, P.6.2 and P.6.3. Potential areas of substitution by solar energy are also marked in the same figures. It is important to note the proposal for redivision of activities between rural and urban regions as marked in the figure P.6.2 and P.6.3 so that more value added activities are shifted to villages for rural prosperity.

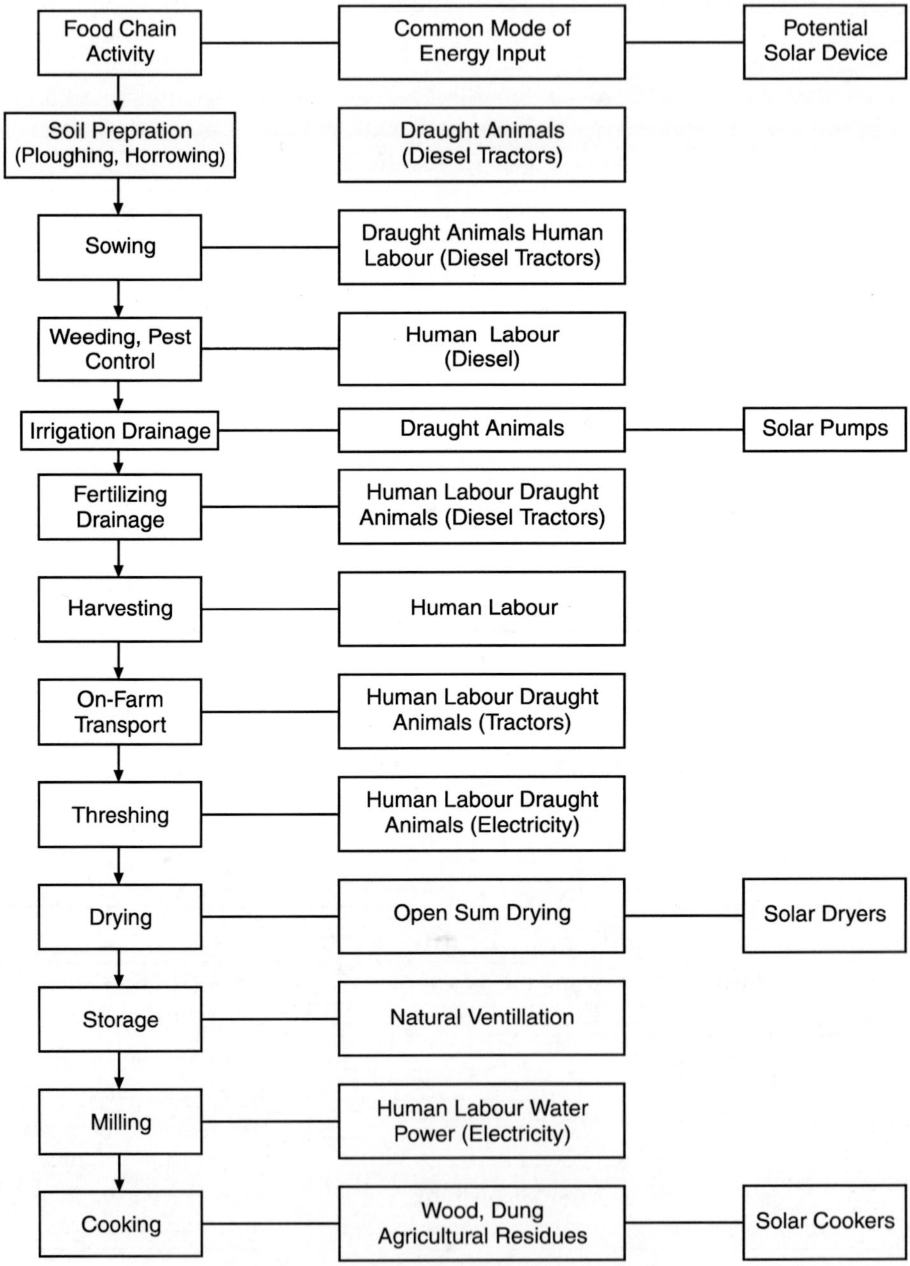

Fig. P.6.1. Imperishable Food Supply System

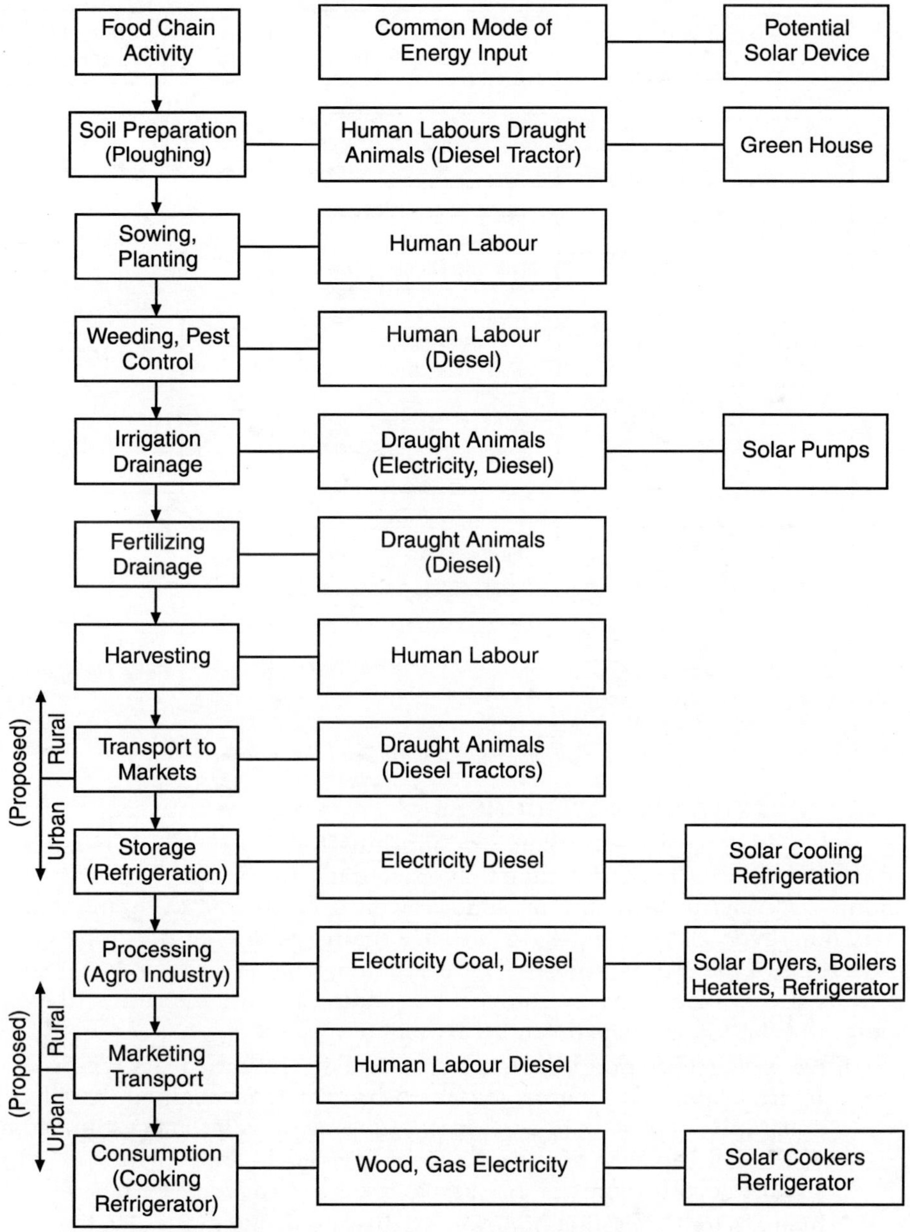

Fig. P.6.2. Perishable Food Supply System

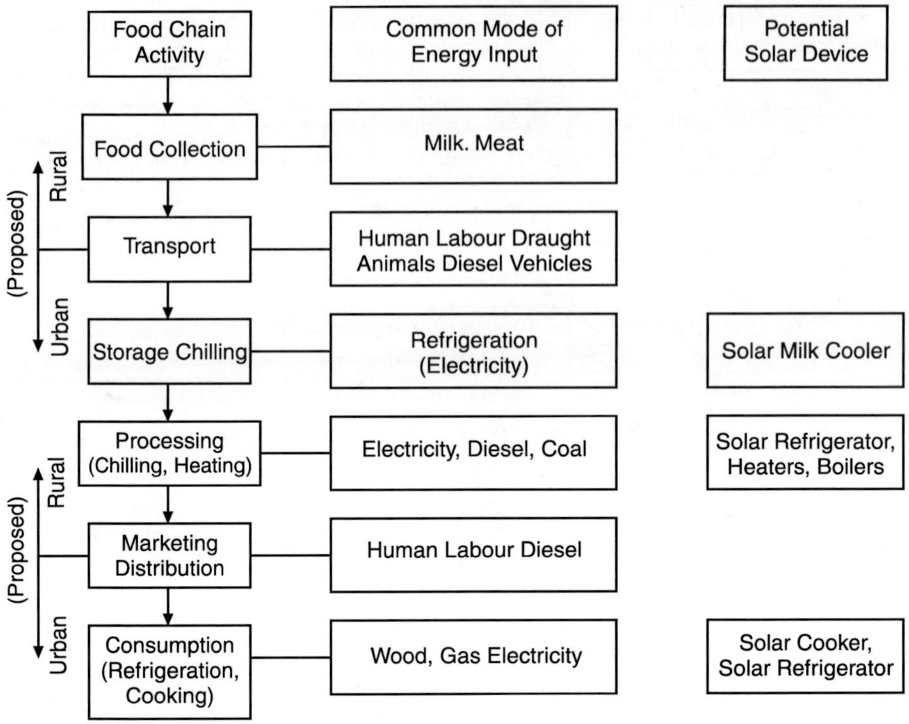

Fig. P.6.3. Animal Husbandary Food Supply System

3. ALTERNATIVE ENERGY SOURCES

India is bestowed with plenty of sunshine for 250-300 days per year. In South and Central India the annual average daily incidence of solar energy is about 4.5 kWh/m^2 with the maximum even exceeding 7 kWh/m^2 with the maximum even exceeding 7 kWh/m^2. Therefore, solar energy can be used to solve some of the energy problems in rural India. Solar energy can be integrated with locally available energy sources. Depending upon the location, the other sources of the village which can be exploited are :

(*i*) water current energy in the rivers and irrigation canals where micro hydro plants working on low/zero head turbines can be installed for generation of electricity and direct pumping of water. It is estimated that 5000 MW capacity of power can be created in India with the help of small hydro turbines of 200 kW to 1000 kW capacity each.

(*ii*) biomass for generation of power by direct burning in fluidised-bed combustors for generation of steam for a thermal power plant or production of fuel gas by gasification and running an I.C. Engine plant. The estimated production of agricultural resisdues in India is 200 million tonnes per year and that of wood is 130 million tonnes. At an energy conversion rate of 35%, total useful potential is about 75000 MW. The fluidised bed units can be used in conjunction with various configurations of power cycles using steam turbines, gas turbines, combined cycles with and without cogeneration.

(*iii*) animal wastes for generation of biogas for running an I.C. Engine or cooking gas. The biogas consumption is about 0.46m^3 per kWh which can be obtained from 13.5 kg of cattledung i.e., daily produce of one cattle.

(*iv*) wind energy for water pumping and production of electricity. Depending upon local wind conditions annual average energy per m^2 of swept surface may be 100 kWh for calm areas and 500 kWh for windy areas.

4. COOKING ENERGY

Eighty percent of the basic energy needs of rural India is provision of cooking energy, which amounts to 0.5 to 1 kWh/person/day with a peak of 0.2 to 1.2 kW per capita it is mostly met from traditional fuels like firewood and cowdung cakes. Cooking usually takes place on open-fire places which have thermal efficiency of the order of 5 percent. In case where firewood is scarce, dung and agricultural residues are used. If too much dung and residues are taken off agricultural land the soil may become slowly depleted of essential nutrients, causing gradually decreasing yields or an increasing demand for synthetic fertilizers

5. HOT PLATE TYPE SOLAR COOKER

The Government of India is trying to introduce box type solar cookers with financial subsidies but with poor success due to many technical drawbacks in the design. The temperature obtained is less than 100°C and is insufficient for chapati making and trying which are very important cooking processes. There is no provision for storage of heat and cooking has to be carried out while the sun is shinning which is an odd time for preparing dinner and breakfast. Cooking has to be carried out in the open without privacy and in the sun which is very inconvenient to the housewife. In order to overcome all these difficulties, a small family size solar cooker was designed and fabricated at REC. Kurukshetra. It consists of a flat plate with mirror boosters, an oil storage tank, a thermosyphon system, a heat pipe and a control valve. It would just replace a bottled gas cooking range. There are no operating and maintenance expenses.

6. BIOGAS PLATE

The technology for conversion of animal wastes into biogas via anaerobic digestion is well established and quite a large number of biogas plants both community type and individual type are already in operation. The biogas obtained from 13.5 kg of cattle dung, i.e., daily produce of one cattle, is sufficient to meet the cooking energy requirements of two persons and surprisingly, the ratio of human beings to cattle in an Indian village is two. The gas produced can be easily transported to specially designed gas plates through network of HDPE pipes

7. WATER SYSTEM

Drip irrigation of corn, wheat, cotton, millets requires 25 m^3 of water per hectare-day whereas flood irrigation of rice, sugarcane, etc., would require 50 to 60 m^3 of water per hectare day. In addition, water requirements for live stock is 40 to 50 liters per head-day. The lifting height of water varies from a few meters to 30 m. The equivalent energy requirements would be 135 W per hectare for surface water and 400 W per hectare for ground water. Out of the total 143 million hectares of land under cultivation in the country, hardly 43 million hectares (a barely 25% of the land under cultivation) is irrigated. Out of this irrigated area, 23 million hectares are served by tube-wells ponds or such similar minor irrigation schemes. There are about 2 to 3 million open wells in our country which are motorised Various modes of irrigation and water supply schemes based on non-conventional energy sources have been developed.

8. SOLAR WATER PUMPS

The National Physical Laboratory, New Delhi has developed, Abhimanyu Solar Pump of 1 kW rating using R114. The total cost of the components (condenser, boiler, expansion motor, storage tank, collector assembly and pump) and development charges is of the order of Rs 37.000.

9. SOLAR PHOTOVOLTAIC WATER PUMPING SYSTEM

The Central electronics Ltd. Sahibabad (UP) has developed a photovoltaic system at a cost of Rs 40,000 which can give an average water output of 30 to 40m^3 per day with a maximum total head of 10 m. It is highly susidised and installed at a cost of only Rs, 5,000

10. WIND MILL

The specifications of a propeller type wind mill-pump set has been standardised for installation in India are :

Tower Height	7m
No. of Propellers	12
Rotor Diameters	5 m
Machine Power Ratio	0.36
Water Depth	20 m

maximum daily output 23 m^3 at a wind speed of 10 m/s

Total cost of machine including transportation and erection is Rs, 11.750 and a similar expenditure is to be incurred on well boring and construction of storage tank. The Government of India provides a subsidy of Rs 7500.

11. BIOGAS ENGINES

The biogas engines for pumpsets produced in India are essentially diesel engines where air intake is connected to biogas supply. The diesel is required for starting the engine and 20% diesel and 80% biogas are mixed automatically to produce power. The biogas consumption is about 0.46 m^3/kWh which can be obtained from 13.5 kg of cattle dung. The price of a typical 3.7kw. 1500 rpm biogas engine with centrifugal pump of 100 × 100mm including accessories and trolley is about Rs 6500. It is subsidised by Government agencies. The cost of a typical gobar gas plant of 85 m^3 capacity is about one lac rupees.

12. BIOMASS GASIFIERS

The agricultural and forest residues can be first dried in solar driers to restrict the moisture content below 15%. These are pressed in compaction briquetting machines to prepare feed size of 50 to 75 mm for the gasifier. On an average 1kg of biosmass produces 2.5 nm^3 of fuel gas and 1kwh of shaft energy to drive irrigation pump. The total cost of a biomass gasifier with gas cooler and scrubber of 15m^3/hr capacity. 5kw/1500 rpm diesel engine and 100 × 100 mm centrifugal pump with accessories and controls is about Rs 37,000.

13. WATER MILLS

Poncet water wheel or a Darrieus turbine rotor can be submerged in a river, water stream or canal to tap current energy from water speed as low as 1m/s. The mechanical energy thus obtained is used for driving irrigation pumps. Such floating water mills can be used in conjunction with Himalayan Mills and Hydraulic rams to meet the total energy requirements of a village in mountaneous region for irrigation and drinking water and all the electrical power for domestic, community and industrial purposes.

14. OTHER SOLAR GADGETS

Photovoltaic solar systems for street lighting, solar TV sets, solar driers for agricultural products, solar kiln, distillation plants, air and water heaters have been developed and are used to provide rural amenities.

14.1. Solar Refrigerator

At the 1975 UNESCO Conference on Solar Energy, several developing countries stated that small solar driven units are urgently needed for preservation of food in rural areas where little electric power is available. A study and design project was undertaken at REC, Kurukshetra for the development of a small solar refrigerator. Solid absorption zeolite refrigeration cycle was adopted because of absence of moving parts, toxic and corrosive chemicals and expectation of long and trouble-free operation. It was designed for fitting onto the cabinet of the present 165L capacity domestic refrigerator. It consists of a $1.5m^2$ flat plate collector filled with 50 mm thick zeolite and can produce 9 kg of ice in 8 hours. The project experience can be used for the development of small solar cold storages for preservation of food in rural areas.

14.2. Solar Milk Cooler

Nearly 500/600 liters of milk per village is sold as liquid milk per day and its chilling is done by the agencies procuring it. A village level chilling/cooling facility will make possible the collection once in a day resulting in large savings. For example, reduction in route kilometerage alone can result in savings of about 12.5% of procurement price. Rural processing of milk will also add value to the produce and increase rural income. The zeolite solar refrigerator can be attached to cooling-cum-storage tank. The water refrigerant circulates by thermosyphon effect in the tank walls and acts as a very efficient heat exchanger, reducing the temperature difference between milk and water vapour to less than 0.5°C.

15. CONCLUSIONS

Geographical conditions dictate the exploitation of locally available energy sources for providing maximum amenities for the rural community. Integrated rural services can be desiged from local resources available, for example, a hydro village in a hilly area, a solar village in south India or a biomass village in the plains of north India and hybrid village elsewhere. The study material of this paper can be used by the village planners as guidelines. The scientists and academicians can benefit from the proposals made for taking up research and technology development projects.

Paper-7

DESIGN PROPOSAL FOR A HOT PLATE TYPE SOLAR COOKER

1. INTRODUCTION

Eighty per cent of the basic energy needs of rural India is the provision of cooking energy, which amounts to 0.5 to 1 kWh/person/day with a peak power of 0.2 to 1.2 kW per capita. It is mostly (for 83% of rural household) met from traditional fuels like fire-wood, agricultural wastes and cowdung cakes. Cooking with fire-wood and cowdung cakes poses a number of problems for the users and for the society as a whole. The serious implications of deforestation and subsequent soil erosion, pollution and lack of recycling of the natural fertilisers have a significant effect on the overall productivity of these areas. Therefore, it becomes essential to find out an alternate source of cooking fuel.

India is bestowed with abundant supply of solar energy and hence it can be a very promising source of cooking energy. Large scale solar cooking technology must be attempted to alleviate the demand for cooking fuel. Other non-conventional energy sources such as bio-gas, would be best used to increase the industrial and agricultural productivity of rural areas, while allowing solar energy to provide the bulk of cooking requirements.

Efforts to design and introduce solar cookers have for many years concentrated on two main types: the cooker with a plane collector (box type) and cooker with special/parabolic mirrors of average precision. The cost of second type is too high to be used economically in rural areas. The Government of India is trying to introduce box-type solar cookers with finanical subsidies but with poor success.

2. NON-ACCEPTABILITY OF PRESENT SOLAR COOKERS

Despite Government efforts to introduce box-type solar cookers under subsidy programmes, they have failed to find a place in kitchens of the rural household contrary to the expectations. The beneficiaries donot accept solar cookers in daily cooking and their possession has made no impact on their conventional fuel consumption pattern.

2.1. Socio-economic Factors

Among the socio-economic factors inhibiting the large scale utilization of solar cookers, initial investment cost is one obstacle to the purchase of solar cooker. Families in rural areas can gather wood and cowdung for fuel for

Electrical India, Mumbai

conventional cooking. Furthermore, solar cooking involves alteration in the household routine, since the housewife must prepare the evening meals during day time, well before the traditional time, which increases the additional problem of keeping the food hot. Secondly the housewife must remain in the hot sun while cooking. On rainy days or when there is insufficient sunshine, conventional fuels have to be used. There is lack of privacy as cooking has to be carried out in the open.

2.2. Technical Drawbacks

The technical drawbacks of the present box-type solar cooker patronised by the Government may be summarised as follows :

(*i*) The temperatures obtained in the cooking vessel (food) are less than 100°C. It does not ensure hygienic cooking as at some stage of cooking cycle, the temperature of the food must exceed 100°C to kill bacteria.

(*ii*) It takes more time than conventional cooking as it is only a low-temperature, low-energy flux device.

(*iii*) Solar cooking cannot be employed for all modes of cooking processes. For example, chapati making and frying which are very important cooking processes in most part of the country especially north India cannot be carried out in a box-type solar cooker. The practice of conventional cooking is very different in India and is a big hinderance to the social acceptability of solar cooking. Solar cooking is very slow and time consuming and some foods have to be pre-cooked before keeping them in the cooker box for final cooking.

(*iv*) There is no provision for storage or heat energy. Cooking has to be carried out while the sun is shining which is an odd time for preparing dinner and breakfast. Also on rainy days and cloudy days when sufficient insolation is not available, auxiliary fuel has to be used.

(*v*) Cooking has to be carried out in the open without privacy and in the sun which is very inconvenient to the housewife. The glare of the mirror/glasses can be injurious to the eyes of the housewife and the accompanying children.

2.3. Study of M.S. University Baroda

A study was conducted by Dr. (Mrs.) Rachel Georage, Reader, Home Management Department, Faculty of Home Science, M.S. University, Baroda, recently on reasons for non-acceptability of solar cookers and her findings are reported as follows :

(*i*)	Lack of Privacy in Cooking	: 100%
(*ii*)	Mirror and glass need special care	: 92%
(*iii*)	Solar cooking should take place during specific periods of the day.	: 92%
(*iv*)	Cooking does not get over in time for meals	: 75%
(*v*)	Solar cooking is too time consuming	: 72%
(*vi*)	Solar cooker is too heavy	: 72%
(*vii*)	Food need to be precooked	: 50%
(*viii*)	Cleaning of utensils (blackened from outside) cannot be entrusted to children.	: 35%

(*ix*) Smell remains in the food and cooker : 25%

(*x*) Paint on the cooking containers wears off fast. : 20%

The figures indicate the per cent response to an opinion poll conducted in rural areas of Gujarat.

The effectiveness of solar cooker can be enhanced by the design of a system which can overcome the above technical and socio-economical problems. The cooker has to be coupled to a heat storage system. It must be capable of storing sufficient heat at a temperature high enough to permit effective cooking later on. It is also desirable to be able to regulate the rate of heat release for cooking.

3. DESIGN PROPOSAL

In order to develop a small family size (four persons) solar cooker, a project was undertaken at Regional Engineering College, Kurukshetra. A solar cooker of hot-plate type was designed and successfully constructed. It consists of a flat plate type collector with mirror boosters. It is coupled to an oil storage insulated tank placed under the cooking platfrom inside the kitchen through a thermosyphon to the outside collector.

In can provide sufficient flux of heat energy at high enough temperature to a hot plate through a heat pipe. The system would be suitable for cooking of conventional foods at convenient time for breakfast, lunch and dinner inside the kitchen. It would just replace a bottled gas cooking range and would be fitted permanently. It will fit in with the age-old habits and practices of rural households and should be acceptable to the user without reluctance.

3.1. Design Data

The following data has been assumed for the design of the cooker after carrying out detailed survey of heating habits, normalised menu for break-fast, lunch and dinner, cooking times and temperatures for different dishes, energy requirements and burner ratings of bottled gas cooking range. Average sunshine hours and solar radiations are taken for Kurukshetra from the handbooks.

Cooking energy requirements (4 persons, three meals per day)	:	2 kWh/day
Hot Plate Rating	:	2 kW.
Cooking Temperature Range	:	150-175°C
Storage Capacity	:	3 kWh.
Annual average solar radiations	:	650 W/m^2
Annual average sunshine hours/day	:	8 hours
Annual average wind velocity	:	10 m/s
Annual average ambient temperature	:	25°C.
Design value of latitude	:	30°N

3.2. Component Sizing

Assuming overall efficiency of 15%, specific heat of oil as 2.63 kJ/kg-K and oil density at 175°C as 0.905 kg/m^3, the following sizes are worked out;

Total collection area	:	2.5 m^2.
Storage volume of hot oil	:	100 litres.

4. SYSTEM DESIGN

The basic system design should respond to local climate: sunshine hours, solar radiation level, wind speed, ambient temperature and to energy demand characteristics. The functional flow diagram of the system is shown in fig. P.7.1. It consists of the following sub-systems.

- Solar collection and conversion.
- Thermal storage.
- Fluid circulation.
- Energy distribution and control.

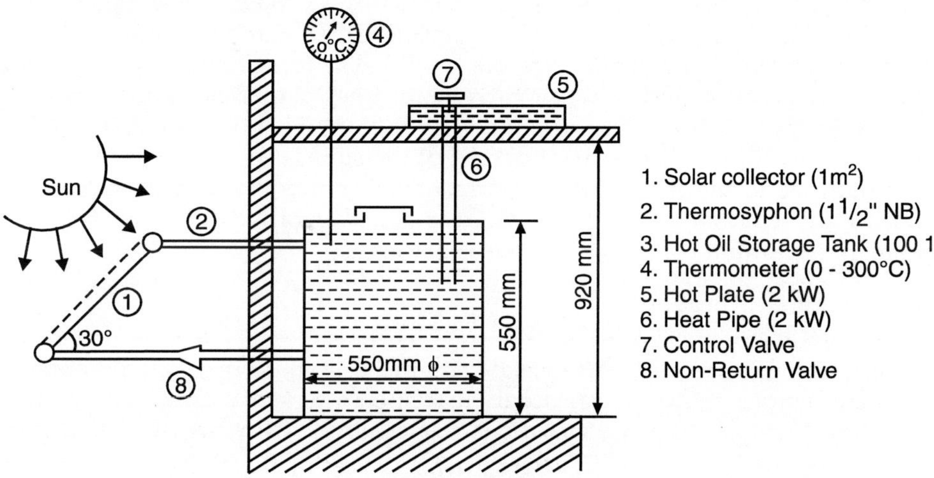

Fig. P.7.1.

4.1. Solar collection and Conversion System

The flat plate collectors are easy to fabricate and cheaper than the concentrated type. In addition, flat plate collectors can absorb all the solar insolation, i.e., beam and diffused radiations whereas concentrated type are not sensitive to diffuse radiations. This factor can be very important during cloudy days especially for Indian regions with long monsoon months. The problem of orientation and tracking can be easily handled in flat plate collectors. Concentrating parabolic collectors with special mirros are not considered on economic and above mentioned technical factors.

The technology of flat plate collector is sufficiently developed for other applications such as space heating of small and large buildings, water heating, space cooling, process heat, crop drying, distillation, etc. It will be, therefore, easy to adopt it for solar cooking.

A temperature in the region of 200°C has been obtained in flat plate collector with selective coating of absorber plate and energy reception augmentation with the help of mirror boosters. In order to obtain a temperature of 475K, we can use augmented flat plates with a concentration ratio of nearly 4 and a/E ratio of 10. Refer to fig. 3 of Author's paper on 'Design criteria for Solar Energy Systems' published in 15th may Issue of Electrical India.

In order to satisfy the conditions of concentration ratio, four booster plain mirrors are used to get the necessary augmentation. The selective coating can be either chrome black or nickle black with the properties given in Table P.7.1.

Table P.7.1. Properties of Selective Coatings

Selective Coating	Shortwave *Absorptance, a*	Longwave Emittance, *E*	Ratio *a/E*
Black Chrome	0.87	0.09	9.8
Black Nickle Oxide	0.85-0.93	0.06-0.1	14.5-15.5

The arrangement consists of flat plate collector of $1m^2$ surface made from Derco Roll Bond Aluminium panels. Three panels of size 915mm × 356 mm × 1.5 mm are arranged in parallel and housed in a properly insulated wooden box. The panels are coated with black chorme. Three window glass sheets 3mm thick and 1000mm × 1150mm size are provided as top glazing to cut off losses due to re-radiation and convection from absorber plate. Four plain mirrors of size 1068mm × 915 mm are arranged at the four sides of the panel box with hinges to manuplate necessary tracking of the sun. The collector is installed at 30° (latitude of the site) with the horizontal. The top and bottom mirrors will need only seasonal adjustments whereas side mirrors can be adjusted every hour for optimum collection of solar radiations.

4.2. Thermal Storage

Thermal storage is used in most solar systems to provide a thermal capacitance effect. Thermal capacitance is required to damp out the fluctuations of collector energy delivery that occurs in response to short-term fluctuations in sunshine level and diurnal variations related to the 24 hours day-night cycle. In our case, we have to collect solar energy during day time only. On a clear day the energy delivered by the solar system is approximately a sinusoidal function of time with a maximum at noon and minimum near sunrise and sunset. The demand on the thermal storage system will be variable and energy will be extracted three times a day for preparation of breakfast, lunch and dinner. The temperature history of storage for a day will therefore, vary and condition of minimum temperature required for cooking must be met. Also on rainy and cloudy days when sufficient insolation is not available, stored heat should be available for cooking.

The amount of thermal storage is usually subject to economic constraints, since very large storage is generally prohibitively expensive. As a compromise, a storage tank of 100 litres capacity is provided which is equivalent to 3 kWh of heat at 175°C. The hot oil used as working fluid is also the sensible heat storage fluid as it has favourable heat capacity properties.

In order to minimize heat loss and tank material requirements, a right circular cylinder of size 55cm diameter and 55cm high fabricated from mild steel of 2mm thickness has been provided. It can be directly placed on the concrete floor. It is properly insulated with 75mm thick glass wool sheathed in aluminium sheet. A thermometer of range 0-300°C is provided in the thermowell welded into the storage tank.

4.3. Fluid Circulation

The practical use of solar energy requires a means to transport energy from the solar collector to the storage subsystem and in turn to the load or demand points. Any fluid flow is subject to viscous and turbulent resistance to flow that must be overcome by a motive force. The most common motive force is pressure generated by a pump which requires an external source of electrical power.

The natural tendency of a less dense fluid to rise above a more dense fluid can be used to cause fluid motion through a collector. The density difference is created within the solar collector where heat is added to the liquid. This thermosyphon system does not need any pump or electrical power supply.

Since the driving force in a thermosyphon system is only a small density difference and not a pump, larger-than-normal plumbing fixtures must be used to reduce pipe friction losses. We have used $1^1/_2$" NB pipes. The flow rate through a thermosyphon system is about 40 litres/m^2 in bright sunshine. After sunset, a thrmosyphon system can reverse its flow direction and lose heat to the environment during night. A gravity-type non-return valve is provided in the return cold leg of thermosyphon to avoid reverse circulation of oil. The heat loss from the piping are controlled by proper insulation with asbestos rope.

The working fluid has to be stable at 200°C and should have favourable heat transfer properties. It has to be compatible with aluminium, copper and carbon steel, the materials of construction of solar thermal system. One hundred and ten litres of oil is required for initial fill of the system. Grade Servotherm Medium of Indian Oil Corporation or Hytherm 500 of Bharat Petroleum are suitable oils.

4.4. Energy Distribution and Control

The heat pipe is relatively new device for transporting large amounts of heat with a very small temperature drop between heat sourcc and sink. A gravity-assisted heat pipe is used to distribute heat energy from storage tank to a cooking hot plate. The evaporate section of heat pipe is 5 cm long and is inserted vertically from top of storage tank. A condenser length of 10 cm is inserted inside the hot plate. The adiabatic length to suit site conditions is properly insulated with asbestos rope. The heat pipe is made from 32 × 2 mm copper tube of 1 m length. The working fluid is Dowtherm grade *E* filled at atmospheric pressure after proper evacuation of air. The boiling point of Dowtherm E is 150°C. There are no wicks inside and liquid oil flows down to evaporator section along axial grooves (2mm × 1mm deep) cut all along inside of heat pipe.

Much controls and instrumentations are not needed. In addition to a thermometer provided in the oil storage tank, a piston-type valve is provided at the top of heat pipe in the condenser section. The rotation of a small wheel controls the travel of piston in the heat pipe and length of condenser taking part in heat transfer. The heat flow rate from the storage tank through the heat pipe will be proportional to effective condenser length not cut off by piston valve.

The hot plate 200mm diameter × 125mm height is made from copper sheet and filled with hot oil for proper heat transfer from heat pipe. It is insulated from all sides except the top cooking surface. The heat pipe has been designed for axial heat flux of 2400 × 10^3K cal/m^2 and radial heat flux of 1200 × 10^3/ Kcal/m^2.

The total cost of materials including hot oil charge and fabrication expenses has been Rs. 3000.

5. CONCLUSION

The proposed design of hot-plate-type solar cooker overcomes most of the technical and socio-economic drawbacks of the present box-type solar cooker. It fits well with the age-old habits of conventional cooking and should be accepted by the rural population in India without any reservations. It is a permanent fixture comparable in performance to the bottled gas cooking burner in heat flux rates and can provide controlled temperatures required for all types of cooking processes and at suitable timing for preparation of breakfast, lunch and dinner. She can cook in all privacy of her kitchen. It is coupled to a thermal storage system and can last for a day without sunshine. The heat flux rate can be conveniently controlled by turning a small wheel suitably placed at the hot plate. There are no operating and maintenance expenses and life long oil fill is provided.

The prototype cost of Rs. 3000 can be substantially reduced during commercial design and manufacture. With suitable Government subsidies as applicable to other solar devices, this is an ideal design for a low cost solar cooker for adoption in rural and muffasal areas.

REFERENCES

1. Singal, R.K., *'Design Criteria for Solar Energy Systems'*, Electrical India, 15th May, 1986
2. The Hindustan Times dated 21st October, 1985.
3. R.L. Loftness, 'energy Handbook'. Van Nostrand Reinhold Company, New York, 1978.
4. A Eggers-Lura, 'Solar Energy in Developing Countries, Vol., 11, Pergamon Press, Oxford.

❑❑❑

Paper-8

DEVELOPMENT OF A SOLAR MILK COOLER

1. INTRODUCTION

India is essentially an agricultural country with 78 per cent of its population living in rural villages. Population-wise the distribution of villages is as follows:

Population	No. of Villages	Per cent of total
Upto 499	349,995	62.0
500 to 999	118,602	21.0
1,000 to 1,999	65,101	11.6
2,000 to 4,999	26,400	4.6
5,000 to 9,999	3,404	0.6
10,000 and above	774	0.2
Total	564,276	100.0

About 95 per cent of the total are small villages with population less than 2000. On an average, the cattle population is 50 per cent of human population in Indian villages. The total milk production per village per day on an average may be taken as 2000 liters; out of which the milk producers consume 25 per cent for self use and convert about 45 per cent into milk products like ghee, paneer and khoa. Therefore, 500-600 litres of milk per village is available for sale as liquid milk per day.

Normally, at present this milk is sold as such and its chilling is done by the agencies procuring it. A village level chilling/cooling facility will surely be advantageous as it will make possible collection of evening milk also once in a day which is more economical than twice a day collection requiring use of available collection, transport and chilling facilities twice each time operating much below their rated capacities. There will thus be large savings through once-a-day collection. For example reduction in route kilometerage only can result in savings of about 12 1/2 per cent of the procurement cost.

Rural processing of milk can add value to the product and increase rural income. The fluid milk can be cooled to the prescribed temperatures by the producers or their co-operatives and keep the milk fresh till the time it is boiled by the consumers for use. Pasteurisation of milk is not necessary as consumers even other-wise boil the pasteurised milk and wastage of energy for double heating can be avoided.

Electrical India, Mumbai

At present, some manufactures have developed milk cooling and storing cabinets suitable for rural areas. Typical specifications of such chillers produced by Larsen & Toubro are given below :

Capacity (Litres)	Connected Load (kW)	Budgeted Price (Rs. in Lacs)
500	2.5	1.32
1000	3.7	2.50
2000	5.0	3.00

The milk is quickly cooled from ambient temperature to below critical temperature and further cooled and stored at 4°C. The cooling is achieved by mechanical refrigeration and therefore, electricity is required for its operation. Unfortunately electrical power is not available in most of the villages and rural electrification is very expensive and practically difficult in far flung areas in a vast country of India's size. On the other hand, India is bestowed with plenty of sunshine all the year round, and development of a suitable solar milk cooler to meet the above requirements of cooling and storage of fresh milk can be a boon to the rural economy and white revolution. Few components, no moving parts, operational simplicity, long repair-free life are essential requisites of a good design of a milk cooler for rural adoption.

2. SOLID ABSORPTION-ZEOLITE REFRIGERATION: BASIC OPERATING CYCLE

Zeolite is a mineral found in many parts of the world and has special adsorption properties. It can adsorb water vapour, ammonia, carbon oxide and freons, all with 30 per cent by weight. The zeolite-water vapour combination is the most efficient system as it requires minimum amount of zeolite due to largest heat of vapurisation of water. The adsorption of water in zeolite is a function of temperature and pressure and nonlinear dependence on pressure makes zeolite very favourable for solar applications.

The zeolite is sealed in an air-tight container irradiated by the sun. During the day cycle, the zeolite is heated to a temperature of about 120°C. At about 40°C, water vapour starts desorbing from the zeolite, and its partial pressure rises. At condensing temperature of 40°C, the partial pressure is about 55 mm Hg and vapour begins to liquify as heat is rejected to the environment. The liquid water is stored in a tank. During the night cycle, the zeolite is cooled by convection to ambient temperature and sky radiation.

Liquid water from the storage tank is introduced into the evaporator, where it absorbs heat from the space to be cooled and is converted into water vapour. If partial pressure is 5 mm Hg, the water in the evaporator boils at 2°C. The zeolite adsorbs the water vapour produced by the evaporator and a partial pressure is maintained below 5 mm Hg. The heat of adsorption is rejected to the atmosphere. The zeolite is fully loaded corresponding to the partial pressure and is ready for next day cycle.

$$Na_2\ Al_2\ SiO_2 + x.\ H_2O = Na_2\ Al_2\ SiO_2.\ \times\ H_2O$$

The above cycle can be utilized for the development of zeolite solar cooler.

3. ZEOLITE SOLAR COOLER

Zeolite solar cooler can be used to produce cooling effect and manufacture ice using sun's rays without any external power and without moving parts. A solar cooler having an efficiency of 15 per cent can produce about 0.9 kWh of cooling for each square meter of collector area receiving a solar input of 6 kWh. This amount of cooling corresponds to about 9 kg of ice manufactured per day for each square meter of collector. This principle can be used to produce ice, refrigerators to store vaccine, domestic rfrigerators, on-site refrigeration of perishable produce and tropical fruits, in-the-field refrigeration of high-cost perishable products etc. Its major applications can be in developing countries like India where energy costs are high and electricity unavailable, unreliable and expensive. It can be of great importance to the fishing industry, especially for refrigeration of fresh fish and other seafood in the remote fishing villages in tropical countries. However, it has a great potential for dairy farmers and rural milk producers.

4. MILK COOLER DESIGN

Possible design of a solar milk cooler can be as per the following specifications :

4.1. Solar Panel

The zeolite solar panel resembles a conventional flat-plate collector. After the zeolite is charged, the whole system is outgassed, evacuated and hermetically sealed. The panels have a zeolite capacity of about 50 kg/m^2 and requires a depth of 50 mm. The upper surface of the panel is painted black for maximum solar absorption. The water loading is 5 to 7 per cent by weight of zeolite for a temperature range of 40°C to 120°C. The inlet and outlet pipes are connected at the bottom of the panel to permit water vapour to enter and leave. The panel is leak-tested prior to use.

The system is ideally suited to local manufacture, but the zeolite will have to be imported and skill is required to prepare zeolite-filled collector and evacute the system. In order to avoid the import of zeolites, synthetic zeolites can be tried. It may be interesting to try strong Acid Cation Exchange resins in either Na or K form as desicants in place of natural zeolites, which are indigenously manufactured.

4.2. Condenser

The air-cooled condenser fin-pipes can be fixed at the back of sloping panels. The condenser pressure will be 55 mm Hg corresponding to ambient air temperature of 40°C. The condensed vapours during the desorption cycle run by gravity into the flooded evaporator.

4.3. Cooling-Cum-Storing TANK

The cooling-cum-storing tank can be made from single-embossed panel coils of stainless steel, the inside being plain sheet and outside made of embossed sheet. This design establishes thermosyphon circulation in the walls during the cooling cycle. The design acts as a very efficient heat exchanger thus reducing the temperature difference between milk and water vapour to less than 0.5°C. As water storage is in the bottom, it makes possible to cool even small quantities of milk. At a pressure of 5 mm Hg, water will boiled at 2°C in the evaporator coils and this low pressure is maintained due to adsorption of water vapour by zeolites.

The walls and roof of the tank are covered with 150mm thick polyurethane insulation having a loss coefficient of 0.2 W/m^2. °C.

Unlike conventional milk coolers using freons or ammonia as refrigerant and brine as secondary fluid, solar milk cooler works with only one fluid. As there are no moving parts or corrosive chemicals and there is no reduction in the adsorption ability of zeolite, it is estimated, the system would work for more than ten years.

4.4. Auxiliary Heat

The cooler uses only solar energy and can be used in rural areas where electricity is unavailable, unreliable or expensive. The collector panel can be provided with pipes that run through the zeolite panel to provide auxiliary heat whenever the sun is not shining. The farmer could then start a small fire from straw, cowdung or other local fuels and let the hot gases from the fire pass through the pipes in the zeolite panel, thereby heating the zeolite and causing desorption.

5. REFERENCES

1. D.I. Tcherves, *'Use of Natural Zeolites in Solar Regrigeration'*, ASSET, Vol. 6, No. 5, 1984, Tokyo.
2. *'Solar Refrigeration for Vaccine Storage and Ice Making'*, ASSET, Tokyo, Vol. 6., No. 1, 1984.
3. D.O. No. DS/84-85/4/731 dated 24th May, 1984 from Dairy Superintendent, National Dairy Research Institute, Karnal.

❑❑❑

Paper-9

DOMESTIC SOLAR REFRIGERATOR

1. INTRODUCTION

Solar energy is better suited to space cooling and refrigeration than to space heating, but until recently this application of solar energy has received little attention. However, energy for air conditioning and refrigeration has been the faster growing segment of energy consumption due to increasing standard of living and population in India and other developing countries. At the 1975 UNESCO conference on Solar Energy several developing countries stated that small solar driven refrigeration units are urgently needed for preservation of food in rural areas where little electric power is available. Consequently, accelerated effort is needed to effect progress in the application of solar energy to space cooling and refrigeration.

The seasonal variations of solar energy are extremely well suited to the space cooling and refrigeration requirements. Since warmest seasons of the year correspond to periods of high incidence of solar energy, it is most available when comfort cooling and refrigeration are most needed. Moreover, the efficiency of solar collector increases with increase in incidence in solar energy and increasing environmental temperature. Consequently, in summer the amount of energy delivered per unit surface area of collector can be larger than that in winter.

Solar cooling and refrigeration are ideally adaptable in India. India is bestowed with high solar insolation and long sunshine hours. Refrigeration units are most needed in rural areas for food preservation where it is difficult to reach with electric power. Therefore, development of small solar-driven refrigeration units is urgently needed for India.

In the light of the above situation, a project was undertaken for the design and development of a small domestic solar refrigerator. The capability of the existing conventional refrigerators was studied and a survey of literature was made to find out a suitable technology for a solar refrigerator to meet the requirements of conventional domestic refrigerator. It was found that a solid absorption-Zeolite refrigerator is most suitable because of absence of moving parts, toxic and corrosive chemicals and expectation of long trouble free operation life. The system is ideally suited to local environment. Heat load and temperature requirements can be met by a convenient and compact sized solar system at competitive price.

There is limited operating experience with solar cooling system. Therefore,

Electrical India, Mumbai

the system and components have been designed from the fundamental principles taking into account the operational experience of a conventional refrigeration cycle and special features of the components in a solar system. The project experience can be used to develop small cold storages for preservation of food in rural areas and commercialization of domestic solar refrigerators.

2. CONVENTIONAL DOMESTIC REFRIGERATOR

It has been found that the present 165L capacity refrigerator working on vapour compression cycle using Freon-12 as refrigerant is most suitable and popular size in India. The design of the solar domestic refrigerator has therefore, been based on the main specifications of 165L capacity refrigerator. A suitable solar refrigeration system has been sized for fitting onto the cabinet of the present domestic refrigerator of this capacity.

The present conventional refrigerator consists of a well insulated cabinet to maintain a temperature of 8-10°C in the refrigerated space and –4 °C in the freezer. The main dimensions of the cabinet are as follows:

Total volume	165L
Food storage capacity	130L
Freezer capacity	35L
Shelf area	$0.5 m^2$
Door shelf area	$0.12 m^2$
Insulation	5 cm Polyurethane Foam

An air-cooled extended surface type condenser of sufficient surface area is mounted on the back of the refrigerator. Plate type or wire finned tubular condensers are normally used with 6.35/4.8 mm copper or steel tubes of total length of about 7m.

The evaporators are housed in the upper space of the cabinet and consists of 7.9/6.35 mm copper tubing of about 5m length soldered to the plate.

An hermetically sealed compressor of 1/8hp is spring mounted to isolate vibrations. A filter and capillary tube acts as division point between high pressure and low pressure sides of the cycle and flow of liquid refrigerant is regulated into the evaporator by the capillary tube.

A maximum cooling load of 317W can be produced by circulation of 125.5 g/min of refrigerant at 1.2 kgf/cm^2 and – 25°C in the evaporator and 12.25kgf | cm^2 and 50°C in the condenser. The maximum coefficient of performance achieved is 4.3.

An experiment was conducted to verify the actual cooling load and temperatures obtained in a conventional domestic refrigerator under normal working conditions. A 165L Kelvinator refrigerator was switched off from the power supply and was loaded with usual eatables. The freezer box was filled with 8kg of ice. The temperature of the cooled space and amount of ice melted were periodically measured. It was found that similar temperature of 8-10°C was obtained as produced by the refrigerator working on compression cycle. The total ice required for 24 hours operation was 7.5 kg. for extreme environmental conditions. In order to take care of a non-sunny day, the solar refrigerator was designed for production of 15kg of ice by the solar system.

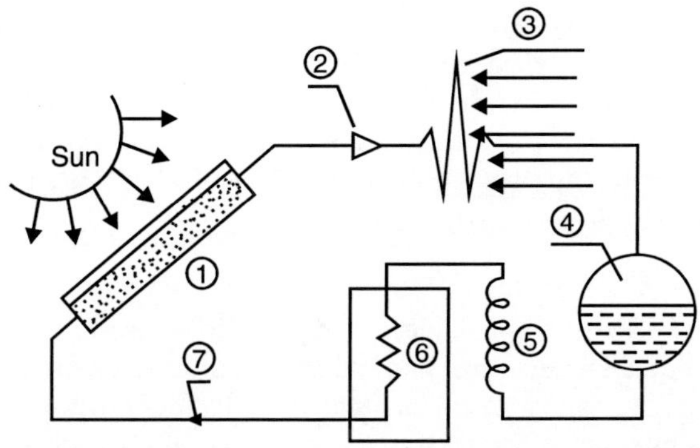

Fig. P.9.1. Flow Diagram

3. SELECTION OF SOLAR REFRIGERATION PROCESS

The main requirements of a solar powered domestic refrigerator can be summarized as follows :

1. An ability to keep refrigerated space at 8 to 10°C in an external temperature of + 43°C during day and + 32°C at night. The system should be compact and suitable for attachment to the cabinet of an existing 165L domestic refrigerator.
2. The system should be free of corrosive, toxic and irritating chemicals.
3. The system should have few components, no moving parts, operational simplicity and long trouble-free life.
4. The physical size, weight should be small for ease of transportation and installation.
5. It should be possible to manufacture the system and components from local materials and skills.

The following processes can be used for the solar domestic refrigerator;

3.1. Photovoltaic-Compression System

These systems use the photovoltaic cells which produce electricity when exposed to light. The electricity that is generated is used to charge a battery that operates a 12 or 24 volt d.c. compressor. It provides cooling through an ordinary freon cycle as in conventional domestic refrigerators. Cooling can be provided during period without sun by a store of ice in the refrigerator or by stored surplus power in batteries. The electric power is controlled automatically by a regulator. The system is the most advanced solar cooling technology and many coolers have been marketed as such.

The photovoltaic panels and batteries are very costly, hence initial cost will be very high. Batteries have to be replaced every two to three years making high operating cost. Reliability of photovoltaic panels is unknown at this stage. The major weakness is the complexity and a number of electrical components.

3.2. Photovoltaic Thermoelectric Refrigeration System

This method uses the Peltier Effect according to which one end of the element is kept at a high temperature and other at low temperature then emf is produced. In this case electricity is produced using photovoltaic panels exposed to sun. The electricity is used to produce a heat flow through a bimetallic (peltier) solid state element. The cold side of this element is exposed to the inside of the refrigerator and the hot side to outside air. A single element can produce a temperature difference of 27°C between inside and outside faces. The elements can be cascaded together to obtain higher performance. A fan is usually required to remove the accumulated heat from the outside surface. This system has same limitations as photovoltaic-compression system.

3.3. Ammonia-Water System

Ammonia gas is boiled out of a solution of water and liquid ammonia in a solar collector and is condensed and stored until the collector cools at night and the pressure in the system falls. The liquid ammonia passes into the evaporator and boils at – 12°C. The frozen surface of evaporator can make 5kg of ice in 8 hours. The ammonia gas returns to the collector where it is reabsorbed into a solution with water ready for the next day. The ice can be removed for use or left to cool the refrigerator in times of no sun.

Ammonia has pungent and irritating smell and is not used in conventional domestic refrigerator. It has leakage problem. The complexity of system and welding of pressure parts require special skills. The system development at present is at university level.

3.4. Lithium Bromide-Water System

This is a similar system as ammonia-water system with the difference that steam is boiled out of the solution of lithium bromide and water. Steam is condensed and its evaporation at low pressure produces cooling.

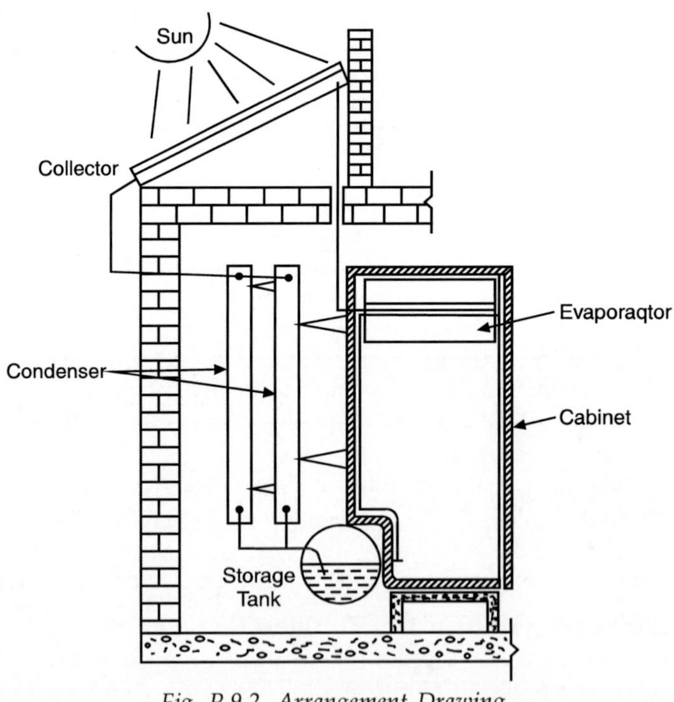

Fig. P.9.2. Arrangement Drawing

In this system the temperature required in the solar collector is in the range of 120°C which cannot be obtained in a flat plate collector. The concentrating type collectors are needed to get the necessary temperature which makes economy unfavourable. It is difficult to obtain adequate circulation expecially during regeneration cycle without a pump. The risk exists that liquid lithium bromide will also vaporise.

3.5. Calclium Chloride-Ammonia System

Calcium chloride is solid at room temperature and it works as solid absorption system. During day, ammonia gas is driven out of solar collector filled with calcium chloride and cement. The gas is condensed by cooling in a water tank to liquid ammonia and passes into the evaporator. At night, as the collector cools and pressure falls, the liquid ammonia passes into the evaporator and boils at – 12°C. The frozen surface of evaporator can make 9kg of ice in 8 hours. The ammonia gas then returns to the collector to be reabsorbed into calcium chloride for the next day. The ice can be removed for use or left to keep the refrigerator cool at times of no sun.

Same objections to the use of ammonia as refrigerant are applicable as were discussed in Water-ammonia system. This system also works at high temperatures of 120°C which cannot be easily obtained in a flat-plate collector. It is difficult to keep solid absorbent as porous and to prevent its clogging.

3.6. Solid Absorption Zeolite Refrigeration System

The solar collector is filled with zeolite which is a sandy material available in nature as lava of volcanoes or manufactured as molecular sieve. Water vapour is driven out of the collector by solar heating during the day. The vapours are condensed and stored in a tank until evening. As the collector cools at night due to sky radiation, a vacuum in the system causes the water in the evaporator to boil constantly at a temperature of – 2°C at a pressure of 4mm of Hg. The frozen surface of evaporator can make 9kg of ice in 8 hours. The water vapours return to the collector for reabsorption into zeolite and the system is ready for the next day. The ice can be removed for use or left to cool the refrigerator at times of no sunshine.

This system has no moving parts, it is quite compact and has a long operational life. No chemicals are used. Water being the refrigerant, there is no leakage problem. Due to high heat of vaporization of water, the amount of refrigerant circulated is very small as compared to other refrigerants. It can run at collector temperatures of 50°C which are easily obtained in ordinary flat-plate solar collectors. Some manufacturing experience is available in USA for zeolite solar refrigeration. This system meets most of the requirements for a domestic refrigerator as enumerated above and hence has been selected for the project.

4. ZEOLITE

It is possible to use a number of liquid or solid desicants for solar refrigeration. Any material that has high affinity for water vapour can be used. It is regenerated by heating with solar energy. It has been found that zeolites have an unusual absorption property that they can absorb about 30% by weight of water at all relative humidities and have been successfully tried in ice-making and solar refrigerators. It is an aluminium-silica complex which has a unique property of absorbing water at low temperatures and giving out at higher temperatures on heating. It can provide refrigeration with excellent engineering efficiency due to their extremely nonlinear absorption isothermals. Since the heat of vaporization of water is largest of any common refrigerant, being ten times larger than that of freons, the zeolite-water vapour combination can provide the most efficient system requiring the smallest quantity of zeolite for its operation.

The adsorption of water zeolite depends exponentially on atlest the second to the fifth power of $\Delta H/RT$, *i.e.*, exp $(\Delta H/RT)^2$ to exp $(\Delta H/RT)^5$. The adsorption of ammonia in water or water vapour in LiBr or in Silica Gel or activated alumina is directly proportional to exp $(\delta H/RT)$ where δH = Heat of Adsorption and T is the absolute temperature. It is this extreme non-linearity of thermal activation that makes zeolite so well suited for cooling.

The zeolite is sealed in an air-tight container that is irradiated by sun. During the day cycle, the zeolite and its container is heated to high temperature. At about 40°C water vapour starts desorbing from the zeolite and partial pressure begins to rise. When the pressure reaches the value determined by the condenser temperature, *i.e.*, 55mm of Hg at 40°C, the vapours begin to liquify. Heat is rejected to the outside and liquid water is stored in a storage tank.

During night cycle, the zeolite is cooled by sky radiation and is ready to adsorb water vapour even at low pressure, liquid water absorbs heat from the space to be cooled and is converted into water vapours. If the partial pressure can be maintained at 4 mm of Hg, the water in the evaporator will boil at – 2°C. The function of zeolite is to adsorb the water vapour produced by the evaporator maintaining the partial pressure below 4mm of Hg.

Zeolite system should be free of air, therefore the zeolite is placed in hermetically sealed metal panel. The panel is painted black for maximum solar absorption. It is connected through ordinary plumbing to the condenser, water storage tank and evaporator. After the zeolite is charged with water vapour, the whole system is outgassed, evacuated and sealed off.

The zeolite solar panel consists of a flat-plate conventional collector. For the extreme condensation and evaporation pressures of 55mm and 4mm of Hg respectively for most zeolites the differential water loading is about 55% by weight between ambient temperature and 120°C which is the maximum temperature attainable with flat-plate solar collectors. In most of the practical uses, it is closer to 3% by weight. After considerable experimentation and computer analysis, it is found that optimum results are obtained with zeolite capacity of 50 kg/m^2 with a depth of 5cm. The upper surface of the panel is painted black and pipes are provided at the bottom to permit the water vapours to enter and leave. Each panel has to be leak tested prior to use.

Zeolites can provide refrigeration from the sun without any external power and without any moving parts. The solar refrigerator with an efficiency of 15% produces about 900Wh of cooling per square meter of collector area for a solar input of 6 kWh. This corresponds to about 9Kg of ice manufactured per day per square meter of collector area.

5. SYSTEM DESIGN

The design data for the solar domestic refrigerator has been assumed as follows:

Cooling load	:	15 Kg ice
Refrigerant	:	Water
Evaporator pressure and temperature	:	4mm Hg and – 2°C
Condenser pressure and temperature	:	55mm Hg and 40°C
Zeolite absorption capacity for water	:	3% by weight
Average annual solar and sky radiations	:	650 $Cal/cm^2/day$
Average annual sunshine hours	:	250
Average annual day temperature (design)	:	+ 43°C
Average annual night temperature (design)	:	+ 32°C

The expected performance of the refrigerator is assumed as follows:

Total refrigerated volume	:	165L
Evaporator Temperature	:	– 2°C
Refrigerator space temperature	:	8-10°C
Daily ice requirement	:	7.5 Kg
Total design capacity	:	15Kg ice (1350 K Cal)
Storage capacity	:	Two days of bad weather.
Heating per day at 55°C @ 15% collector efficiency	:	1460 KCal.
Cooling by sky radiations at 32°C	:	1460 KCal.
Daily average C.O.P.	:	1350|1460 = 0.92
Total cost of one unit	:	Rs. 4500

6. Specifications

The main specifications of the solar refrigerator are worked out as follows:

1. Collector

Collector Type	:	Flat Plate
Collector Area	:	$1.5 m^2$
Dimensions	:	0.75 × 2.0 m
Absorber panel material	:	All copper, 0.5 mm thick.
Surface treatment	:	Copper black on copper.
Glazing	:	Double 3mm low iron glass.
Case material	:	Aluminium extrusions 2.15m × 0.9m × 0.15 m
Insulation back, edge	:	Glass wool 50 mm thick
Sealants and gaskets	:	rubber
Method of installation	:	Wall mounted at an angle of latitude of the site.

2. Evaporator

Daily refrigeration effect	:	1350 KCal.
Design heat rate	:	1350|8 = 168.75 KCal/hr.
Heat transfer coefficient	:	342 KCal|hr-m^2-°C (water to water)
Temperature terminal	:	2°C
Heat transfer area	:	$0.247 m^2$

Type	:	Panel coil, double embossed *P* type evaporator immersed in water.
Size	:	30 cm × 30 cm.
3. Condenser		
Refrigerant flow rate	:	0.251 Kg | hr.
Heat rate of rejection	:	168.25 KCa | hr.
Heat transfer coefficient from steam to air	:	8.52W/m^2.°C
Temperature terminal	:	5°C
Heat transfer area	:	3.95m^2
Panel size	:	75 cm × 115cm. Two nos.
4. Capillary Tube.		
Material	:	Copper
Size	:	1.5 | 0.5mm × 1.47m length.
5. Miscellaneous		
Water Storage tank	:	MS tank 20L capacity. 50 cm diameter × 25cm length.
Low pressure Piping	:	32NB copper NRV ball type 4cm LDPE
High pressure piping	:	10 NB copper NRV 1-5cm aluminium.

7. REFERENCES

1. Frank Kreith and J.F. Kreider, '*Principles of Solar Engineering*', Hemisphere Publishing Corporation, Washington.
2. Solar Refrigeration for Vaccine storage and Ice-making, Selected Article from ASSET, Vol. 6, No. 1, 1984.
3. Tchernev, D.I., '*Use of Natural Zeolite in Solar Refrigeration*' Asset, Vol. 6, No. 5, 1984.
4. Tchernev, D.I., '*The Use of Zeolites for Solar Cooling*', The Zeopower Company, USA.
5. Trade literature of Zeopower Company, USA.
6. Trade literature of Kelvinator India Ltd.

❑❑❑

Paper-10

FLOATING TYPE MICRO HYDRO POWER PLANTS

INTRODUCTION

India has made great strides in industrialisation during post independence years. This has brought prosperity to a selected few and created concentrated centres of employment near the industrial complexes and big cities where economic activities have flourished. The material and energy resources have been diverted to feed these industrial and commercial 'gluttons'. The high technology industry could not penetrate the traditionally agricultural rural and remote areas where nearly 78% of India's population live. The rural population felt neglected and many of them are flocking to the new economy centres in search of their livelihood. The continuous migration of population and uneven distribution of wealth, material and energy resources have created many social and political problems which can lead to very serious consequences, if not further checked.

The solution of the problem lies in concerted efforts for the development, welfare and prosperity of the rural areas which is turn depend upon the availability of cheap and abundant energy for domestic cooking, heating and lighting, community development, irrigation water, mechanised agriculture and small industries.

Transportation of commercial fuels like coal, diesel and patrol over long distances to remote and rural areas is very difficult and transmission and distribution of electricity through power grid to far flung areas are prohibitively expensive. The farmers and craftsmen are not in a position to pay for high costs of energy, thus small scale industries in these areas are reduced and there are limited irrigation possibilities in agriculture.

The modern high technologies do not adopt to rural ecology. An energy production technique based on renewable and appropriate technology which can be taken care by trained local people for its operation and maintenance problems is ideally suited. Only appropriate technology involving local resources, manpower and raw materials and environment can ensure environmental harmony and sustainability of the development. The utilisation of energy in flowing water through micro hydro schemes is a good example of appropriate technology for rural development.

SMALL HYDEL PROJECTS IN INDIA

Hydropower, available free of cost in the form of flowing water, is an important, pollution free renewable source of energy. The large potential of hydroenergy available from flowing water in the rivers and irrigation canals, however, still remains untapped, since generating power on a small scale from low and ultra low head drops of 3 to 20 metres has not been an economic proposition because of the high civil and equipment costs as well as high operating cost to be incurred per unit of power generated. The situation is further deteriorated by very low load factors in rural areas.

Chinese have developed 1200 MW of power by small hydel projects by creating water heads of about 5m along the rivers. It is estimated that 5000 MW capacity of power can be created in India with the help of small hydro turbines of 200 kW to 1000 kW capacity each. M/s. Jyoti Limited, Baroda have the necessary know-how and capacity to manufacture these small hydro turbines, see Table P.101. There are also other organisations showing interest in the manufacture of such small sets. Bharat Heavy Electricals Limited also has the necessary know-how which can be passed on to some ancillary unit for quick manufacture of small hydro turbines without disturbing their main manufacturing plan for large sets.

Table P.10.1

Manufacturing Range of Joyti Ltd. For Small Hydel Sets.

Type of Turbine	Operating Head range (m)	Output Range (kW)
Pelton	100 to 400	50 to 6000
Turgo Impulse	50 to 200	25 to 3000
Francis	20 to 200	50 to 3000
Propeller Kaplan	2 to 30	10 to 6000
Horizontal Tubular	2 to 25	10 to 6000

Due to civil costs of dams, water conduits and diversion channels, independent small hydel projects become very costly. It may cost Rs. 18000 per kW installed. Such projects should be planned on the basis of multiple purpose use of water resources such as irrigation and power generation. Even though these schemes are costlier when compared with the cost of unit power generation from major hydel projects, these are encouraging keeping long term energy availability in view. Also hydropower is renewable in nature, once built, there is built-in protection against inflation of cost, water is available essentially free, it can be instantaneously switched on, it need not run continuously.

According to the size of hydrotubine, the small hydel projects may be classified as :

Micro Hydel Projects : 50 to 100 kW
Mini Hydel Projects : 100 to 1000 kW
Small Hydel Projects: 2 MW to 15 MW.

Small hydel projects can be completed in four to five years, mini and micro projects in three years provided civil works of irrigation projects are already completed. Here water released for irrigation only is used and power is therefore, generated for 6 to 7 months a year.

In Maharashtra, 130 schemes of various categories of small hydro projects of about 11000 kW capacity for a layout of Rs. 2000 millions have been planned. A seperate circle has been created in UP State Electricity Board with headquarters at

Nainital for small hydel projects. Sixteen micro hydel schemes are already functioning in eight hill districts of UP with total installed capacity of 7730 kW. Four schemes are under construction for a total cost of Rs. 3.37 crores. The total installed capacity after completion of these schemes will be 8830 kW. Other 32 schemes of 20,000 kW are in the pipeline.

Other hilly states are also going for small and micro hydel schemes in a big way.

RIVER CURRENT ENERGY

Western Ghats of Maharashtra and regions of Uttar Pradesh, Himachal Pradesh, Jammu & Kashmir, Sikkim and North Eastern States located in the Himalayas are suitable sites for small hydel projects where medium to high heads can be available. Equipment costs are low in the high head schemes. As we come down the hills, the available head decreases and equipment costs go up. For example, the cost of one hydro set of horizontal tubular design of 3.2 kW capacity operating at a head of 3m may be Rs. 1.28 lacs, *i.e.*, Rs. 40,000 per kW installed. The civil costs are site-specific but taking 30% of equipment costs, the total installed cost will be a high value of Rs. 52,000 per kW installed.

Most of our rivers flow through flat terrains which can be harnessed to produce power much in excess of 5000 MW as estimated above. Chinese model is difficult to develop due to different topography in India. Special 'zero' head turbines operating on kinetic energy of flowing water should be developed to harness huge river current energy. The civil costs or micro hydel power projects can be drastically cut by using floating type plants. There is no need of diversion channels, deep excavations and massive foundations. The floating plant consists of two floats with a vertical (or horizontal) runner in between, which rotates by the velocity of water. The capacity developed is a function of the size of the runner and cube of water current speed. The rotational speed of runner is rather low and electric generator can be operated via a transmission gear. The floating plant moves up and down with water level and gives constant performance all the year round which is a big advantage over firmly installed water turbines especially during monsoons.

The current energy of flowing water which can be exploited by floating type power plant may be expressed as :

$$P = 1/2\, A V^3 E \text{ (kW)}$$

Where P = Power generated

A = Embrassing area of runner (m^2)

V = Water current speed (m/s)

E = Efficiency of runner and depends upon the type of the runner and varies between 0.5 to 0.85.

Although hydraulic wheels will operate at low water velocities, it is unlikely to be an economic proposition in current speeds below 1 m/s (2 knots).

PONCELET WATER WHEEL

In ancient times, floating mills were used in Germany on large rivers such as the Rhine, the Weser and the Elbe to drive electric generators and machine tools. A vertical undershot water wheel consisting of a horizontal shaft and radial blades fixed to it, was fitted on the side of a barge which was either moored to the bank or firmly anchored in the stream. A smaller barge was used to support the other end of the shaft. The hulls of the barges were built with local materials. The efficiencies realised were very low, in the range of 20 to 30% because of bad geometry (straight radial blades).

Poncelet water wheel is an improvement on the straight blades of undershot water wheel, which have suitable curvature. Water strikes the vanes practically without shock (inlet vane angle 15°) and drives it by impulse. Water is discharged from the blades almost vertically downwards. The efficiencies achieved are 55 to 65%.

The main drawback of such machines is low power output for a given size and weight because at any instant only a small part of it is actually in the water being driven by the current while the rest of the machine is idle.

DARRIEUS TURBINE ROTOR

Darrieus turbine rotor is similar to modern vertical-axis turbine which was first introduced in France by G.J.M. Darrieus in 1920's. Its configuration consists of four hydrofoil blades which rotate at much higher speed than the water current speed. The speed of rotation is primarily dependent upon the mchine overall dimensions. It has relatively low starting torque. Because of simplicity of the blade design and because they are relatively thin blade fabrication costs are low. They require no pitch control for synchronous applications.

Two prototypes have been developed by Intermediate Technology Development Group, London and are shown in Fig. P.10.1 and P.10.2. The blades can be made out of timber or fabricated from steel, ferrocement or glass fibre. The expected performance from these models are plotted in Fig. P.10.4.

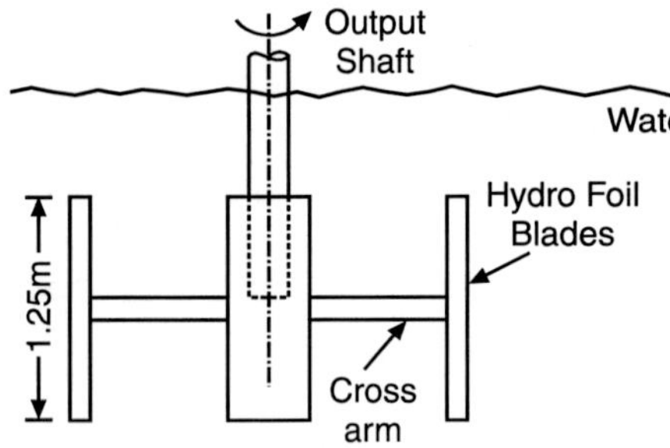

Fig. P.10.1. Vertical Axis Rotor 13.5 RPM in 1m/s Current Power Takeoff Above Surface Swept Area 37.5m²

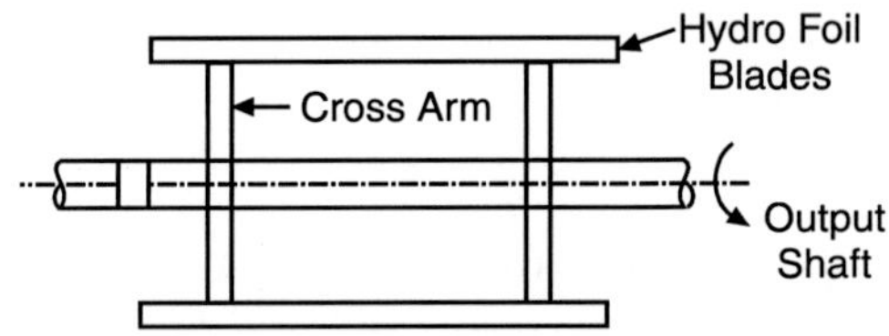

Fig. P.10.2. Horizontal Axis Rotor 32 RPM in 1 m/s Current Power Take offs under Water Swept Area 3.75 m²

INSTALLATION

Floating type micro hydro power plants are suitable for use in situations where there is no 'head' of water available and is designed to extract kinetic energy from a river or canal current. Therefore appropriate sites are along the banks of large rivers or on irrigation canals.

As the energy flux depends upon the cube of water velocity, the latter must be as large as possible. The depth of water required for a $4m^2$ runner is atleast 2.5m.

The Darrieus turbine rotor is a low speed machine. It is low solidity device which is completely submerged during operation and this, together with its high efficiency (above 50%), means that a rotor of moderate size and modest materials content can extract useful amounts of energy from current speeds as low as 1m/s (2 knots).

Irrigation canals are expected to be an important area of application for floating type micro hydro power plants. In a large canal, the turbine rotor could be mounted on a pontoon as in a river situation, whereas on a small canal it might be cheaper to mount the rotor on a bridge or beam over the canal. The power output from the rotor could be increased by shaping the canal banks to form a venturi to increase the water velocity through the rotor.

The mooring details and site requirements are shown in Fig. P.10.3 and P.10.5 respectively. The mooring cable arrangement protects the rotor against submerged or semi-submerged objects striking the blades.

The pontoon which carries the turbine, generator and the transmission system, can float on two rows of barges or oil drums under which wooden keels are affixed. The mooring cable can be adjusted with a winch to ensure sufficient angle of the keels to the current direction so that water side thrust on the keels keeps the pontoon at a distance from the bank.

AUXILIARIES

The rotational speed of Darrieus turbine is very low: 13.5 rpm for vertical axis and 32 rpm for horizontal axis turbines as shown in Fig. P.10.1 & P.10.2. For Poncelet turbine, the peripheral velocity of runner is about half of water current speed. Therefore multi-stage belt drives or gear boxes are needed to increase the speed of the generator shaft to about 950 rpm. Permanent magnet alternators or induction generators are preferred over synchronous generators. To reduce equipment costs, the conventional mechanical or hydromechanical speed governor should be dispensed with. The intertia of rotor being small, there is a problem of instability in power supply quality.

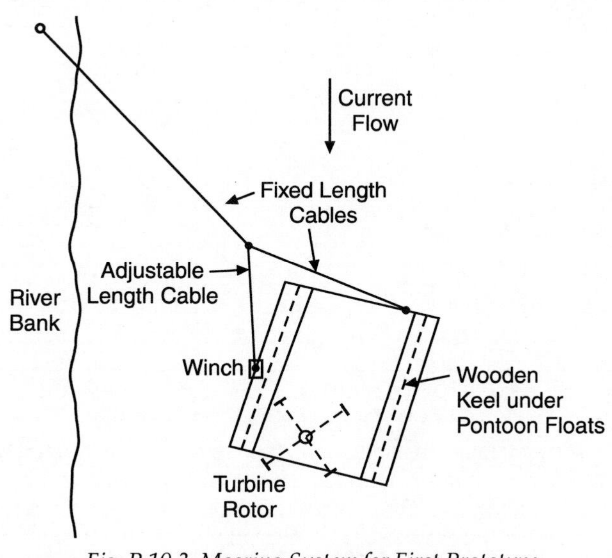

Fig. P.10.3. Mooring System for First Prototype.

The above problems can be solved by use of the flywheel which can ensure stable power supply and can shave-off power demand fluctuations. A flywheel designed for this purpose can be accelerated to extremely high speeds without any risk of breaking apart. It can be made from concentric rings of quartz fibre as windings, the rings being fitted over one another and close gaps filled with an elastic substance to keep the plies of the rims together. The flywheel is coupled to a generator and enclosed in a sealed evacuated casing to reduce windage losses. The device operates as a generator when power demand on the system grows, and as electric motor, when it is time to accumulate energy. According to some calculations, the cost of 1 kW will be less than that with pumped hydroelectric storage. A project has been described contemplating a 5m diameter flywheel of 1.96 MN weight that could be charged to 20 MWh. The operating speed is 3500 rpm.

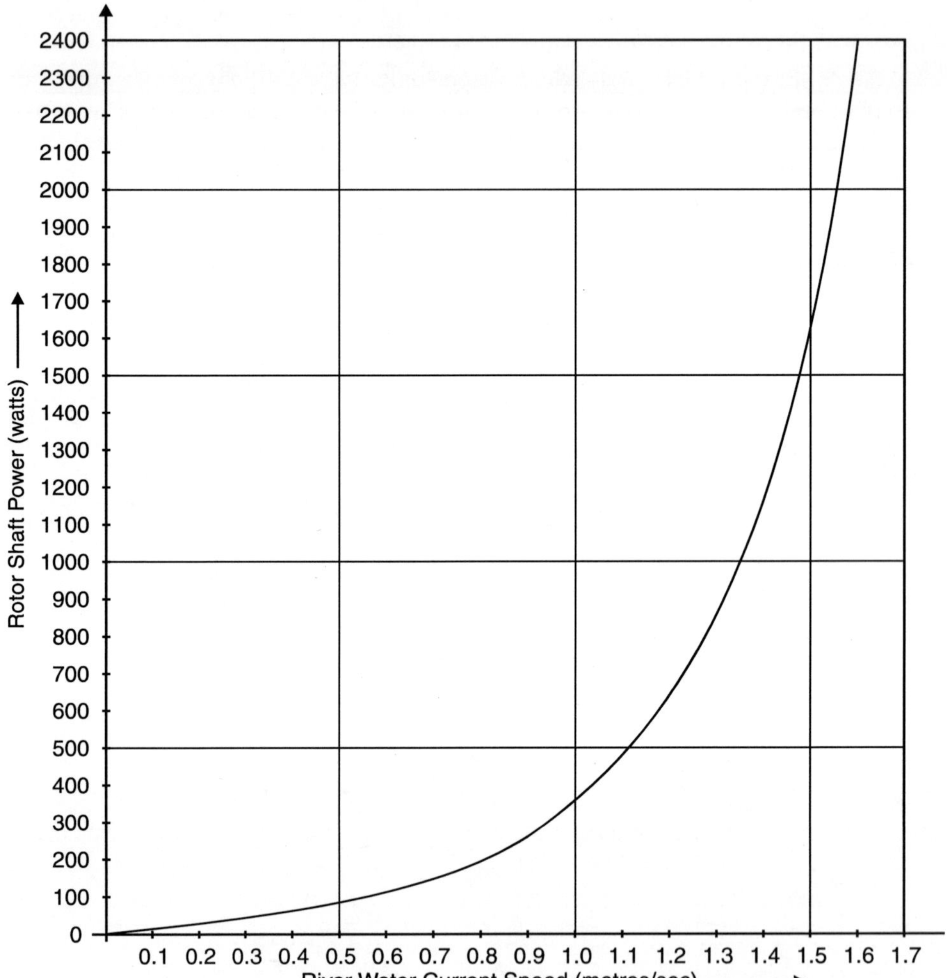

Fig. P.10.4. Graph showing how turbine rotor power varies with river current speed for a turbine with swept areas $4m^2$ and rotor efficiency 25%.

The mechanical energy storage system as described above can fit into appropriate technology level suitable for rural environment. The load governors based on electronic circutory or microprocessor technology may be able to slove the problem to some extent but may not be easily acceptable in rural areas because of higher levels of technology involved. Further, these load governors donot operate happily against unstable operation of turbine rotor. Load governors invariably require an auxiliary load or another hybrid system to store surplus electric power generated by hydel power plant during low demand period. This is not always a good decision thermodynamically, as electrical energy is the highest form of energy and hybrid systems help to degrade the energy form and in addition there are heavy conversion losses.

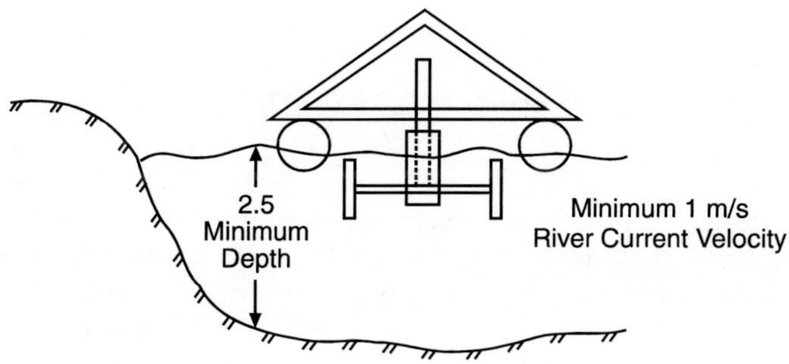

Fig. P.10.5. Site Requirement for Turbine with 4m^2 Swept Area.

CONCLUSION

Floating type micro hydro power plants can be a very attractive proposition to harness enormous amounts of hydroenergy available as river current energy in the flat country side. Civil costs are reduced and cheap and reliable equipment can be developed for tapping the flow energy in thc irrigation canals by constructing the banks into venturi shapes. Research and development efforts are needed to develop efficient turbine rotors working on the principle of wind turbines or old time water wheels followed by intensive field trials. The problem of stability and energy storage to improve load factors can be solved by the development of suitable flywheels. The development efforts must be constrained keeping in view the principles of appropriate technology suitable for harmonious development of rural areas involving local people, local materials and local resources.

REFERENCES

1. Peter German, *'The Development of a Turbine for Tapping River Current Energy'*, Intermediate Energy Technology Development Group, London.
2. *'Upgrading and Further Development of Traditional Devices for the Utilisation of Hydro-energy'*, Bremen Overseas Research and Development Association, Bremen.
3. Jagdish Lal, *'Hydraulic Machines'*, Metropolitan Book Co, Pvt. Ltd. Delhi.
4. N.P. Cheremisinoff, *'Fundamentals of Wind Energy'*, Ann Arbor Science Publishers Inc., Michigon, 1978.
5. P.d. Mohindra, *'Current from Currents'* The Tribune, 14th June, 1982.
6. Brij Bhardwaj, *'Surplus Power by 1986 Envisaged'* Hindustan Times dated 5th April, 1984.
7. *'UP submites Five more Micro Hydel Schemes,'* Hindustan times dated 30th Sept., 83.
8. Venikov & Putyatin, *'Introduction to Energy Technology'* Mir Publishers, Moscow, 1981.

Paper-11

SOLAR "AHARA" FOR VILLAGES

The quality of rural life depends upon the availability of cheap and abundant energy for socio- economic activities. Plenty of sunshine available in the village offers an attractive solution to some of the energy problems. In rural areas, households are the major consumers of energy followed by agriculture and cottage industries. It is mostly met from traditional fuels like wood, dung and agricultural residues. Cooking and domestic heating usually take place on open fire, which has low thermal efficiency of the order of 5 per cent.

Cooking with firewood and cowdung cakes poses a number of problems for the users and for society as a whole. The serious implications of deforestation and subsequent soil erosion, pollution and lack of recycling of natural fertilisers have a significant effect on the overall productivity of rural areas. Therefore, it becomes essential to find an alternative source of domestic fuel. Other non-conventional energy sources such as biogas, wind and micro-hydro, would be best used to increase the industrial and agricultural productivity of rural areas, while allowing solar energy to provide the bulk of energy requirements for cooking and domestic heating.

Efforts to design and introduce solar cookers have for many years concentrated on two main types; the cooker with a plane collector (box type) and cooker with special parabolic mirrors of average precision. The Government of India is trying to introduce box type solar cookers with subsidies but with poor success. This type of cooker has certain technical drawbacks due to very low temperature (less than 90° C) obtained. The cost of second type of solar cookers is too high to be used economically in rural areas. The main reasons are high cost of parabolic mirrors and complicated tracking system required with these types of cookers.

An "AHARA" is a must for every household in the Indian villages in which a slow fire is maintained for almost the whole day with the help of cowdung cakes and firewood. This is multi-purpose cooker-cum-heater with low heat flux inputs. A versatile, multi-purpose, low-cost solar ahara has been developed at Regional Engineering College, Kurukshetra, which can be used as small solar water heater, oil heater and ghee boiler in addition to solar cooker for rural delicacies of India such as rice, dals, sarson-ka-sag, palak and other vegetables and cereals requiring moderate cooking temperatures and low heat flux inputs to preserve taste, flavour and nutritional values of food.

The design of the ahara is based on the use of plane mirrors cut to special shape and size and arranged at suitable angles to approach a paraboloid. The tracking is manually achieved in two axes by the use of simple pipes and joint threads. Only

---

The Tribune, Chandigarh

those materials have been used which are easily available in the villages and can be assembled locally without the need of any special skills.

The total cost of the prototype, including materials and manufacturing labour, is Rs 500 only. The cost can be easily reduced further during commercial production. The marketing price of such as heater will be much less than box type solar cookers patronised by the Government of India. Technically, it is far superior in versatility of applications and a temperature of 125° C is easily obtained for more than four hours on a normal sunny day. The cooking efficiency and cooking time are very favourable as a pressure cooker can also be used with the device. This is not possible in box-type solar cookers.

The solar "ahara" is a light-weight, portable device and can be placed near the south facing wall of the house and tracking can be effected through a window from inside the house. This is a very effective and ideal design for a low-cost solar cooker for adoption in rural and muffasil areas.

Paper-12

RURAL POWER

FLUIDISED BED TECHNOLOGY COULD HELP MEET ALL POWER NEEDS OF A VILLAGE

A recent study conducted by scientists from the Regional Engineering College (REC), Kurukshetra, reveals that all energy needs of village can be met locally from the available biomass, using fluidised bed technology-an industrial technique for efficient combustion.

The technology involves suspending the material to be burnt by blowing an upward stream of air through it in a "reactor". The material in the resulting fluidised bed, which resembles a boiling viscous mass, is more acessible to chemical reactions than the same material in a solid, static sate. Up to 98 per cent combustion is achieved through this process. Further, the process is environmentally benign since the emissions of nitrogen oxides are negligible.

A survey conducted by the REC in Birohar village in Kurukshetra district revealed that the biomass required for a plant of 3 megawatt capacity based on 35 per cent overall efficiency is 25 tonnes per day.

Rural energy requirements can be met with smaller plants based on fluid bed technology. Small steam generator or package boilers, for instance, can be designed to produce steam for power generation. Similarly, standalone external furnaces or thermal fluid heaters can be set up to produce the hot water to meet domestic and village industry requirements.

Meanwhile, the Union ministry of human resource development has also started projects to build pilot plants based on fluidised bed systems for rural developement.

❑❑❑

Down to Earth, New Delhi

PART VII

APPENDIX

APPENDIX

Table 1. Basic and Derived Units.

Type	Unit	Symbol	Formula
Length	meter	m	
Mass	kilogram	kg	
Time	second	s	
Temperature	kelvin	K	
Force	newton	N	$(kg)\ (m)/s^2$
Pressure	pascal	P_a	N/m^2
Energy, Work	joule	J	Nm
Power	watt	W	J/s

Table-2. Prefixes of Multiples and Submultiples.

Factor	Prefix	Symbol
10^{-18}	atto	a
10^{-15}	femto	f
10^{-12}	pico	p
10^{-9}	nano	n
10^{-6}	micro	m
10^{-3}	milli	m
10^{-2}	centi	c
10	deka	da
10^{2}	hecto	h
10^{3}	kilo	k
10^{6}	mega	M
10^{9}	giga	G
10^{12}	tera	T
10^{15}	peta	P
10^{18}	exa	E

Table-3. Units for Energy and Power.

Unit	Description
1 joule [J]	Basic unit for energy, work, heat
1 watt [W]	Basic unit for power, energy flow, energy flowrate
1 kWh	3600 kJ
1 B T U	British Thermal Unit = 1,05504 J.
1 joule	1 newton × 1 meter [Nm] 1 watt × 1 second [Ws]
1 W/m2	Energy flux.
1 kWh/kg	Specific energy 4.166667×10^{-2} MW. day/tonne. 3.6×10^{6} J/kg 3.6×10^{6} Ws/kg.
1 W	Power 1 [J/s]; 10^{-3} kW 1.341020×10^{-3} hp 6.241457×10^{6} MeV/(s) (cm^{3})
1 W/kg	Specific Power 1×10^{-3} [kW/kg]

Table-4. Conversion Factors for Energy Units.

	kJ	kcal	kWh	CEU	OEU	m^3 (Natural gas)
kilojoule (kJ)	1	0.2388	0.000278	0.000034	0.000024	0.000032
kilocalorie (kcal)	4.1868	1	0.001163	0.000143	0.0001	0.00013
kilowatt hour (kWh)	3600	860	1	1.123	0.086	0.113
Coal equivalent unit (CEU)	29 308	7000	8.14	1	0.7	0.923
Oil equivalent unit (OEU)	41868	10000	11.63	1.486	1	1.319
m^3 (natural gas)	31736	7580	8.816	1.083	0.758	1

Table-5. Pressure and Temperature Units

Pressure	1 Pa = $1N/m^2$ 1 bar = 105 Pa = 0.1 mPa = 750 mm Hg = 1.02×10^4 mm H_2O
Temperature	T[K] = t[°C] + 273.15 t[°C] = **5/9 t[°F] – 32)** $T[R]$ = t[°F] + 459.67 $T[k]$ = 1.8 [R]
Standard	0°C and 1.013 bar

Table-6. Lower Heating value of Fuels.

Fuel	Quantity unit	Lower heating value [kg/kJ] or [kJ/m^3]	CEU-factor
Hard coal	kg	29809	1.017
Hard coal coke	kg	28650	0.978
Hard coal briquettes	kg	31401	1.071
Lignite	kg	8303	0.283
Lignite briquettes	kg	19259	0.657
Lignite (hard)	kg	14980	0.511
Fuel wood (1m^3 = 0.7t)	kg	14654	0.500
Peat	kg	14235	0.486
Mineral oil (crude)	kg	42622	1.454
Engine gaso line, benzene	kg	43543	1.486
Crude gaso line	kg	43543	1.486
Aviatron gaso line, light Fuel for aviation turbines	kg	43543	1.486
Heavy fuel for aviation turbines, petroleum	kg	42705	1.457
Diesel oil	kg	42705	1.457
Fuel oil (light)	kg	42705	1.457
Fuel lil (heavy)	kg	41031	1.400
Petroleum coke	kg	29308	1.000
Liquified gas	kg	45887	1.566
Refined gas	kg	48358	1.650
Coal gas, city gas	m^3	15994	0.546
Blast furnace gas	m^3	4187	0.143
Natural gas	m^3	31736	1.083
Crude bar	kg	37681	1.268
Pitch (asthalt)	kg	37681	1.286
Load	kg	10,000	0.340
Crop residues	kg	14,000	0.478
Biogas	kg	28,000	0.955
Methanal	kg	23,000	0.785
Ethanal	kg	30,000	1.023

$$\text{CEU factor} = \frac{LHV}{29308}$$

Table-7. Current Cost of Energy

Power Plant Type	Installation cost [¢/kW]	Electircity generation cost [¢/kWh]	Remark
1. Nuclear	2500 – 3500	0.015 – 0.02	P = 5MW
2. Hydro	2500	0.07 – 0.13	P = 200 MW
3. Tharmal	2800 – 3000	0.015 – 0.02	coal-fuel
4. Combined cycle	470 – 1200	0.015 – 0.02	
5. Wind power	2500	0.04 – 0.15	P = 2.5 MW
6. Solar Thermal	3500	0.1 – 0.13	
7. Polar thermal Hybrid	3500	0.04 – 0.1	
8. Photorolataic	7000 0.35 – 0.5	0.6 – 0.9 Africa	Europe
¢ = U. S. cents.			

Table-8. The elements

Element	Symbol	Atomic number, Z	Element	symbol	Atomic number, Z
Actinium	Ac	89	Mercury	Hg	80
Aluminum	Al	13	Molybdenum	Mo	42
Americium	Am	95	Neodymium	Nd	60
Antimony	Sb	51	Neon	Ne	10
Argon	A	18	Neptunium	Np	93
Arsenic	As	33	Nickel	Ni	28
Astatine	At	85	Niobium	Nb	41
Barium	Ba	56	Nitrogen	N	7
Berkelium	Bk	97	Nobelium	No	102
Beryllium	Be	4	Osmium	Os	76
Bismuth	Bi	83	Oxygen	O	8
Boron	B	5	Palladium	Pd	46
Bromine	Br	35	Phosphorous	P	15
Cadmium	Cd	48	Platinum	Pt	78
Calcium	Ca	20	Plutonium	Pu	94
Californium	Cf	98	Polonium	Po	84
Carbon	C	6	Potassium	K	19
Cerium	Ce	58	Protactinium	Pr	59

Contd....

Element	Symbol	Atomic number, Z	Element	symbol	Atomic number, Z
Cesium	Cs	55	Promethium	Pm	61
Clorine	Cl	17	Protactinium	Pa	91
Chromium	Cr	24	Radium	Ra	88
Cobalt	Co	27	Radon	Rn	86
Copper	Cu	29	Rhenium	Re	75
Curium	Cm	96	Rhodium	Rh	45
Dysprosium	Dy	66	Rubidium	Rb	37
Einsteinium	Es	99	Ruthenium	Ru	44
Erbium	Er	68	Samarium	Sm	62
Europium	Eu	63	Scamdium	Sc	21
Fermium	Fm	100	Selenium	Se	34
Fluorine	F	9	Silicon	Si	14
Francium	Fr	87	Silver	Ag	47
Gadolinium	Gd	64	Sodium	Na	11
Galium	Ga	31	Strontium	Sr	38
Germanium	Ge	32	Sulfur	S	16
Gold	Au	79	Tantalum	Ta	73
Hafniurm	Hf	72	Technetium	Tc	43
Helium	He	2	Tellurium	Te	52
Holmium	Ho	67	Terbium	Tb	65
Hydrogen	H	1	Thallium	Tl	81
Indium	In	49	Thorium	Th	90
Iodine	I	53	Thulium	Tm	69
Iridium	Ir	77	Tin	Sn	50
Iron	Fu	26	Titanium	Ti	22
Krypton	Kr	36	Tungsten (Wolfram)	W	74
Lanthanum	La	37	Uranium	U	92
Lead	Pd	82	Vanadium	V	23
Lichium	Li	3	Xenon	Xe	54
Lutecium	Lu	71	Ytterbium	Yb	70
Magnesium	Mg	12	Yttrium	Y	39
Managanese	Mn	25	Zinc	Zn	30
Mendelevium	Md	101	Zirconium	Zr	40

Table-9. Physical Properties of Water (Saturated Water at 298 K)

Property	Symbol	Value
Molecular Weight	M	18.02
Gas constant	R	461.4J/kg-K
Density	ρ	999 kg/m^3
Specific heat	C	4181 J/kg-K
Dynamic viscosity	μ	9.239×10^{-4} kg/m-s
Kinematic viscosity	ν	9.248×100^{-7} m^2/s
Thermal conductivity	K	0.604 W/m-K.
Prandtle number	P_r	6.395
Heat of vaporization	h_{fg}	2257 kJ/kg

Table 10. Physical Properties of Air (Standard conditions)

Property	Symbol	Value
Molecular weight	M	28.97
Gas constant	R	287 J/kg-K
Specific heats	C_p	1005 J/kg-K
	C_v	718 J/kg-K
Density	ρ	1.185 kg/m^3
Dynamic viscosity	μ	1.84×100^{-5} m^2/s.
Kinematic viscosity	n	1.55×10^{-5} m^2/s
Thermal conductivity	K	0.0257 W/m-k
Ratio of Specific heats $\left(\frac{C_p}{C_v}\right)$	γ	1.3997
Prandtle number	Pr	0.720

Table-11. Common Constants

Constant	Symbol	Value
Acceleration due to gravity	g	9.80665 m/s^2
Velocity of sound in air	C	344 m/s
Boltzmann Constant	k	1.380×10^{-23} J/K
Faradays constant	F	9.65×10^{7} c/kgmal.
Planck's constant	h	6.626×10^{-34} Js
Stefan-Baltzmann constant	e	5.67×10^{-8} J/m^2-s-K
universal gas constant	$\overline{R}$	8314.4 J/mol-K.
NTP	To	273.15 K
Solar constant	S	1353 W/m^2

Table-12. Solar Radiation data for Indian cities. (MJ m^{-2} day^{-1})

Location		Months											
		Jan.	Feb.	March	April	May	June	July	August	Sept.	Oct.	Nov.	Dec.
1. Ahmedabad	H_b	18.72	20.01	19.50	17.21	15.70	9.11	3.96	4.35	11.11	18.38	19.80	19.83
tilt = 30°	H_d	4.85	5.58	6.18	7.35	7.32	10.15	11.38	10.53	8.48	5.50	4.09	4.31
2. Calcutta	H_b	13.68	14.86	14.19	12.45	10.25	5.07	4.15	4.82	6.81	10.88	14.28	14.80
tilt = 30°	H_d	5.87	6.50	7.59	8.80	9.84	10.20	10.10	9.58	8.80	7.19	5.76	5.30
3.Madras	H_b	14.69	18.71	18.52	15.99	11.94	8.49	8.94	8.42	10.82	10.17	9.67	10.14
tilt = 20	H_d	7.01	6.28	6.50	7.12	8.53	9.58	10.27	10.03	8.90	8.02	7.49	7.56
4. New Delhi	H_b	16.15	17.65	17.42	14.99	12.14	7.72	6.10	7.18	12.62	17.80	19.07	16.96
tilt = 30°	H_d	4.94	5.61	6.52	7.74	8.76	10.48	9.31	8.38	7.11	5.34	4.42	4.60
5. Poona	H_b	19.93	21.52	19.40	16.47	13.89	6.99	2.94	3.29	8.16	15.74	19.26	18.74
tilt = 40°	H_d	4.31	4.45	5.51	6.67	6.98	10.27	11.04	11.33	9.74	6.24	4.58	4.47
6. Visakha patnam	H_b	20.65	20.00	17.25	14.41	11.63	5.65	4.36	6.11	8.74	13.53	17.34	18.32
tilt = 30°	H_d	5.46	5.43	6.36	7.18	8.10	9.30	9.80	9.82	8.65	6.62	5.79	5.34

REFERENCE BOOKS

1. A. Eggers-Lura, *"Solar Energy in Developing Countries"*, Vol. II, Pergamon Press, Oxford.
2. Ashok V. Desai, *"Non-conventional Energy"*, New Age International Limited, Tokyo (2001).
3. B.T. Nijaguna, *"Biogas Technology"*, New Age International Publishers, New Delhi.
4. C.P. Gupta and Rajendra Prakash, *"Engineering Thermodynamics"*, New Chand & Bros., Roorkee (1981).
5. Daniel M. Simmons, "Wind Power", Noyes Data Corporation, Park Ridge, New Jersey (1975).
6. Frank Kreith and J.F. Kreider, *"Principles of Solar Engineering,"* Hemisphere Publishing Corporation, Washington.
7. Jagdish Lal, *"Hydraulic Machines"*, Metropolitian Book Co. Pvt. Ltd., Delhi.
8. *K.H.* Spring, *"Direct Generation of Electricity,"* Academic Press, London (1965).
9. M.M. El-Wakil, *"Power Plant Technology"* Mc-Graw–Hill Book Company, New York (1985).
10. N.P. Cheremisinoff, *"Fundamentals of Wind Energy"*, Ann Arbor Science Publishers Inc., Micligon (1978).
11. P.K. Nag, *"Power Plant Engineering"* Tata Mc-Graw–Hill Publishing Company Limited, New Delhi (2003).
12. Nikolai V. Khartchenko, *"Green Power: The Eco-Friendly Energy Engineering"*, Tech Book International, New Delhi (2003).
13. Robert Noyes, *"Fuel Cells for Public Utility and Industrial Power"*, Noyes Data Corporation, New Jersey (1977).
14. S.L. Soo, *"Direct Energy Conversion"*, Prentice-Hall (1968).
15. Technorama, The Institution of Engineers, Kolkatta (2004).

INDEX

INDEX

A...

B...

C...

D...

E...

F...

G...

H...

□□□

Win Prizes **ATTENTION READERS** **Win Prizes**

We request for your frank assessment regarding various aspects of the book as given below :

NON-CONVENTIONAL ENERGY RESOURCES

(Alternative Energy Sources and Systems)

By: Dr. R.K. Singal

Please Fill up the Sheet in Neat and Clean Writing

(i) What topics of your syllabus, important for your examination point of view, are not covered in this book ?

....

....

....

....

(ii) What are the chapters wherein the treatment of the subject matter is not systematic or organised or updated ?

....

....

....

....

(iii) Have you come across misprints/mistakes/factual inaccuracies in the book? Please specify the chapters and the page numbers.

....

....

....

....

(iv) What other books on the same subject have you found/heard better than the present book? Please specify in terms of price and quality.

....

....

....

....

(v) Further suggestions and comments for the improvement of this book.

....

....

....

....

(PS: If need be please attach a separate sheet to write your views)

OTHER DETAILS

(i) Who recommended you this book? (Please tick (✓) in the box near the option relevant to you.

☐ Teacher ☐ Friends ☐ Bookseller

....
....
....
....

(ii) Name of the recommending teacher, his designation and address :

....
....
....
....

(iii) Name and address of the bookseller you purchased the book from :

....
....
....
....

(iv) Name and address of your institution (Please mention the University or Board as the case may be)

....
....
....
....

(v) Your name and complete postal address :

....
....
....
....

(vi) Course you are enrolled in :

....
....
....
....

The best assessment will be awarded each month: The award will be in the form of our publications as decided by the Editorial Board.

Please mail the filled up coupon at your earliest to :

S.K. KATARIA & SONS®

4885/109, Prakash Mahal, Dr. Subhash Bhargav Lane,

Opposite Delhi Medical Association, Daryaganj, New Delhi–110002 (INDIA)

Phone: +91-11-43551243, *Mobile:* +91-9871775858, 7678425209

e-mail: katariabook@gmail.com, katariabooks@yahoo.com

Website: www.skkatariaandsons.com